Bennet Woodcroft

Patents for Inventions

Abridgements of specifications relating to sugar. A.D. 1663-1866

Bennet Woodcroft

Patents for Inventions
Abridgements of specifications relating to sugar. A.D. 1663-1866

ISBN/EAN: 9783337128999

Printed in Europe, USA, Canada, Australia, Japan

Cover: Foto ©berggeist007 / pixelio.de

More available books at **www.hansebooks.com**

PATENTS FOR INVENTIONS.

ABRIDGMENTS

OF

Specifications

RELATING TO

SUGAR.

A.D. 1663–1866.

PRINTED BY ORDER OF THE COMMISSIONERS OF PATENTS.

LONDON:

PRINTED BY GEORGE E. EYRE AND WILLIAM SPOTTISWOODE,
PRINTERS TO THE QUEEN'S MOST EXCELLENT MAJESTY.

PUBLISHED AT THE
OFFICE OF THE COMMISSIONERS OF PATENTS FOR INVENTIONS,
25, SOUTHAMPTON BUILDINGS, HOLBORN.

1871.

PREFACE.

THE Indexes to Patents are now so numerous and costly as to render their purchase inconvenient to a large number of inventors and others, to whom they have become indispensable.

To obviate this difficulty, short abstracts or abridgments of the Specifications of Patents under each head of invention have been prepared for publication separately, and so arranged as to form at once a Chronological, Alphabetical, Subject-matter, and Reference Index to the class to which they relate. As these publications do not supersede the necessity for consulting the Specifications, the prices at which the printed copies of the latter are sold have been added.

The number of Specifications from the earliest period to the end of the year 1866 amounts to 59,222. A large proportion of the Specifications enrolled under the old law, previous to 1852, embrace several distinct inventions, and many of those filed under the new law of 1852 indicate various applications of the single invention to which the Patent is limited. Considering, therefore, the large number of inventions and applications of inventions to be separately dealt with, it cannot be doubted that several properly belonging to the group which forms the subject of this volume have been overlooked. In the progress of the whole work such omissions will, from time to time, become apparent, and be supplied in second or supplemental editions.

This volume contains Abridgments of Specifications to the end of the year 1866. From that date the Abridgments have not been published in classes, but will be found in chronological order in the " Chronological and Descriptive Index " (*see* List of Works at the end of this book). It is intended, however, to publish these Abridgments in classes as soon as the Abridgments of all the Specifications from the earliest period to the end of 1866 have appeared in a classified form. Until that takes place, the reader (by the aid of the Subject-matter Index for each year) can continue his examination of the Abridgments relating to the subject of his search in the Chronological and Descriptive Index.

This series of Abridgments embraces not only the inventions which relate to the making, purifying, and refining of all kinds of sugar commercially in use, but also those which relate to the preparation and revivication of animal charcoal and its substitutes. The apparatus required for nipping or otherwise breaking up sugars are likewise included in this series, but not inventions which relate to the use of sugar in manufacturing lozenges and other artiles of confectionary or for medicinal purposes.

The Abridgments marked thus (* *) in the following pages were prepared for another series or class, and have been transferred therefrom to this volume.

B. WOODCROFT.

August, 1871.

INDEX OF NAMES.

INTRODUCTION.

SUGAR is a substance which extensively abounds, but it is found more especially secreted in different parts of a variety of plants,— in some plants in their stems, in some in their fruit, and in others in their roots ; indeed, there are but few vegetable substances, which do not contain it in more or less quantity. Still there are only a very few substances, and these are vegetable, which have hitherto been found to yield it in so large a quantity as to make its manufacture from them remunerative.

Although the qualities of sugar are very numerous and varied, according to their mode of production and to the nature of the substance from which they are derived, still, it may be said that, practically, there are only two kinds of sugar which have commercially any great importance, and these are commonly known under the names of cane sugar and grape sugar, more recently called glucose, but the latter sugar, although science has now taught us of its contemporaneous existence with the former, is a substance of a comparatively modern manufacture.

In the first century, sugar (no doubt the sugar from the cane) was described by Dioscorides and Pliny as "resembling salt, and " only used in medicine," whilst Strabo, Theophrastus, Arrian, and others speak of it as a sort of honey produced from a reed growing in Arabia and India, but although sugar must have been known in the East from a very early date, it is said that it was not known in Europe as an article of food until the time of the Crusades, sugar canes having been found in abundance by the Crusaders when in Syria.

The sugar cane appears to have been transplanted from Cyprus to Madeira and from thence to the West Indies in 1506, where, the climate and soil both proving favourable to its cultivation, in the same century a great number of works were put in operation for making sugar; and from that time, and for some con-

siderable time afterwards, the West India Islands were the chief
sources of supply of sugar to Europe and afterwards to America,
for it was not until towards the middle of the eighteenth century,
about 1751, that the sugar cane was brought into and cultivated
in America upon the banks of the Mississipi near New Orleans.

The sugar cane alone was known as the source for obtaining
this description of sugar until the year 1747, when M. Margraff,
of Berlin, demonstrated that cane sugar was present in the juice
of the birch, also in the parsnip, carrot, and white and red beet ;
and in 1796 M. Achard, of Berlin, obtained from beet root 5 per
cent. of white sugar and 3 per cent. of molasses. The price of
cane sugar being high, towards the end of the eighteenth or
early in the beginning of the nineteenth century, a factory was
established at Cunern, in Lower Silesia, for extracting sugar from
beet root, but no other attempt was made to prosecute this manufacture on a large scale until the time of the first Napoleon, who,
bent upon ruining the colonial trade of this country, excluded
our colonial products from France, and offered at the same time
a premium of one million of francs for the best method of producing sugar from native products. The effect of this policy
was that very soon factories were established in France for making
sugar from beets, and these gradually extended to Belgium,
Germany, Austria, and Russia, in all of which countries the
manufacturing of beet-root sugar is now prosecuted on a most
gigantic scale, the manufacture having been encouraged and fos-
tered from its commencement in each of these countries by
a protective duty.

Many endeavours have from time to time been made in this
country to manufacture sugar from the beet, but it would appear
that, up to a few years ago, not much success attended any of
them. However, from a letter from Mr. James Caird to the *Times*,
which was afterwards copied into the Journal of the Society of
Arts of 25th November 1870, it would appear that the manufacture of sugar from beet is now being worked successfully by
Mr. Duncan at Lavenham in Suffolk, where, it is said, about
400 tons of roots are weekly converted into sugar, and it is
further added, that " this is now the third year of the Lavenham
" factory and of the growth and manufacture of English beet
" root sugar on a considerable scale—a scale equal in extent to
" that of continental factories." It may likewise be here stated
that the cultivation of the beet for the manufacture of sugar

appears now to be seriously engaging the attention of a number of the farmers of this country, and a most useful and instructive paper upon " The cultivation and use of sugar beet in England " has lately been read before the Society of Arts by Dr. Voelcker, and will be found in the Society's Journal, 10th of March 1871.

Dr. Voelcker has evidently bestowed much thought and labour upon this subject which appears to have puzzled the British nation more than it seems to have done our continental neighbours, for that gentleman says, "that there are at present over " 500 beetroot sugar factories and distilleries in France, nearly " 200 in Belgium, about 300 in Prussia, and a goodly number " in Austria, Russia, and other continental states," and finally he concludes by saying that "probably the number of continental " beet root factories and distilleries does not fall much short of " 2,000, and according to reliable reports most of them do a " lucrative business."

It may be said that the processes for the production of sugar from the juice of beets are materially the same as those for producing sugar from the juice of the sugar cane.

Grape sugar, or as it is now frequently called *glucose,* although it is also known under a variety of other names, such as starch or fruit sugar, &c., &c., was first recognized in 1792 by Lowitz as a distinct species of sugar, differing in many of its properties from cane sugar, as for instance, it does not so readily crystallize as cane sugar, and when it does crystallize, it does so in confused masses ; it is much less soluble in water, cane sugar requiring for its solution only about one-third of its weight of cold water, whereas grape sugar requires about one and a half times its weight, and it is also much less sweet to the taste. Cane sugar likewise differs in other respects from grape sugar, as for instance, in forming salts with the alkalies and alkaline earths, and also as will be shewn further on.

Grape sugar although found in a great number of substances, especially fruits, is not known to exist in any one of them in such quantity as to render its manufacture from fruits of any value; but chemistry has stepped in, and supplied this deficiency, and the production of grape sugar or glucose, by acting upon starch or such like substances with a dilute acid and heat, has now become a manufacture of some considerable importance, especially on the continent, and this has arisen partly from the fact that this article can be brought into this market (as nearly as may be) colorless and without any flavour but its sweet taste, and these properties make it highly valuable for confectionary purposes, and likewise from the important fact, that, malt liquors to which a portion of cane sugar has been added do not keep so well as those to which a like portion of grape or glucose sugar has been added; consequently it has now come into considerable favor with brewers who are in the habit of using sugar along with malt in the brewing of their beers.

One of the best modes of distinguishing grape sugar from cane sugar is by adding to the solution of sugar to be tested in a flask a few drops of an alkaline solution of tartrate of copper and gently boiling; if any grape sugar is present a bright red metallic looking precipitate of suboxide of copper is shortly thrown down from the liquid, if no grape sugar is present no red precipitate takes place, and the solution remains clear with a slight blue tinge from the addition of the copper solution.

In a solution of a mixture of the two sugars a quantitative examination may thus be made; the grape sugar is first estimated in one portion of the solution of the sugar by a standard solution of the alkaline tartate of copper, and the quantity noted; then another portion of the original solution of the two sugars is boiled for a short time after having added to it a few drops of an acid, say sulphuric acid; this boiling with the acid converts the cane sugar that may have been present into grape sugar, when the whole of the sugar is again estimated, by adding the copper solution from time to time to the boiling solution so long as a precipitate takes place, the increased quantity of the copper solution required over the first experiment will give the amount of cane sugar in the mixture.

The value of this test will be better understood when it is stated that, owing to the mode in which sugars, but especially those known as low sugars, are manufactured, they contain in many instances grape sugar; and likewise from the fact, that large quantities of grape sugar or glucose are used for the purpose of adulterating brown cane sugars, thus improving their color, but at the same time deteriorating their sweetening or saccharine properties to a material degree.

In the process given above for testing the value of samples containing both cane and grape sugar it will be noticed that cane sugar is converted into grape sugar by boiling its solution with a few drops of an acid, sulphuric acid recommended, and this action of acids upon cane sugar has formed the subject of a patent which will be found abridged in this volume (Garton's patent). The product so obtained is manufactured on a large scale by Messrs. Hill & Co., of Southampton, and sold under the name of "saccharum;" it has found considerable favour with a large number of brewers in England, who employ a portion of it along with the malt, and by so doing are said to obtain from this mixture a malt liquor of a more permanent character than when brewed with malt alone or a mixture of malt and any cane sugar, however pure the cane sugar may have been.

If the process has been properly conducted the product should be grape sugar or glucose, with a certain per-centage of water, but in many samples which have been examined from time to time very few have been found in which the cane sugar has been perfectly converted, and most of the samples have exhibited a considerable per-centage of unconverted sugar.

SUGAR.

SUGAR.

A.D. 1663, February 4.—N° 141.

WILLOUGHBY, Francis, Lord, HYDE, Lawrence, and
" DE MARCATO, David.—Makeinge and frameinge of sugar
" mille."

[No Specification enrolled. Letters Patent printed, price 4d.]

A.D. 1691, August 22.—N° 271.

TIZACK, John.—"A way by an engine to be worked by one
" or more men for the well and more easy oyling and dressing of
" leather and cloath." This apparatus, it is said, "may be of
" great use to all such as doe worke at those and some other
" trades, and may be also applicable to the raiseing of water,
" washing of cloathes, milling of sugar canes, pounding of,
" mineralf, and pounding and bruising of all sortf of seedf,
" pounding charcoale to make powder of, and pounding and
" making raggf fitt to make paper and the like."

[No Specification enrolled. Letters Patent printed, price 4d.]

A.D. 1692, March 24.—N° 310.

SMITH, Anthony Forester.—"Boyling and heating waters
" and all other liquors, as alsoe for melting and refineing sugars
" and all other things that are done by fire, with lesse charge
" and greater expedition than is now used in our dominions,"
and also "communicating the heat of a small quantity of any
" hot or boyleing liquors in a short time to any greater
" quantities."

[No Specification enrolled. Letters Patent printed, price 4d.]

A.D. 1721, July 12.—N° 433.

HARDING, William.—"Sugar mills, engines, and wormes."
These, it is said, have hitherto suffered by imperfections from
" being chiefly made with large timber and wooden coggs, only

S.   A

" having a case of iron on the timber and an iron gudgeon
" through them, which often breaks by the imperfeĉcon of the
" rollers." Now models have been made of " sugar mills, engines,
" and wormes," the "rollers and coggs, and gudgeons, whereof
" are all iron, cast and wrought in a different manner and from
" all those now in vsd, and such equal proporĉons and dimensions
" that if by any accident a gudgeon or any other part of the
" rollers should be disordered it may be repaired without the help
" of any other workmen than those about the mill, for one-eighth
" part of the charge it vsed to cost, and be set to work againe
" in four or five houres time; the engine for supplying the said
" mills with water having an iron pinion or small wheel of a
" peculiar form and manner, with works circular within an iron
" double or endless rack, toothed all round, so contrived within
" as to command one sucker or forcer, performing both offices at
" the same in a single barrel or cilinder, making a purchase of
" any length required, which being wrought by the power of the
" mill, or otherwise, will supply all the works with a continual
" stream of water without intermission, and the wormes being
" cast and made on a core of mettle, so particularly contrived as
" to be taken out after the worms or pipes are cast thereon, with
" much more ease and conveniency than any ever before made
" vse of by any artificer whatsoever, by which meanes the wormes
" and water pipes will be compleatly finished without sorder or
" being bent after casting, as those now in vse generally are, and
" being kept cool by a constant supply of water will prevent
" yearly the loss or spoiling many thousand gallons of rum, and
" by removeing the waters from those places where it super-
" aboundẹ to places where it is extremely wanted will greatly
" encrease the value of many hundred thousand acres of land in
" all those our dominions ih America."

[No Specification enrolled. Letters Patent printed, price 4d.]

A.D. 1766, November 21.—N° 862.

BARCLAY, ALEXANDER, and YONGE, JOHN GREENHILL.—
" A new method of constructing sugar mills by the application of
" friction wheels to diminish the resistance arising from friction."
This consists in " applying the said friction wheels in such manner
" as that the axis of one wheel revolves on the periphery of
" another."

[Printed, 4d. No Drawings.]

A.D. 1773, August 5.—Nº 1051.

MELVILL, John.—"A new machine or stove engine, which " can be made of any form whatever, for the more easy, cheap, " and expeditious mode of either boiling sugar, soap, or other " articles which require to be boiled in large vessels, or distilling " all kinds of liquors." No description is given of the above apparatus, but there is a drawing, in which are set forth a number of different shaped vessels, and in which three of these are united together and appear to be bowl-shaped, having each a jacket; these jackets communicate the one with the other, and contain in their lower part hot water; the water in the first jacket is heated by a fire underneath, any steam escapes by a pipe fixed close to the top of the jacket of the last vessel.

[Printed, 10d. Drawing.]

A.D. 1773, November 7.—Nº 1056. (* *)

CHRYSEL, Christopher.—"Method of constructing and set- " ting boilers of any dimension for the use of fire engines, salt " works, brewhouses, distillers, soap-houses, sugar-houses, and " sugar works." This consists, first, in changing " the boilers, " coppers, and pans, containing four or more feet deep of water " or liquors," "into shallow boilers of about one foot in depth " and of a convenient length and breadth," to hold the same quantity, " or double or treble the quantity, which will make no " difference in the quantity of fuel required." In fire engines " the shallow boilers must be about two feet and an half, whereof " half a foot is assigned to the depth of the water, and two feet " for the confinement of the steam." Second, " setting the said " shallow boilers upon a small furnace, and serpentine flue, " contrived in such manner that the fire goes out of the furnace " in full flames under the whole bottom of the shallow boiler for " about a hundred feet, ascending to the top of the chimney." There are two regulators, " one for regulating the access of the " free air, and the other for regulating the exit of the fire flames " and smoak."

[Printed, 4d. No Drawings. Rolls Chapel Reports, 6th Report, p. 162.]

A.D. 1773, December 18.—Nº 1057.

FLEMING, John.—"An entirely new and particular kind of " machine or engine for pressing sugar canes and squeezing the

" juices therefrom, which actuated by wheels upon a peculiar and
" much more simple construction than any invention of the like
" nature hitherto found out, and, as I humbly conceive, would
" be much more useful and advantageous to the publick than
" any invention or discovery of the like nature." This consists
of two rollers fixed perpendicular, between them is " the spindle,
" a center roller," these are all cased with iron, "and are turned
" ⅛ of an inch hollow to prevent their touching and wearing
" smooth when the cane is taken in ;" there are "two wheels of
" the same iron fixed on the head of the rollers. The inside that
" goes over the wood is in an octagon form; they have four
" tenons that project two inches from the center of every other
" square of the octagon. These tenons keep the wheels firm
" when let in the wood. There is a set of coggs on each of these
" wheels which connects with a set of wallerers that are placed on
" a wallerer wheel that goes over the spindle, and is brought
" down to answer with the roller wheels." There is a small
roller fixed close to the "wallerer wheel," for " passing the cane
" after it is once squeezed to the roller that squeezes it a second
" time. This small roller is turned by a set of coggs on the lower
" part of the wallerer wheel that is fixed on the spindle. There
" are eight iron plates, two inches shorter than the iron cases,
" fixed in the small roller, which project about three inches,
" The rollers must be placed so as the plates may come within
" four inches of the center roller and within two of the roller that
" passes the cane a second time, there is an iron pivot at each
" end of this roller which revolves in two brasses fixed for that
" purpose." A piece of wood is placed oblique from the roller.
" The cane first passes to the plates in the small roller. This
" piece of wood directs the canes to the plates which takes and
" forces it between the rollers which squeeze it a second time."

[Printed, 8d. Drawing. Rolls Chapel Reports, 6th Report, p. 139.]

A.D. 1774, January 24.—N° 1061.

FORDYCE, George.—" Preparation of blood, which is prepared
" so as to preserve those qualities which will render it useful for
" the making of sugar for a great length of time, and in any
" climate." This consists as follows :—" Take the blood of oxen,
" or any other animals, separate from it the gluten, and to the
" remainder add residuum of ether in greater or smaller quantity,

" according to the state of the blood, so as to check putrefaction
" during the operation. If the blood is in such putrefiscent state
" as to require it, then evaporate the aqueous particles in a regu-
" lated heat, and for use dissolve the preparation in water, and
" apply it in the same manner as fresh blood would be applied."

[Printed, 4d. No Drawings. Rolls Chapel Reports, 6th Report, p. 162.]

A.D. 1784, April 17.—N° 1428.

BOUSIE, WILLIAM.—"The refining of sugar, and the making
" of sugar from the cane juice, and by means of which muscavado
" or brown sugar may be made to yield a considerably greater
" proportion of refined sugar, and better in quality than can be
" obtained by the methods usually practised."

[No Specification enrolled.]

A.D. 1784, August 20.—N° 1448.

MURRAY, ROBERT.—" For the refining of sugar and the
" making of sugar from the cane juice, and by means of which
" muscavado or brown sugar may be made to yield a consider-
" ably greater proportion of refined sugar, and better in quality
" than can be obtained by the method usually practised." For
making sugar from cane juice the juice from the mills is immediately run through linen into two or three copper cauldrons with
flat bottoms, from 60 to 66 inches diameter of a certain weight,
thickness, and height and made of three sheets of copper jointed
and soldered in a certain manner, to a depth of 12 or 13 inches
only, when it is quickly, boiled, skimmed, and concentrated until
drops allowed to fall form small sheets, when the fires under the
cauldrons are put out or suppressed, the liquor from two or three
cauldrons are passed through a woollen strainer fixed on wicker
work into one or more cauldrons, called boilers, about nine inches
deep ; when it is again boiled until it is " ready to crystallize in
" large distinct grains." The fire is then covered or extinguished
and the mass put into a wood cooler five feet square with an
elevation of eight inches lined with pure white metal " similar to
" Smith's patent metal," in which it is frequently stirred with a
wooden spatula until crystallization is completed when the mass
is placed in a draining trough of wood four feet long, three wide,
one high, with three rails in the bottom, on which is placed an
osier hurdle ; under this is a trough for the drainings which are

treated again as above. The sugar in the trough is removed into hogsheads, as raw sugar, or clayed as afterwards described when refining raw sugar. " If any animal syrup can be procured (and " whites of eggs are preferable to any other), an admixture thereof " in the cauldron will assist in getting rid of the extractive part " and render the raw sugar most beautiful."

For refining raw sugar it is put into cases or boxes to the height of about eighteen inches, levelled without any compression, covered with about three quarters of an inch of white clay the consistence of thick pap. When the water from the clay has filtered into the mass of sugar the dry clay is removed from the surface and a second coating of the same clay about an inch thick is applied. When "the white clay begins to retreat and " rack upon the surface, from 100 to 106° F. of heat is intro- " duced into the warehouse," "this dissolution infilters itself " through the mass into a recipient below such as was used in the " former process for boiling the cane juice only deeper. The " clay when dry is removed and the heat continued until the " draining ceases. The purified or bleached sugar is taken from " the case." The black and bitter molasses no longer suscep- tible of chrystallization may be fermented for " spirits fit for " varnishing but which is disagreeable to the palate if drank." " The raw sugar clayed, purged, or purified by this process is " then in a proper state to be refined, an operation which "nearly " resembles that which is commonly practised by refiners; it " differs only in not making use of lime to dissolve the sugar, " according to the usual process, but of common pond or river " water, and in the clarifying with white of eggs in place of " using bullock's blood which is the practise of the present " refiners."

[Printed, 4d. No Drawings. Repertory of Arts, vol. 4, p. 289.]

A.D. 1785, July 27.—Nº 1492. (* *)

WOOD, SUTTON THOMAS.—" Distilling, rectifying, refining, and " preparing spirits, oils, sugars and salts, and other substances " and solutions, by the power and application of steam, and " certain discoveries in the application of steam to the carrying " on and to the various purposes of trades and manufactures." This consists "in making use of steam instead of fire." Some modes of doing this are shown: in one instance a steam jacket

surrounds the still; in another the vessel or still containing the
liquor to be distilled surrounds a boiler. Hot plates and ovens
on the above principles are described, but it is stated it is of no
consequence in what shape or form the vessels, &c. may be made,
" or in what manner the heat of the steam is communicated,"
" provided the effect is produced upon the principle aforesaid."

[Printed, 1s. Drawing. Carpmael's Reports on Patent Cases, vol. 1, p. 182;
Rolls Chapel Report, 6th Report, p. 172.]

A.D. 1786, October 6.—N° 1559.

REEDER, JOHN.—"New invented coppers or vessels." This
consists of a copper vessel for clarifying cane juice with a slightly
convex bottom, a pipe with a tap is fixed just at the bottom "for
" conveying the liquor from the vessel instead of leading it out
" from the surface as heretofore used." Across the part of the
pan from which the pipe is led is "a strainer or door to prevent
" any of the scum or filth from passing through the pipe with
" the pure liquor." In "the granulating copper or vessel in
" which the last operation of boiling the sugar is performed the
" contents are discharged through a pipe fixed in some convenient
" part of the bottom of the copper or vessel in which the same is
" boiled."

[Printed, 6d. Drawing. Rolls Chapel Reports, 6th Report, p. 175.]

A.D. 1791, March 3.—N° 1794.

SHORLAND, WILLIAM.—"New invented mill, machine, or
" machinery calculated to prevent mills and all kinds of works
" drove by water from being flooded or impeded by back or tail
" water, through the operation and effect of double power over-
" shot wheels and floodgates, and for grinding, &c. of every
" species of grain, foreign or domestic produce, and the powder-
" ing and whitening of every sort of sugar, such mill, machine,
" or machinery to be worked either by water, wind, manual
" labour, steam, horses, or any other cattle." These are, first,
" the overshot wheels to be made of any height or width the
" power of the work may require," and made "of wood, iron,
" brass, copper, or any composition of metals," and fixed upon
suitable bearings. Two or more of these wheels are fixed as
required with "spur or cog wheels to be fixed one on each side"
of each water wheel, working "in each other in the usual way

" of wheel turning another." These wheels have a flood hatch
or hatches, gate or gates, " at the head of the thorough or mill
" pond, to let off or discharge the overflow or waste water."
" This water or waste water is let to run under the overshot
" wheels by means of drawing up the flood " hatch, hatches, gate,
or gates.

Second, in grinding using " English, Welsh, or foreign " stones ;
if metal be used, " iron, steel, brass, copper, or lead, or any com-
" position of metal or metals, wood and stone put together in
" the form of stones," shaped and fixed as " for grinding all
" sorts of edge tools." The bed stones are of the above materials, and "made or shaped in a concave form." The face of the
runner for grinding is solid or hollow ; if hollow, of " the form
" or shape of unshrouded water wheel, with a ball or globe of
" iron, or any sort of metal to be put in the inside, between the
" arms, to help and accelerate the motion of the machine."

[Printed, 4d. No Drawings. Rolls Chapel Reports, 6th Report, p. 184.]

A.D. 1794, October 30.—N° 2019.

COLLINGE, JOHN.—" Invention and improvement on sugar
" mills, which will occasion less power to work them and are not
" liable to be out of order." This consists " in constructing the
" step blocks to receive the bottom gudgeons of the mill in any
" manner," so that they shall be immersed in oil or grease, and
" by providing a recess below the bearing points or cappooses, in
" any manner so that the foul oil or grease shall escape beneath
" them." To avoid taking down the mill cases, &c. to cleanse
the stop block an aperture capable of being closed by a plug or
screw is made in the lower part of its recess to let out the foul
oil or grease. "The head blocks which receive the upper
" gudgeons are to be constructed in any manner so that they
" will receive cotton, wool, or any other suitable elastic material "
that will emit oil or grease when supplied either from recesses
around above that part of the gudgeon which works therein.
" The step and head blocks should be fixed in the framing as
" near the ends of the mill cases as possible to render the
" gudgeons shorter, which will prevent in a great degree their
" being broke. These blocks may be made in any external form
" that is most suitable to the framing of any mill to which they
' may be applied."

[Printed, 4d. No Drawings. Rolls Chapel Reports, 6th Report, p. 188.]

A.D. 1801, June 2.—N° 2506.

WAKEFIELD, Thomas.—"A new art or method of refining
" sugar." The sugar to be refined "contained or inclosed in a
" cloth felt vessel or covering either having pores or holes
" therein," is placed in a convenient place and pressure is applied
" by weight or by rollers or by the screw or wedge, with the help
" of a steam engine, mill, or power capable of acting by way of
" pressure." Such sugar previously containing moisture, the
moisture or a part thereof passes from the sugar and out of or
" through the said covering, and thus a part of the colouring
" matter and of the impurities of the sugar will be expelled, and
" the degree of refinement procured will be proportioned to the
" quantity of pressure applied and the quantity of moisture
" expelled or squeezed out of the sugar." " From the moisture
" or substance thus forced out there may be extracted some sugar,
" molasses, and rum or other spirits by the usual process."
" The sugar refined as aforesaid may be further or otherwise
" refined and manufactured by the methods now in use or by
" other means."

[Printed, 4d. No Drawings. Repertory of Arts, vol. 16, p. 1.]

A.D. 1802, August 2.—N° 2639.

WYATT, Charles.—"Certain improvements in the apparatus
" for and mode of distilling, drying coffee and sugar." These
are, in reference to this subject, in drying sugar two cylinders
having handles by which they may be revolved contain the sugar,
&c. These are suspended in an oblong square-shaped box,
" steam from a closed boiler being introduced " through a pipe
into the cavities between the cylinders and the square box, passes
off through a tube at the other side of the box, and in its passage
it warms the air which is supplied into the cavities by openings
in the box, which openings are also for the discharge of the cylinders. " By the revolution of the cylinder placed in the current of
" air thus warmed, the substances within it are dried."

[Printed, 10d. Drawing. Repertory of Arts, vol. 2 (second series), p. 9.]

A.D. 1805, July 8.—N° 2866.

BATLEY, Benjamin.—"A new and improved method of refin-
" ing sugar." First, "charge the sugar pans with the usual

" quantity of lime water, and for each ton weight of sugar to be
" refined, allow at the rate of ten gallons of skimmed milk,
" more or less, as may be necessary, according to the quality of
" the sugar; mix five gallons of such milk with the lime water,
" then skip the sugar, and after the sugar is skipped let it remain
" in the pans till next morning, when the whole, so mixed, is to
" be stirred together, and after taking off the first scum of the
" sugar, add more skimmed milk, and continue to repeat the
" same after each scum of the sugar, until the liquor is perfectly
" cleared."

[Printed, 4d. No Drawings. Repertory of Arts, vol. 8 (*second series*), p. 17;
Rolls Chapel Reports, 7th Report, p. 190.]

A.D. 1805, November 26.—N° 2899.

INGRAM, JAMES.—"A new method of manufacturing powder
" sugar from raw sugar alone, and from syrup of sugar alone, and
" the mixture of raw sugar and syrup of sugar together." First,
" making powder sugar from raw sugar alone." Mixing at the
rate of about seven pounds of raw sugar with one pint of water,
or of lime water, " of the same strength usual in refining lump
" sugar," boiling this in a metal kettle until the stem of a tobacco
pipe, or any other thing that will answer the purpose, dipped
into the boiling sugar, and then put in cold water, the sugar
can be taken off from the pipe, &c., breaking brittle, it is sufficiently boiled; it is thrown on a slab previously greased or oiled,
and in a few minutes, until it begins to harden, it is rolled into
two or more rolls. These "rolls are drawn with your hands
" repeatedly over an erect iron spike or other proper thing " until
the sugar whitens and becomes stiff, when it is taken off the
spikes, and after six hours or more, when quite hard and firm, the
lumps are stoved like lump sugar for two or three days, and when
they are quite dry they are ground in a sugar mill, the powder
passed through a " fine wire or other proper sieve," returned to the
mill and mixed "with bastard or raw sugar, in the proportion
" of fourteen pounds or thereabout to each hundredweight of
" powder sugar or thereabout; to condition it in the usual way
" bastard sugar is conditioned; the powder sugar will then be fit
" for sale."

Second. "Making powder sugar from syrup of sugar alone."
Proceed with it as above, " except as to the mixture of water
" which is unnecessary."

Third. " Making powder sugar from the mixture of raw sugar " and syrup of sugar together." Mixing these substances toge- ther in equal weights, or thereabouts, and proceeding as above.

[Printed, 4d. No Drawings. Repertory of Arts, vol. 9 (*second series*)' p. 170; Rolls Chapel Reports, 7th Report, p. 188.]

A.D. 1806, August 20.—N° 2958.

CURTIS, James, and ROBBINS, Josias.—" Certain improve- " ments in boilers for manufacturing sugar, and in the mode of " fixing the same, whereby much labour and fuel will be saved." These are as follows :—The vessels are of an oblong shape placed in a row; the bottom of the clarifiers is just higher than the skim- ming copper; " the bottom of the skimming boiler is placed a " little above the top of the evaporating boiler, so that by opening " a cock or other proper convenience it will feed the evaporating " boiler. The evaporating boiler is fixed a little above the top of " the taiches, and has two cocks to feed the two taiches." " The " two taiches have a cock or other proper conveyance in each " which discharge the sugar into ye coolers, and under the taiches " are damper plates " to prevent burning of the sugar at time of discharging. All these boilers are fixed in cast-iron or other metal curbs. The fires are " under each taich, and from them conveyed " under the evaporating boiler, from that to the grand or great " skimming boiler, and from that to the clarifiers (if wanted), but " when they are not required the flame is conveyed into or up the " first chimney by means of a damper." There is a " fire-place " for the clarifier only. The fire acts entirely upon the bottom of " these vessels, by which means they are less liable to be burnt or " are worn out and prevented from corroding."

[Printed, 8d. Drawing. Rolls Chapel Reports, 7th Report, p. 192.]

A.D. 1807, March 7.—N° 3021.

NEWMAN, Henry Charles Christian.—" A cattle mill for " expressing the juice of the sugar cane." There is " a ring or " circle of hard wood, stone, or cast-iron, either raised on arches " or otherwise, or sunk below the surface of the ground, and com- " monly called in the West Indies a Pitt Mill, which is to be in " the centre of the aforesaid ring or circle;" there is a cog wheel on the spindle of the mill with a socket on the top of the spindle and precisely in the centre of it. A gudgeon lets down into the socket and turns on a steel plate at the bottom of it; a horizontal

shaft or lever passes through an eye or ring in the gudgeon to the
further extremity of the cog wheel; at the bottom of this horizontal shaft or lever is a heavy wheel. Over the eye or ring in the
gudgeon is a lantern wheel or pinion also firmly fixed on this end
of the lever. To this wheel or pinion is attached "a pole, on each
" side of which one or more horses are harnessed, which pole has
" a collar in which the lever turns."

[Printed, 6d.   Drawing.   Repertory of Arts, vol. 12 (*second series*), p. 348;
Rolls Chapel Reports, 7th Report, p. 196.]

A.D. 1809, September 23.—N° 3261.

VAUGHAN, George.—" Certain improvements in the process
" of refining sugars." There is a floor on which a series of tubes
are placed for a day's work of sugar; in these tubes are placed a
number of thimbles. The moulds of the ordinary sugar loaf
shape, filled with boiled sugar, are placed in these thimbles "with
" a collar of leather on the nose," "which is to keep it air-tight."
The cover of the mould is made to fit in a groove, round which
there is some putty placed to keep it air-tight." The series of
tubes above named are all connected with a horizontal tube which
is connected with receivers, which receivers are again connected
with an air pump, by the working of which and by certain cocks in
the tubes of the apparatus being open and others shut, the syrup is
drawn from the moulds into the receivers. When the sugar has to
be clayed the communication with the air pump is cut off; "take
" off the covers and put on the clays immediately in the usual
" way; let this clay stand on for four days if for double-refined
" sugars, and longer if for singles or others; at the end of which
" time take the clays off and put the covers on and work the air
" pump;" cut off the connection with the air pump, and immediately clay it a second time and proceed as before. Before
claying for the third time, the sugars are knocked "out of the
" moulds to know what kind of clay is necessary to put on," and
the sugar is finally stoved. The moulds, in preference, are of
block tin, but may be of "tin and zinc, or tin, zinc, and lead
mixed;" they "are cast in iron or brass moulds, and may be
" made in the usual form, with a groove or without, and may be
" put to stand in funnels instead of thimbles," but thimbles are
preferred made of " tin, zinc, and lead, or any two of those metals
" in equal proportions, and cast in a mould," or three of tin to
one of zinc, or three of tin, one of zinc, and one of lead. "The
" tubes on and under the floors may be made of ,tinned plates

" or any other metal, as also the receivers, all of which are to be
" painted outside."

[Printed, 10d. Drawing.]

A.D. 1810, May 17.—N° 3338.

BELL, JAMES.—" Certain improvements in the manner of refining
" sugar, and of forming sugar loaves of a certain description."
These are, first, instead of placing the moulds upon pots, "into
" which the syrup drops from the moulds containing the sugar,"
" placing trunks or gutters under the moulds in such manner as
" to receive the syrup dropping from them, and to convey it into
" cisterns, from whence it may be again conveyed into pans."
These trunks, gutters, and pipes, to convey the syrup into cisterns,
may be formed of any substance which will not injure the sugar,
as wood, pottery, artificial stone, metal, or " partly of one and
" partly of another." " They should be placed in an inclining or
" slanting direction with an inclination sufficient to make the
" syrup run into another gutter placed under their lower extremity
" so as to conduct the syrup into pipes communicating with cis-
" terns appropriated to different syrups, which may be conveyed
" with pleasure from these cisterns into the pans by means of
" pipes and stopcocks."

Second, " the manner of forming sugar loaves." " The chief
" difference " in this case between this " method and that now
" followed consists in the difference of the mould." The mould
instead of being plain " is fluted or striped, or has figures, orna-
" ments, or devices in the inside of it." " No particular direc-
" tions are necessary for making the moulds," but in order to
have " the fluting or stripes of the figures, ornaments, or devices
" cut, engraved, or otherwise made and polished in the inside,
" the mould should be made in two pieces," which can be soldered together or may be joined together by " douls or by hinges
" and clasps." The bottom of the loaf has any " mark, letter,
" name, ornament, figure, or device " upon it the sugar refiner
may choose.

[Printed, 6d. Drawing. Repertory of Arts, vol. 19 (second series), p. 70.]

A.D. 1811, March 6.—N° 3407.

WILLIS, ABRAHAM.—" New method of producing steel toys of
" different descriptions, such as barbers' curling irons, sugar
" nippers, snuffers, and other articles."

[No Specification enrolled.]

A.D. 1811, March 26.—N° 3425.

BELL, James.—" Certain improvements in the manner of cut-
" ting, shaping, or scraping sugar loaves, and lumps, and of
" pulverizing and reducing to small grains or powder sugar
" loaves, lumps, and bastard sugar." These are, first, " applying
" to the sugar knives or blades set in a moveable frame, which may
" be put in motion and made to act against the sugar by any of
" the common powers now in use." The frame " containing the
" knives or blades, may, by means of screws, be placed less or
" more obliquely at pleasure, so as to cut the sugar of a smaller
" or larger grain if wanted." This wheel, if moving horizontally
above the sugar " is fixed upon a perpendicular rod, which passes
" upwards and downwards at pleasure through an iron cylinder
" that moves horizontally in the frame that supports the machine.
" This cylinder is turned by an iron tooth bevil wheel, which is
" placed on it, and is acted upon by another iron tooth bevil
" wheel turning vertically, assisted by a fly wheel. The machine
" is constructed on and supported by a frame, in which is a slid-
" ing bar acting upwards and downwards in grooves in the frame,
" assisted by friction wheels at each end." The wheel containing
the blades or knives, suspended by a pulley, is let down on the
sugar, secured on a board by prongs, a central prong in the wheel,
lays hold of the sugar, the fly wheel turned, the sugar is " shaved
" down into grains."

Second, fixing " the knives or blades, either transversely or
" straight, on a moveable cylindrical frame, placed parallel to the
" floor, and moving vertically on pivots, to one of which is fixed
" a cog wheel, acted upon by a lesser cog wheel to which motion
" may be given by a handle or any of the powers now in use."
Fronting the knives or blades, and close to them, is fixed " a
" board or tray, with a steel plate on one edge thereof. At the
" opposite edge of the tray is placed a sliding bar, which by
" means of chains or cords fixed to it and wound round friction
" pullies attached to the spindle or shaft of the small cog wheel,
" pushes the slide forward to the blades, and it stops when it
" reaches them by means of a catch box which disengages the
" friction pullies." The loaves or lumps of sugar " intended to
" be shaved, being placed in the tray are pressed on to the knives
" or blades by the sliding bar, and the sugar is cut in grains,"
smaller or larger by the friction pullies being of a " lesser or

" greater diameter." The same moving cylindrical frame may
be placed perpendicularly or obliquely.

Third, placing " knives or blades on a wheel in a similar manner
" to that described in the first-mentioned method," but moving
" vertically with a shaft on which is fixed a cog wheel acted upon
" by a lesser cog wheel moved by a fly wheel and handle. At
" the side of the wheel on which are fixed the knives or blades
" is a long trough, placed obliquely close to the wheel, and pro-
" ceeding from thence by an angle upwards." The loaves or
lumps of sugar being put into the trough " press forward by their
" own weight on the knives and blades."

[Printed, 10*d*. Drawings.]

A.D. 1812, February 27.—N° 3541.

CONSTANT, Louis Honore Henry Germain.—" A new
" method of refining sugars." First, preparing charcoal of wood
by washing it with water, grinding it up coarsely in a mill or
otherwise with a little water, then very finely in a mustard mill,
&c., washing it by decantation, filtration, or otherwise, forming it
into blocks " about the size of a large cheese," and drying " the
" same in the sun or by a moderate temperature, after which the
" same may be kept for use in casks or other fit packages."

Second, " clarifying or refining muscovado or other clayed or
" soft sugars." The sugar is dissolved by boiling and stirring in
a vessel until the solution marks a certain specific gravity, in preference, by a hydrometer, the graduation of which is described,
then adding to the boiling liquid according to the quality of the
sugar in solution from 5 to 10 lbs. of the above prepared charcoal
for every hundredweight of sugar, stirring and mixing it well in,
allowing the whole to repose for a short time, then boiling up as
speedily as possible, and when near boiling over, adding with
agitation the " usual finings of white of eggs, or blood or other
" albuminous material," boiling up to rise the coagulated albumen in the form of scum, then allow the whole to remain at rest
at a very gentle heat, skim off the charcoal, and carefully filter
the syrup. The charcoal obtained from the skimmings is agitated
in boiling water, thrown on a filter, and washed with hot water,
making " use of this water in the subsequent processes of solution
" and clarifying of sugars." In constructing furnaces for the
above purposes, not only using doors and registers to the grate,

&c., but also constructing and using a plate of metal or other fit material that can be slided or moved in and out of the fire-place, and be suddenly interposed when required between the bottom of the boiler and the fire; this is used when skimming off the charcoal, &c.

Third, in " refining of sugars in lumps, pieces, or loaves, instead " of the old method of clarifying," effecting and performing the same " by gradual percolation of my purified syrup cold through " the said sugars in order to clear out the coloured syrup or " molasses which occupies the interstices between the crystals of " the sugar at the first formation thereof." The quality of the sugar to be treated determines the strength of the syrup to be used, and likewise its quality. In bleaching or refining loaves of sugar, the upper part of the loaf commonly called the fountain is cut off until the sugar appears solid and firm, the loaf is turned upside down with its mould, after an hour or so is struck from the mould, and a piece of rag plugged in the point of the mould. the loaf is returned to the mould point downwards, as fairly up-right as possible, and afterwards pour on a due quantity of the white purified syrup, and in about 24 to 48 hours the plug is taken out, and the yellow or dark coloured syrup flows out and becomes replaced by the white syrup. It is practicable by this method to bleach and refine all sugars without turning or agitating them, or putting "in a plug as before described, but that in this " case there would be danger of spots and irregularity of colour " near the point, and the loaf would be porous. Using & applying this percolated syrup to purify the lump or masses of sugar purified by the prepared charcoal in the boiler as above. Also syrups used in purifying by percolation sugars royal or of the finest quality, to purify in like manner very good sugars, and using again this last syrup in purifying the said lumps. In filtering the syrup before mentioned, it is "very convenient to support " the filter or filtering cloth upon or within a basket" made for the purpose. The first runnings (which are less clear) are returned to the filter again, as is usual "in operations of this or the like " nature."

[Printed, 4d. No Drawings. Repertory of Arts, vol. 21 (second series), p. 207, also vol. 27, p. 252; Webster's Patent Law, p. 24 (also p. 128, case 54); Carpmael's Reports on Patent Cases, vol. 1, p. 294; Parliamentary Report, 1829 (Patent Law), p. 198; Rolls Chapel Reports, 8th Report, p. 89.]

HOWARD, EDWARD CHARLES.—" A process for preparing and
" refining sugar." It is stated that "from the known fact that
" water dissolves the most uncrystallizable in preference to that
" which is the most crystallizable sugar," and also from having
discovered " that no solution (unless highly concentrate) of sugar in
" water can without material injury to its colour and crystallizing
" power, or both, be exposed to its boiling temperature during
" the period required to evaporate such solution to the crystal-
" lizing point," the following operations have been " established
" and adopted." First, raw or muscovado sugar is made into
" a magma of the consistency of well worked mortar," with
water at the temperature of the atmosphere, left for an hour or
more, then heated by preference from 190° to 200° F. in a vessel
with a steam or hot water jacket, more sugar is added, if required,
" in order to put the mass into such a state of imperfect fluidity
" that the same shall readily close behind the stirrer." The
moulds are then filled, and when the sugar has become cold in
the moulds the stoppers are taken out, and when the liquid has
nearly drained out, the upper surface of the lump is pared down,
" until the sugar presents a uniform and firm appearance." The
sugar so pared off is mixed " with cold water to a magma of such
" consistency that the same shall not readily close behind the
" stirrer," although a magma of a thinner consistency is preferred, and placed upon a uniform and firm surface, and when
" the magma becomes moderately dry, there is carefully poured
" on it a " saturated solution of fine sugar in cold water about
" half an inch deep." When the sugar is close grained, and the
surface hard, "an unsaturated solution of sugar, or even water
" itself, may be poured thereon without running it," but it is not
recommended, as it " requires too much nicety for general prac-
" tice." It is not "necessary that the identical sugar taken from
" the surface of the loaf should be used as magma or syrup
" thereon, but on the contrary," and sugar of a finer colour may
be employed.

Second, the lumps, loaves, or masses are drawn from the
moulds, and in the usual manner. "The net or good sugar " is
separated from that which retains molasses, this latter going to
be mixed up with raw sugar for operating with as above. The
former is refined as follows :—" Pouring upon it in any convenient

S.   B

" vessel 6 lbs. of water (by preference boiling) to every 5 lbs. of
" sugar, deducting about 6 per cent. for the moisture " in the
sugar. When dissolved, and the impurities subsided, the solution
is drawn off into a vessel, and further purified by what the
patentee terms "my ordinary or common finings," prepared as
follows :—2½ lbs. of allum for every cwt. of solid sugar are dissolved in about sixteen times its weight of boiling water, adding
to such solution about 70 or 80 grains of whiting for each pound
of alum ; the solution is drawn from the precipitated matter, and
cream of lime (passed through a sieve) is added to it until it shows
by turmeric paper very slightly caustic, the precipitate called the
finings, is collected on a blanket filter, and drained until the mass
begins to crack, when it is fit for use. The solution of sugar is
gradually added to the finings with stirring, until they are brought
to a uniform creamy state, and this mixture is poured into the
whole quantity of the said solution of sugar with agitation. The
whole is then allowed to rest for the night or so, when the clear
liquor is drawn off and evaporated by steam heat, or water at a
temperature of about 200° F., more or less, until the liquor is
about 1·37 sp. gr., when it is transferred to another vessel, stirred
until it becomes granular; it is then filled into the moulds, and
as soon as cold the stoppers of the moulds are removed, and when
the syrup is run the loaves are examined, and if necessary they
are pared down as before described. If the sugar appears sufficiently fine, it is removed from the moulds, the smaller ends
not clear of syrup are cut off, and the remainder dried. If the
loaf is not sufficiently dense or close in its frame, it is re-moulded
as is well known previous to drying by " stamping the grains into
" a metallic or other mould which will immediately re-deliver
" the same." To "keep and retain the point" of the loaf,
" without returning the syrup contained in that point upon the
" body thereof," a pipe is appended to the point of the mould,
and the lower portion of the sugar which will be in the said pipe
along with the redundant syrup, is removed when the pipe is
removed. The liquors left in the two cisterns are next dealt
with. To the one containing the gross dirt and impurities about
its bulk of boiling water is added, the liquor filtered through a
cloth, then add to it the other liquor containing the finings, and
then "abstract therefrom by washing and subsidence, all the
" sweet they contain," which sweet liquor is used for magmas.
" Inferior syrups may be advantageously mixed with muscovado

" sugar instead of water," as directed in the primary operation.
The loaves refined by the use and application of the before-
described finings "may be still farther refined by the use and
" application of other finings " made as follows :—3½ lbs. of
alum are used for every cwt. of solid sugar, and the process is
proceeded with as before, except in place of the milk of lime
being added, a strong solution of caustic soda is employed, the
precipitate washed is named the "finings."

[Printed, 6d. No Drawings. Repertory of Arts, vol. 23 (second series),
p. 129, also vol. 5 (third series), pp. 220 and 271; Webster's Patent Law,
p. 26 (also p. 137, case 125); Carpmael's Reports on Patent Cases, vol. 2.
p. 235; Mylne and Craig's Reports, vol. 1, p. 487.]

A.D. 1813, November 20.—N° 3754.

HOWARD, EDWARD CHARLES.—"A process for preparing and
" refining sugars." The first part of this process is said to be
an improvement upon the former invention in No. 3607 and to
consist "in the application of steam to the sugar at the same
" time that it is exposed to the action of the finings, and in there-
" after causing the impurities to separate by filtration or by
" repose."
Second, this is said to be an improvement upon No. 3607, and
to consist " in the application of the finings completely neutra-
" lized, when the process of filtration is to be employed and that
" when filtration is not employed it does consist in exposing the
" unrefined sugar successively to the two parcels of finings, the
" first whereof, in conjunction with steam," removes "part of
" the impurities chiefly by chemical agency, and the second
" whereof, by mere diffusion without steam, removes the residue
" of the impurities chiefly by mechanical agency, and is capable of
" being afterwards used as a first parcel in repeating this double
" process upon another quantity of sugar." The excess of
caustic lime should be removed when neutralizing the alum by
milk of lime " by the subsequent addition of alum, not more than
" sufficient to leave the colour" of the "turmerick paper"
unchanged when immersed therein, it is also adviseable to add
about 3 oz. of whiting to every 2½ lbs. of alum. The finings thus
prepared will answer the purpose of the finings described in
No. 3607, "for the superior degree of refinement in sugars."
Third, "the evaporation or concentration of the saccharine

" solution by the application of heat to a solution in a vacuum
" and the maintenance of the continued action of a pump or
" other exhausting instrument, although the common gage
" should indicate no change." There is " a syphon gage with
" the intervals of pressure or stations of the mercury set off and
" the corresponding boiling temperatures expressed in numbers,"
without drawing off any of the said solution. Although an
arrangement with a tube is shown by which a portion of the
solution may if required be conveniently drawn out without
forming an immediate communication between the boiler and the
external air. Having brought the syrup to the proper density
it is run " from the boiler into a granulating vessel, in the nature
" of that now commonly used as a cooler capable of being heated
" by steam. The temperature of the liquid is regulated while it
" undergoes the requisite or usual stirring to effect granulation.
" It is said that the crystalline grains have the most tendency to
" arrange themselves between the degrees of 150 and 160 F., but
" in practise it has been found advantageous to heat up the same
" as soon as it has acquired some grain to about 180°, and
" subsequently cool it to about 150°," " either by withdrawing
" the application of heat and stirring out or allowing the escape
" thereof, or by the addition of a due quantity of colder eva-
" porated liquor as it has been usual to give heat by what is
" called skippings, or by both, or by any other fit means." The
alternations of temperature are repeated until the best grain is
formed to the eye; finally before moulding, in preference, the
temperature is brought to nearly 200°, not higher. The moulds
are of a greater length and have not the pipe appended as
described in No. 3607, in order that the loaf may have as much
" of its pointed end cut off as will not part with its syrup."
This is done " by a revolving instrument resembling a chuck in
" the art of turnery, and made concave and provided with a side
" cutter within." The improvement upon the former process
which " relates to the granulation or crystallizing of sugar does
" consist in substituting for the method of skippings" the " ap-
" plication of different degrees of temperature alternately to the
" same mass of evaporated or concentrated saccharine liquor."

[Printed, 8d. Drawing. Repertory of Arts, vol. 25 (second series), p. 257,
also vol. 5 (third series), p. 340; Register of Arts and Sciences, vol. 4 (new
series), p. 41; Webster's Patent Law, p. 26 (also p. 137, case 125); Carpmael's Reports on Patent Cases, vol. 2, p. 235; Mylne and Craig's Reports,
vol. 1, p. 487.]

A.D. 1813, December 20.—N° 3771.

SUTHERLAND, John.—"An improvement in the construction
" of copper and iron sugar pans and sugar boilers, and a new
" method of hanging the same, and also an improvement in the
" construction of the furnace or fire-places in which pans and
" boilers ought to be placed." "The boilers are shallow and
" hung from a cast iron ring built into the brickwork," so
as to expose the whole surface of the boiler to the action of the
fire. Two boilers are shown contiguous to each other, and the
fire at one end of the first passes through flues underneath each.
The division between the boilers through which the fire passes,
is formed " of a hollow brick arch made in one or more pieces.
" The current of air passing through that hollow arch in some
" degree counteracts the action of the fire on the inner surface."

[Printed, 6d. Drawing. Rolls Chapel Reports, 7th Report, p. 115.]

A.D. 1815, May 8.—N° 3912.

MARTINEAU, Peter, the younger, and MARTINEAU, John,
the younger.—"A new method or methods of refining and
" clarifying certain vegetable substances." The articles employed
" for purifying and clarifying sugar," are, firstly, "animal char-
" coal, that is to say, animal substances properly burnt or charred
" or calcined," as "ivory black, bone ash, &c., and afterwards
" reduced into smaller pieces or powder."

Second, " bitumous earths, commonly called coals, either in the
" state in which are mined, or articles of their products after
" fusion, and reduced as afore-mentioned."

Third, " certain argillaceous earths known by the name of
" ochres."

Fourth, " vegetable charcoal, usually called lamp black." The
first articles named, however, are performed.

The manner of applying them " may be greatly varied," but
the following method is preferred :—The pans or boilers are filled
" with sugar and water, adding a little more water or lime water
" than in the common mode of refining." A quantity of the
substances before-mentioned "according to the quality of the
" sugar to be refined and clarified" is added; but preferring
" two to five pounds of charcoals or earths before-mentioned to
and for every hundredweight of sugar to be clarified and refined,"
adding " the usual finings of eggs, blood, or other albuminous

" matter in rather larger quantities than in the usual mode of
" refining," stirring, applying heat until the scum has completely
risen, and pouring " the whole of the liquid sugar and scum
" into and upon the usual or any other known filter." If the
first runnings are not quite clear, they are returned back into the
filter. This clarified solution is evaporated and granulated in the
usual manner.

> [Printed, 4d. No Drawings. Repertory of Arts, vol. 27 (*second series*),
> pp. 193, 252; Webster's Patent Law, p. 24 (also p. 123, case 54) : Carpmael's
> Reports on Patent Cases, vol. 1, p. 299; Parliamentary Report 1829 (*Patent
> Law*), p. 198; Rolls Chapel Reports, 8th Report, p. 105.]

A.D. 1815, June 22.—Nº 3933.

TAYLOR, JOHN.—" A certain method or methods of purifiying
" and refining sugar." This consists as follows :—For purifying
raw sugar, "it must first be brought to a moist state, and if the
" process be employed in the original manufacture in the West
" Indies, the degree of moisture at which the sugar will be upon
" draining a short time after it is taken from the coolers in which
" it is crystallized will be sufficient." But if the process is
" practised in this country on sugars as dry as they usually are
" imported, they will require to be mixed with a certain propor-
" tion of cold water or lime water," varying " according to the
" opinion of the operator and the quality of the sugar, and will
" readily be determined by trial, as no exact rule can be laid
" down for each case." In general, "the proportion of water may
" be from one-eighth to one-tenth the weight of the sugar."
They are well mixed, and subjected to such a degree of pressure
as " to express all the fluid part therefrom, which will be found
" to contain all the molasses and the soluble impurities, and a
" certain quantity of sugar in solution." Any apparatus for
carrying out the above may be used. " The sugar contained in
" the expressed syrups may be obtained therefrom by the usual
" processes of evaporation, and from its not being injured by the
" usual application of heat, is capable of being made into an
" inferior sort of refined sugar."

> [Printed, 4d. No Drawings. Repertory of Arts, vol. 28 (*second series*), p. 70;
> Rolls Chapel Reports, 8th Report, p. 107.]

A.D. 1816, February 3.—Nº 3978.

DRUKE, JOHN GEORGE.—" A certain method of expelling the
" molasses or syrup out of refined sugars in a shorter period than

" is at present practised with pipe clay." This consists as follows:—" Sulphates of lime, natural and artificial combinations
" of calcareous earths, with suitable acids or other fit substances,"
plaster of Paris preferred, is mixed with "any fit liquid, by pre-
" ference cold water," adding one pound of the plaster of Paris
to three pints of water, stirring it well, and when it is brought to
the consistency of a thickish paste, so that a finger dipped into
it to the depth of one or two inches below the surface, and drawn
along, the impression or channel caused by the "finger will re-
" main, or not be closed up readily." This substance is to be
used in the same manner as when using clay. In preference, a
piece of calico or pieces of damp calico are placed between the
upper part of the loaf of sugar and the above clay substitute.
" The operating mass and intermediate substance before-described
" may be applied with equal utility and success to sugars in
" whatever vessels or receptacles they may be placed for the
" purpose of being separated from the molasses."

[Printed, 4d. No Drawings. Repertory of Arts, vol. 29 (second series),
p. 321; Rolls Chapel Reports, 8th Report, p. 112.]

A.D. 1816, July 27.—N° 4048.

HAGUE, JOHN.—" Certain improvements in the method of
" expelling molasses or syrup from sugars." These are " either
" by withdrawing the air from the under surface, or by com-
" pressing the air on the upper surface of the sugar." "A trough
" or box (open at the top) either square, round, or any other
" shape," in preference, larger at top than at bottom, is used.
In this trough is placed a false bottom, (in preference of sheet
copper), perforated with many small holes, on this is a cloth or
web of hair or other material, on which the sugar, previously
regularly moistened " with water, lime water, or some other
" liquid," is placed. To the real bottom or sides is fixed a pipe
which leads to a receiving vessel placed below for the reception of
the molasses. The real bottom is in communication with an air
pump, worked by water, wind, steam, animal, or other power, the
passage of the air through the sugar separates and expels the
molasses, which falls on the real bottom, and runs into the vessel
below. The other method, namely, by compressing the air on the
upper surface of the vessel is effected in a trough or box, covered
on the top and furnished with a false bottom and cloth or web, as
by the first method, and having spread the moistened sugar on

the cloth or web, as by the former method, the air is compressed
" by means of a force pump, bellows, or some other contrivance,
" worked by some power, either natural or artificial," on the
upper surface of the sugar, which produces the same effect on it
as the operation first described."

[Printed, 4d. No Drawings. Repertory of Arts, vol. 31 (*second series*),
    p. 328; Rolls Chapel Reports, 8th Report, p. 116.]

A.D. 1816, November 1.—N° 4072.

VARLEY, WILLIAM, and FURNESS, ROBERT HOPWOOD.—
" An improved method of obtaining or producing saccharine
" matter or substance from wheat, rye, oats, and barley, bear or
" bigg." This consists as follows :—The flour obtained from any
of these serials, but, by preference, barley flour, and, by preference,
separated from the bran by " the well known methods," is
gradually added, with stirring, to water acidulated with sulphuric
acid, 3 lbs. of " concentrated sulphuric acid or good oil of vitriol
" of commerce" to every 45 gallons of water, brought to a temperature of 130°, to one hundredweight of the flour. The vessel,
by preference, of fir timber, has a lid, which is then put on, and
the joints having been made steam tight, a stop cock connecting
it with a steam apparatus is opened and the whole brought to the
temperature of 212° F., and continued (at or as nearly as possible at) for the space of thirty hours or so, the insoluble matter is
then removed by filtration, and the saccharine solution, either
before or after concentration, has the acid nearly neutralized by
lime, and the neutralization finished by the addition of carbonate
of lime or pulverized chalk, limestone, marble, oyster shells, &c.,
and on standing the sulphate of lime precipitates and is removed.
The concentration of the saccharine fluid is conducted in a leaden
vessel surrounded with an iron jacket, into which steam is
admitted. " The syrup" is drawn off for use.

[Printed, 4d. No Drawings. Rolls Chapel Report, 8th Report, p. 120.]

A.D. 1817, January 23.—N° 4093.

DE CAVAILLON, JOSEPH.—" Certain improvements in the
" preparing, clarifying, and refining of sugar, and other vege-
" table, animal, and mineral substances, and in the machinery
" and utensils used therein." These are, first, the several
charcoals, animal or vegetable, such as " bone charcoal, ivory

" black, blood charcoal, vegetable or wood charcoal, or that sub-
" stance known under the name of residue of Prussian blue,
" which are all used in a state of powder for clarifying and bleach-
" ing syrups and other liquors or substances, are generally
" considered of no use or value when they have been once used,
" but are thrown away as completely refuse." It is here proposed
to collect them, and burn them in close vessels, except one small
orifice for the escape of any elastic or other vapour. They "must
" be exposed in proper furnaces to a white heat, until no further
" steam or vapour" arises from them. When cold, they are taken
out of their vessels and they are carefully sifted, washed, and
filtered in preference making the funnel or support of the " fil-
" tering bag square like the hopper of a mill, and with a flat
" bottom." " The hopper or funnell of the filter may be of wood,
" iron, copper, tin, or other metal, and of any convenient size,"
in which is applied " the wicker basket, and linen, woollen, or
cotton bags usually made use of."

Second, " in the boiling and boiler, which instead of making a
" fixture over the fire," the boiler is " to be so hung or suspended
" in three or more chains or slings," so as to be raised or lowered "
from the fire that is " contained in a fixed furnace with a flue
" beneath it," the pan can be thus adjusted 'with the greatest
nicety; the pan has a lip.

[Printed, 4d. No Drawings. Rolls Chapel Reports, 8th Report, p. 117.]

A.D. 1817, January 23.—N° 4095.

WILSON, DANIEL.—" Certain improvements in the process of
" boiling and refining sugar." These are, in place of a vessel
heated by an open fire, applying the heat necessary for the boiling
or evaporating of syrup " by means of a current of heated fluid,
" which passes under and round the sides of the pan containing
" the sugar to be so boiled or evaporated, and the fluids which
" are thus used " are " above the heat of boiling sugar, but below
" the degree of heat necessary for burning or decomposing it, so
" that it can never be injured from an excess of heat." The pans
are double, with a space between them; in the inner is the syrup,
and the outer pan communicates with a vessel in which "the
" fluid (forced by a pump from the boiler) for conveying the heat
" is heated." The substances employed as the medium for conveying the heat are "whale, spermacetti, cod, seal, herring,
" pilchard, or other fish oil, linseed, rape, hemp, olive, nut, palm,

" sunflower seed, poppy seed, and castor oils, tallow, lard, butter,
" grease, animal fat, and wax."

[Printed,. 6d. Drawing. Repertory of Arts, vol. 32 (second series), p. 69 ;
Rolls Chapel Reports, 8th Report, p. 117.]

A.D. 1817, June 3.—N° 4130.

WYATT, CHARLES.—"Method or methods for preventing any
" disadvantageous accumulation of heat in manufacturing and
" refining sugar." This consists " in bringing the heated particles
" of cane juice or of a solution of sugar into contact with the air
" in a very extended surface," so as to promote "a copious
" evaporation " at a "temperature below the boiling point of
" the liquor." In deep vessels the operation is performed "by a
" number of thin sheets of metal or other convenient substance
" attached to a handle, and moved up and down by the hand, or
" by any other power." " But when shallow, and particularly
" when hemispherical vessels are used,". " a number of discs or
" circular plates placed on a horizontal axis " are 'preferred ; and
the exposure of the liquor to the air " is performed by the revolu-
" tion of the said discs." Or a number of tubes are " bent to the
" shape of the internal part in the vessel, and fastened by their
" extremities to an axis or handle ; " and an extensive surface is
produced "by means of spheres or segments of spheres, cones,
" frustums of cones, prisms, ellipsoids, cylinders, or an endless
" series of plates or sheets loosely connected together, suspended
" and acting round two axes like the common jack chain, the
" lower part always passing through the liquor." Copper is
preferred "to any other substance for the execution of these
" contrivances, although cloth, silk, linen, cotton, canvas, earthen-
" ware, wood, and other bodies may be used with inferior effect."

[Printed, 4d. No Drawings. Rolls Chapel Report, 8th Report, p. 118.]

A.D. 1817, November 28.—N° 4181.

HAGUE, JOHN.—-" Certain improvements in the method of ex-
" pelling molasses and syrup from sugars, and also in the refining
" of sugars."

[No Specification enrolled.]

A.D. 1818, February 3.—N° 4220.

WILSON, DANIEL.—"Certain improvements in the process of
" boiling and refining sugar." These are, first, separating from

juice or raw sugar solutions their chemical impurities, the most abundant of which "is called by chemists extractive matter, the " others are tannin and gallic acid." "These substances possess " the property of forming insoluble compounds with the salts and " oxides of tin and zinc." The sulphate of zinc is preferred. In refining of sugar the pan is charged with strong lime water and the sugar added in the usual manner, and the fire set; to every hundredweight of sugar a solution of four ounces of sulphate of zinc in a minimum of water is added, and the whole well stirred; a precipitate of sulphate of lime and tannate and gallate of zinc takes place. "There is in the lime water employed for dissolving " the sugar a quantity of lime more than sufficient to neutralize " the sulphuric acid of the sulphate of zinc: but when the raw " sugar employed contains much acid, and a strong grain is " required," one ounce of lime in powder is made into milk of lime and well mixed "with the sugar in the pan about five " minutes after the addition of the solution of zinc." The solution, after being treated, is brought to the boiling point, run through a filter, and boiled to a proof. "Where the use of lime " water is exceptionable," precipitated oxide of zinc, prepared by adding lime water or lime in substance to a solution of sulphate of zinc, is preferred; the precipitated oxide' is used in the same manner as the sulphate. For every hundred gallons of cane juice eight ounces of sulphate of zinc and two ounces of lime are employed.

Second, the improvements in apparatus, described in No. 4095, consist in the employment of the medium (one of the oils, fats, or waxes) there named. The sugar is contained in a single pan, on the bottom of the inside of which is a coil of pipe; this pipe communicates at both ends with the boiler in which the medium is heated, having a force pump attached to one of them, so as to drive the heated oil through the coil of pipe.

[Printed, 4d. No Drawings. Repertory of Arts, vol. 34 (second series), p. 134; Rolls Chapel Reports, 8th Report, p. 125.]

A.D. 1818, July 11.—N° 4276.

BAIRD, JOHN.—"New invented boiler for evaporating the juice " of the sugar cane or syrup derived from thence." This consists in making or casting cast-iron "sugar boilers thin; and by the " process of annealing, which was formerly applied only to small " vessels, they are rendered less brittle and more durable." This

annealing is accomplished in a kiln or furnace as follows :—
The outer walls of a large round furnace are made of fire-brick,
surrounded and bound together by a case or ring of cast-iron; in
the inside of this furnace is a chest, also made of fire-brick, in
which the boilers or pans meant to be annealed are placed. There
is a moveable dome arched upon a cast-iron ring, to which eight
chains are attached. The dome is raised when required by a
crane; around the kiln are eight fire-places, having flues underneath and surrounding the chest. There is a passage "from the
" furnace to the chimney, part of which, by being built on a plate
" protected by bricks against the fire, is so constructed as to be
" moveable, and is taken away before the dome is removed."
The crane is used also for putting in and taking out the boilers
or pans. These boilers are cast of the usual shape, "but only
" about one half the usual thickness."

[Printed, 10d. Drawing. Rolls Chapel Reports, 8th Report, p. 125.]

A.D. 1819, February 23.—N° 4344.

BROCKSOPP, Thomas.—" The application of certain ma-
" chinery to the purpose of breaking or crushing of sugar."
This consists of a frame with standards supporting a pair of
rollers, the sugar in its rough state is fed into a hopper at the top
and falls down between the rollers which "devolve against each
" other by means of the pinion and cog wheel" attached to their
axles being put in action; "but as the pinion and cog wheel are
" of [different diameter, the rollers will necessarily turn with
" different velocities, and consequently rub against each other.
" There are also grooves in one roller, and pins in the other."
A box covers the rollers and supports the hopper. A knife or
scraper is placed under the roller that receives the cogs. There
" is a fly wheel attached to the axle of the pinion, and turned by
" by the handle designed to regulate the motion." There is a
" box under the rollers which receives the sugar after it has
" passed through the rollers and becomes broken or powdered."

[Printed, 6d. Drawing. London Journal (Newton's), vol. 1, p. 13; Rolls
Chapel Reports, 8th Report, p. 131.]

A.D. 1820, April 15.—N° 4447.

ROHDE, Major.—" A method of separating or extracting the
" molasses or syrup from muscovado or other sugar." This is

effected by absorbing these substances into "linen or some other
" substance of absorbing quality, assisted by mechanical or manual
" motion and friction," as follows :—Any lumps are broken, so as
to admit of the sugar being passed " through a sieve of sufficient
" texture or size without breaking the grains or chrystals ;" the
sugar is spread in thin layers on the absorbing substance, and
" having folded it, place it in bags or other packages, and apply
" manual or other mechanical power to put it in motion, so as to
" afford the friction necessary to separate the molasses or syrup
" from the sugar." The molasses or syrup is absorbed into the
material used, and " the purer chrystals remain on the surface, and
" are separated by brushing, shaking, or scraping them off." The
molasses or syrup is afterwards extracted from the absorbing substance by means of water or steam. Or any means are employed
" by which an absorbing substance " comes in contact with the
sugar, " so as to allow of its absorbing the molasses or syrup,
" whilst it leaves the chrystals on its surface."

[Printed, 4d. No Drawings. Repertory of Arts, vol. 40 (*second series*),
p. 78 ; London Journal (*Newton's*), vol. 1, p. 413.]

A.D. 1821, January 15.—N° 4528.

DANIELL, JOHN FREDERICK.—" Certain improvements in
" clarifying and refining sugar."
[No Specification enrolled.]

A.D. 1821, August 14.—N° 4583.

COLLINGE, JOHN.—"An improvement in cast-iron rollers for
" sugar mills, by more permanently fixing them to their gud-
" geons." This consists, " in casting the rollers or cases of sugar
" mills upon wrought-iron axes or gudgeons, by which means
" they will be more permanently fixed than can be effected by
" any method hitherto known or made use of for this purpose."
It is stated that permanent tightness and fitting can never be
obtained by driving wedges into the bearings. In carrying out
the above it is stated " that the gudgeons at each end of the axis,
" being properly and truly turned in a lathe, care must be taken
" to place it exactly perpendicular in the mould, that the case
" or roller, when cast, may be as true to its centres as possible."
When quite cold it is taken from the mould and " the gudgeons
" are again placed in the lathe, and the cylindrical surface of the

" roller or case is turned in the usual way until it is perfectly
" true with its bearings."

[Printed, 4d. No Drawings. London Journal (*Newton's*), vol. 4, p. 303.]

A.D. 1822, May 9.—N° 4674.

KNIGHT, RICHARD, and KIRK, RUPERT.—"A process for the
" more rapid crystallization, and for the evaporation of fluids, at
" comparatively low temperatures, by a peculiar mechanical
" application of air." This consists as follows :—"A quantity of
" air is propelled by means of a blowing engine, bellows, or other
" machine used for propelling air through a pipe or pipes made
" of lead, copper, iron, or other fit material into the lower part of
" the copper, pan, or vessel containing the heated syrup," " coiled
" or otherwise shaped " at the bottom of the vessel " being per-
" forated with a number of small holes." " In lieu of the perforated
" pipe, a shallow metallic vessel of the nature of a cullender within
" the boiler may be connected, the air pipe and the cullender
" being perforated with small holes, the heated air may be driven
" through this perforated cullender." The air "may be heated
" by forcing it through metallic or other pipes surrounding the
" inside or bottom of the same vessel which contains the matter
" to be acted on or by an arrangement of pipes included between
" an inner and outer vessel, and heated by fire, water, steam, or
" otherwise, or by causing the air to pass through heated iron,
" lead, copper, or other pipes or tubes, or by means of flues,
" ovens, chambers, retorts, or other similar apparatus."

[Printed, 4d. No Drawings. London Journal (*Newton's*), vol. 5, p. 295 ;
Carpmael's Reports on Patent Cases, vol. 1, p. 504 ; Webster's Letters
Patent, p. 55.]

A.D. 1823, June 19.—N° 4805.

SMITH, JAMES.—"An apparatus for the applying of steam to the
" boiling and concentration of solutions in general, crystallizing
" the muriat of soda from brines containing that salt, melting
" and refining of tallow and oil, boiling of sugar, distilling, and
" other similar purposes." This consists of an oblong steam boiler
having " a flat top or upper surface " which forms " the bottom
" of the pan or vessel required to perform the various operations
" set forth." Fire is applied to the bottom of the boiler " in any
" of the ordinary ways now in use." Each end of the pan
projects over the boiler. At the upper part of one end of the

boiler is a funnel and feed pipe by which water is introduced into
it to the depth of about two inches; at the other end of the
boiler "is a gage cock by which to ascertain when the water
" attains that depth." Below the feed pipe is "a cock for giving
" vent to the rarified air when necessary," and below that again
is another "cock for the purpose of drawing off the water again
" when occasion shall require."

[Printed, 6d. Drawing. Repertory of Arts, vol. 44 (second series),
p. 74; London Journal (Newton's), vol. 7, p. 190; Mechanics' Magazine,
vol. 2, pp. 39 and 104; also vol. 3, p. 58; Register of Arts and Sciences,
vol. 2, pp. 10, 17, and 89; Engineers and Mechanics' Encyclopædia, vol. 1,
p. 212.]

A.D. 1824, May 6.—No 4949.

CLELAND, WILLIAM.—"An improvement in the process of
" manufacturing of sugar from cane juice, and in the refining
" of sugar and other substances." This consists in the use as
filters of long narrow bags "about six feet in length (more or
" less) and from five to six inches in breadth when flat and
" empty, and when full and extended from three to four inches
" in diameter," made of any fit material, but preferring "that
" kind of linen cloth called English duck;" the tops are cased
for "two or three inches downwards with stout woollen cloth,"
and likewise "affix a fillet or edge of double woollen cloth about
" half an inch broad round the top of the mouths of the bags, to
" form a sort of flanch or rim to them." Fitting inside the
mouths of these bags and "somewhat less at the bottom than the
" top so as to correspond with holes bored or made in the bottom
" of a cistern or head," are mouth-pieces or funnels with a bow or
handle fixed to and over them, open both at the top and bottom
fixed perfectly tight. "The bags are open at the bottom for
" the convenience of cleaning, and are tied close with twine
" about an inch from the bottom when used, for which purpose
" they are let down through the said holes formed in the bottom
" of the cistern or head." The funnels or mouth-pieces "are
" forced into the said holes by means of a key or winch, so as to
" render the said cistern or head tight and prevent the liquor
" under operation escaping anywhere except through the funnels
" or mouth-pieces." The cistern or head is kept filled with the
liquor to be filtered, and the liquor filtered through the bags
passes into a proper receptacle beneath. The above mode of
fixing the bags to the bottom of the cistern is preferred, although

they may be fixed by screws or other means. The head or cistern
is supported by some means; when it is necessary to retain the
heat such head or cistern is supported over a close case of non-
conducting material.

[Printed, 6d. Drawing. Repertory of Arts (*third series*), p. 139; London
Journal (*Newton's*), vol. 9, p. 81; Register of Arts and Sciences, vol. 2,
p. 212.]

A.D. 1825, March 15.—N° 5127.

BARLOW, Joseph.—"A method or process for bleaching and
" clarifying, and for improving the quality and colour of sugars,
" known by the name of bastard and piece sugars." This consists
in the employment for the above purpose " of molasses instead of
" clays or other contrivances which have been heretofore employed
" for that purpose," as follows :—Instead of purifying these sugars
by the usual practice by claying, by which process a part of the
sugar itself becomes dissolved by the water passing from the clay
into and through the sugar, pouring " upon the top of the sugar
" in the mould a quantity of the ordinary West Indian molasses,
" which after a few hours will have filtered through the sugar
" in the mould, and ultimately run off at the bottom, carrying
" the colouring matter with it." The quantity of molasses used
varies with the quality of the sugar and " of the molasses itself,
" which if too thick a consistency, may be diluted with water."

[Printed, 4d. No Drawings. Repertory of Arts, vol. 3 (*third series*), p. 65;
London Journal (*Newton's*), vol. 10, p. 190; Register of Arts and Sciences,
vol. 3, p. 156.]

A.D. 1825, July 26.—N° 5222.

FREUND, Charles.—"An improvement or improvements in
" the process of refining sugar." This, it is said, " consists, not
" in the form or construction of any of the different cisterns or
" vessels " used, " but entirely in the application of any suitable
" vegetable alkali, combined with fuller's earth, in proper pro-
" portions, by the use of which in the manner and way " afterwards described, " a considerable improvement will be produced
" in the quality of the sugar so refined, and a great saving will
" also be effected when compared with the usual method of
" refining with blood or other albuminous articles." While the
pan is hot, "84 gallons of water and 15 lbs. of American pearl or
" potash " are put into it, and when the alkali is thoroughly

dissolved 18 cwts. of raw or Muscovado sugar, which should be free from lumps, are added, and the whole thoroughly mixed and then allowed to rest "from two to four hours, as may be most " convenient." 20 to 30 lbs. of the lightest coloured fullers' earth previously made dry are made into a cream with water, and added to the ingredients in the pan, and the whole well mixed, after which the contents are left at rest till next morning, which is the usual practice, and the pan cleared the following morning, " or the fire may be applied and the pan cleared without further " loss of time." The fire being lighted the syrup is brought by water to a proper consistency and the scum removed in the usual manner of the old process of refining with blood, when it is transferred to a settling cistern, which is a vessel having three cocks at different heights from the bottom, the clear liquid is drawn off at different periods from the different taps, and the dirt in the bottom is taken out through an aperture secured by a screw cap. The dirt or sediment, scums, &c. are afterwards dealt with.

[Printed, 6d. Drawing. London Journal (Newton's), vol. 13, p. 327; Register of Arts and Sciences, vol. 4, p. 30, also vol. 1 (new series), p. 148.]

A.D. 1825, October 22.—N° 5272.

JENNINGS, HENRY CONSTANTINE.—"Certain improvements " in the process of refining sugar." These are, "the application " of rectified spirits, being principally alcohol," for the refining of sugar. In carrying out this, raw or Muscovado sugar is put into a " conical vessel holding from 500 lbs. to 1,000 lbs., having " a wire copper gauze or perforated bottom," and the process is assisted "by using all and every of the well known means, whereby " the liquids are made speedily to percolate through solid sub- " stances, whose parts are not in actual contact; these means " are hydrostatic, hydraulic, or hydropneumatic." The spirits used may be "rectified spirits of wine, rum, brandy, or any liquor, " being principally alcohol, which has very little affinity for " saccharine matter or sugar, and a great affinity for coloring " matter, water treacle, &c." "When any spirit is passed through " the mass of sugar so as to drop no more," to remove all or " nearly all the spirit," "about 30 gallons of saturated syrup" is passed through the mass, this " leaves the sugar only moistened " by the syrup and ready for putting into the hogshead." The

s.   c

spirit or wine, &c. percolated "may be used again over inferior
" sugars, and after it is very thick it may be rectified, and
" the spirit re-obtained in an uncombined state without much
" loss."

[Printed, 4d. No Drawings. Repertory of Arts, vol. 6 (*third series*),
    p. 335; London Journal (*Newton's*), vol. 11, p. 370; Register of Arts and
    Sciences, vol. 4, p. 118.]

A.D. 1827, April 28.—N° 5488.

LAWRENCE, Morton William. — "Improvement in the
" process of refining sugar."

[No Specification enrolled.]

A.D. 1827, July 4.—N° 5520.

CLELAND, William.—" Certain improvements in the process
" of preparing, refining, and evaporating sugar." These are said
" to consist in the exposing of fresh surfaces of fluid in continual
" succession to the evaporating power of air by means of a
" revolving coil of metal pipe heated with steam." Beneath the
pan or boiler containing the fluid to be evaporated is a boiler for
water which is heated by fire "applied in any of the ordinary
" manners." There is "a hollow tube formed of pewter, copper,
" tin, or tinned copper or iron, or other suitable metal bent or
" wound in a screw-like form, so as to have the appearance
" of a cylinder." This cylinder formed of tube as aforesaid is
supported, mounted, or fixed upon a central axis "above the
" pan or boiler containing the flued to be evaporated," by means
" of arms or radii placed at proper distances upon such central
" axis to render the whole stiff and firm, and the central axis
" has proper turned bearings at two ends, upon which that axis
" with its cylinder of tube fixed as aforesaid may revolve on its
" proper centre." The two ends of the central axis are made
hollow for a few inches at each end, and the two ends or
terminations of the tube cylinder are "introduced and fixed
" by soldering or otherwise into such tubular parts of the axis."
One of the bearers or supports of the axis is also tubular and
opens into the upper part of the boiler below the pan, while its
upper end is connected with the hollow part of the axis of the
cylinder by means of a steam joint so that the steam from the
boiler passes through the coiled tube cylinder and out by the

opening at the other end of the axis. The cylinder is rotated by a hand lever a strap or a pulley fixed upon the axis. The boiler for producing the steam may be detached.

[Printed, 6d. Drawing. London Journal (*Newton's*), vol. 1 (*second series*), p. 359; Register of Arts and Sciences, vol. 2 (*second series*), p. 311.]

A.D. 1827, October 11.—N° 5555.

STOKES, JAMES.—"Certain improvements in making, boiling, " curing, clarifying, or preparing raw or muscavado sugar, " bastard sugar, and molasses." These are, first, to the liquor or cane juice placed in the purifier add "about fourteen pounds of " charcoal, about seven pounds of bark of the wild elm tree and " about one pound of lime, and proceed to clear the liquor, take " off the scum, &c., in the usual way, and when sufficiently " clear" "pass it through a blanket into a clarifying cistern or " other vessel, and pump it or otherwise convey it into the pans " or teaches for the purpose of boiling or evaporation." Previous to the sugar being put into hogsheads it is put into boxes or vessels constructed for the purpose and mixed in the proportion of one gallon of "brandy, geneva, rum, or any other spirit or " spirits" to "every hundredweight of sugar," and then submit it to hydraulic or other pressure "which forces out the moisture " and leaves the sugar sufficiently dry to put it into the hogsheads " and much improved in quality and colour. Or,"

Second, "take raw sugar or bastard sugar in its finished or " manufactured state, and when boiled by this or any other " method" put "it into boxes as above mentioned, proceed to " mix, press, &c. in the manner as before described and pro- " ducing the same effect."

The quantities or apparatus may be varied as necessary, "intro- " ducing any other article or articles of similar chemical powers " and affinity."

[Printed, 4d. No Drawings. Repertory of Arts, vol. 8 (*third series*), p. 278; London Journal (*Newton's*), vol. 3 (*second series*), p. 259; Register of Arts and Sciences, vol. 2 (*new series*), p. 276.]

A.D. 1827, December 4.—N° 5572.

FAWCETT, WILLIAM, and CLARK, MATTHEW. — "An im- " proved apparatus for the better manufacture of sugar from the " cane." This consists as follows:—There is a round boiler for the production of high pressure steam of sufficient temperature to boil cane juice to granulate it into sugar, it may be made " of malleable iron, of copper, or of cast iron." In the interior of the boiler are two wrought iron or copper tubes of sufficient capacity to contain the fire, flame, and smoke necessary to produce the high pressure steam, and these tubes are wholly immersed in the water; they terminate in smoke flues regulated by a damper or dampers. The sugar pans are above the boiler, additional heat is afforded by copper tubes into the pans supplied by steam from the generator, but to this application an exclusive right is not claimed. There are " stop valves to shut off the " communication with the boiler when the tubes are taken out, " pipes for discharging the syrup from the pans, a safety valve " on the boiler. From the upper part of the boiler is a pipe for " supplying an engine for a cane mill or other object with steam. " There is a feed pipe into the bottom of the boiler for supplying " it with water and a float and levers for regulating the supply of " water."

[Printed, 1s. Drawings. Repertory of Arts, vol. 7 (third series), p. 267; London Journal (Newton's), vol. 6 (second series), p. 92; Register of Arts and Sciences, vol. 3 (new series), p. 115.]

A.D. 1828, March 29.—N° 5635.

DAVIS, JOHN. — (A communication.) — " An improvement in ". boiling or evaporating solutions of sugar and other liquids." This consists of an apparatus attached by means of a pipe from the top (in which there is a tap) to a sugar pan, " whereby the " operation of boiling in vacuo may be performed without the use " of air pumps as heretofore by means of a vacuum formed by " the admission and eduction of water and a Torricellian column, " and in such manner that the steam " which is evolved from the sugar pan is drawn therefrom and divides itself into two portions, one portion passing downwards into a condenser, from the bottom of which is the Torricellian column (above a well of waste water) and the other portion passes upwards into the upper part of a double vessel containing water, acting with pressure upon the surface of the water and by that means forcing a portion of the water in the inner vessel through a pipe leading from the bottom of the vessel into the condenser to assist in condensing the first-mentioned portion of steam. The double vessel above referred to consists of an outer square vat filled with water in which is placed lengthways and fixed a stout cask; an agitator is inside this cask which is turned by a handle working in stuffing boxes to prevent the admission of air. A glass tube communicates at each end with the inside of the cask and serves as a guage to show the quantity of water in the cask. There are pipes and taps for supplying and discharging water or air to or from the vat and cask. The agitator in the cask is for disengaging any air in the water.

[Printed, 10d. Drawing. Repertory of Arts, vol. 8 (*third series*), p. 456; London Journal (*Newton's*), vol. 5 (*second series*), p. 79; Register of Arts and Sciences, vol. 4 (*new series*), p. 132.]

A.D. 1828, May 17.—N° 5657.

POWELL, Thomas, POWELL, William, and POWELL, John.—"Certain improvements in the process, machinery, or " apparatus for forming, making, or producing moulds or vessels " for refining sugar, and in the application of materials hitherto " unused in making the said moulds." These are, first, a block the shape of a loaf of sugar, in the centre of the broad end is a hole about half of an inch in diameter, and in the other or small end is a spindle projecting about three inches " in diameter, the " size of the hole to be in the small end " of the mould.

Second, "an upright spindle in a frame and turned with a " large vertical wheel and string in a similar way to that of " turning potters' wheels for throwing or making ware." The block above is put on the spindle. The spindle in the small end of the block aforesaid works in a box of brass or other material so that when the upright spindle is turned the block " which is " then standing perpendicular upon its broad end will also " turn."

Third, for preparing the clay, is a tub over a table, upon which is a sliding board, and upon which board is a flat mould about half an inch deep, " and of the shape and size of the clay required " to make one sugar mould or vessel. The tub is filled with clay " which is pressed through the tub by a screw upon a strong " board or flat plate of iron, when the mould is filled a wire is " passed between the bottom of the tub and flat mould by which " means a cake of clay is produced of the thickness and size " required for making one sugar mould or vessel, and this is " done on the machine described under the second head." Another vessel for preparing the clay is a tub made of thin frames

of wood braced together, when this tub is filled the braces are
slackened and wires are passed and drawn between the frames,
and a cake of clay is cut off the size of and shape of one mould
or vessel for refining sugar.

Fourth, a kiln of an oblong form, the inside is divided by
arches one over the other, there are holes made in the arches to
admit the fire through the kiln. Every mould is set on its broad
end, the kiln gear is so arranged as to give free circulation to the
fire within as well as without the moulds or vessels. The moulds
are made of stone ware clay, the same being glazed inside and
outside with a salt glaze.

[Printed, 10d. Drawings. London Journal (*Newton's*), vol. 8 (*second
series*), p. 191; Register of Arts and Sciences, vol. 3 (*new series*), p. 35;
Engineers and Mechanics' Encyclopædia, vol. 2, p. 328.]

A.D. 1828, November 27.—N° 5718.

KNELLER, WILLIAM GODFREY.—"Certain improvements in
" evaporating sugar, which improvements are also applicable to
" other purposes." These are, in reference to this subject,
" forcing by means of bellows or any other blowing apparatus
" atmospheric or any other air in a hot or cold state through the
" liquid or solution subjected to evaporation," by means of pipes
reaching nearly to the bottom of the pan. The pan may be heated
" by a naked fire, steam, or hot air." The cane juice or syrup is
brought "to the proof or chrystallizing points" by keeping it
beeween 140° and 170° F., but preferring it between 160° and
170°. To more quickly "remove the steam or vapour from the
" surface of the liquid or solution and thereby to favour the
" evaporation, I sometimes, particularly when I use hot air for
" heating the pan or boiler, conduct the hot air, after it has given
" out part of the heat to the bottom of the boiler, to the surface
" of the liquor or solution; but I do not consider this contrivance
" necessary in any nor adviseable in all cases." The depth of
the solution is from 4 to 6 inches, and in preference, equal in
depth in every part and the bottom of the evaporator perfectly
level.

[Printed, 8d. Drawing. Repertory of Arts, vol. 9 (*third series*), p. 69;
London Journal (*Newton's*), vol. 4 (*second series*), p. 321; Mechanics'
Magazine, vol. 28, p. 325; Register of Arts and Sciences, vol. 4 (*new
series*), p. 298; Webster's Patent Law, pages 11 and 133 (case 97); Barne-
wall and Adolphus's Reports, vol. 2, p. 370; Carpmael's Reports on Patent
Cases, vol. 1, p. 501: Law Journal, vol. 9 (*King's Bench*), p. 242; Patentees
Manual, p. 107; Billing on Patents. p. 42.]

A.D. 1828, December 6.—N⁰ 5725.

HAGUE, JOHN.—" Certain improvements in the method of
" expelling the mollasses or syrup from sugar." These are,
" by occasioning a pressure of the atmosphere on a surface of
" the sugar," " either by withdrawing the air from the under
" surface, or by compressing this air on the upper surface of the
" sugar." " A trough or box (open at the top), either square
" round, or any other shape," in preference, larger at the top
than at the bottom is used. In this trough is placed a false
bottom (in preference, of sheet copper) perforated thickly with
small holes, on this is placed " a cloth or web made of hair or
" some other material." To the real bottom or sides is fixed a
pipe which leads to a receiving vessel placed below for the reception of the molasses. The real bottom is in communication
with an air pump. The sugar is spread " about three inches
" deep, or any other proper depth, all over the cloth or web
" that covers the false bottom of the trough or box ;" the air
pump set to work, the passage of the air through the sugar
separates and expels the molasses and ultimately it falls into the
receiving vessel. This operation is continued " by keeping the
" air pump at work until the molasses or syrup is sufficiently
" expelled from the sugar." Before or during the process,
sometimes it is " desirable to moisten the sugar with water, lime
" water, or some other liquid." The other method by compressing the air on the surface of the sugar by means of force
pump, bellows, &c., is conducted in much the same apparatus as
the above.

[Printed, 4d. No Drawings. Repertory of Arts, vol. 10 (*third series*),
pp. 212, 221; London Journal (*Newton's*), vol. 3 (*second series*), p. 303.]

A.D. 1829, April 28.—N⁰ 5785.

DAVIS, JOHN.—(*A communication.*)—" A certain improvement
" in the condenser used with my apparatus for boiling sugar in
" vacuo." This is said to be improvements in respect of the
apparatus described in No. 5635, as follows, namely :—" The
" substitution of the common waste pipe to the bottom of the con-
" denser instead of the Torricellian column, and an enlargement
" of the condenser to about six times the size of that required "
in the former apparatus described in No. 5635; by which improvement the required vacuum can be formed by the introduction

of steam from the boiler or pan into the condenser. When the
steam in the boiling pan has been got up to a pressure of more
than fifteen pounds to the square inch, the water in the condenser
must be drawn off by the waste pipe, the cock on same being
open, and also the cock in the pipe from the boiler to the condenser being open, the steam from the pan will rush into the
condenser and expel what atmospheric air may be in it, and rush
out of the waste pipe. When this is accomplished, the cock on
the waste pipe is shut and a communication is opened to the
double vessel above, a portion of the steam causes a pressure on
the surface of the water, while another portion of steam rushes
down into the condenser, striking upon the perforated plate and
distributing itself for condensation.

[Printed, 8d. Drawing. Repertory of Arts, vol. 9 (*third series*), p. 1 ;
London Journal (*Newton's*), vol. 5 (*second series*), p. 83.]

A.D. 1829, August 1.—No 5824.

BATES, JOSHUA.—"A new process or method for whitening
" sugars." This consists in the use of "a separate vessel or pan
" for holding the water employed in whitening sugar, such vessel
" or pan being sufficiently porous to allow the water to percolate
" through and to be distributed on the sugar in a state of minute
" division. Also "a vessel (divided within into several compart-
" ments, each having its separate drip hole) for holding raw or
" Muscovado sugar during the process of whitening the same
" by the percolation of water," whether by the common process
of claying or by means of a pan or vessel " sufficiently porous to
" allow the water to percolate through and to be distributed on
" the sugar in a state of minute division." The pans may be
of earthenware or of filtering stone, or pans of wood or any other
material with a sieve-like bottom covered with sand or any other
suitable substance to retard the percolation of water. In the case
of loaf sugar these pans are " placed on the top of the sugar in
" the inverted conical pots."

[Printed, 6d. Drawing. Repertory of Arts, vol. 9 (*third series*), p. 146 ;
London Journal (*Newton's*), vol. 9 (*second series*), p. 176.]

A.D. 1829, September 15.—No 5848.

AITCHISON, JOHN. — " Certain improvements in the con-
" centrating and evaporating of cane juice, solutions of sugar,
" and other fluids." These are as follows : — An oblong or

square frame supports a pan made of stout sheet copper and in the form of a segment of a cylinder. This pan is surrounded with a jacket, and the space between is filled with steam for heating the syrup or other fluid contained in the pan. A hollow cylinder revolves by means of tooth gear in the pan, "dipping " into the pan to within about two or three inches of the " bottom." This cylinder is formed of two concentric cylinders, so as to leave a space between them of about two inches for a steam chamber; and by an arrangement of hollow axles, &c. steam is introduced into the space or chamber of the double cylinder, and from thence to the jacket of the pan, and ultimately escapes with the air and condensed water by an open cock at the bottom. There are scrapers arranged for acting upon the different surfaces as the syrup or other fluids thicken. Although only one rotatory drum has been described in this apparatus, on a large scale there may be introduced "two, three, or more revolving " drums into one pan," and sometimes the drums or rotatory cylinders are constructed " with four, six, or more surfaces by " connecting several concentric cylinders one within the other, " and conducting the steam into their several chambers by hollow " arms leading from the central shaft, similar to those already " described."

[Printed, 8d. Drawing. Repertory of Arts, vol. 10 (*third series*), p. 34; London Journal (*Newton's*), vol. 7 (*second series*), p. 220; Mechanics' Magazine, vol. 21, p. 417; Register of Arts and Sciences, vol. 4 (*new series*), p. 262; Rolls Chapel Reports, 7th Report, p. 131.]

A.D. 1829, December 14.—N° 5877.

GOULSON, BENJAMIN.—" Certain improvements in the manu- " facture of farina and sugar from vegetable productions." These are as follows :—The roots " of dahlia, carrots, turnips, beet, " mangel wurtzel, and potatoes are well washed or deprived of " their peel;" "they are then steeped, either whole or cut into " slices, in pure water and an acid." "Any acid substance or " acid salt will answer the purpose," but sulphuric acid is preferred. To every hundred pounds weight of roots employing two pounds of acid for dahlia, and the quantity of acid required varies in the order in which the names of the roots are placed " up to ten pounds for every hundred weight of potatoes." When by steeping the root can be rubbed by the hand into a paste, the liquor is run off and the roots washed until they cease to have an acid taste, when they are dried in baskets in the sun or in a stove, and "ground into powder become farina." This farina is converted into "saccharine matter or sugar" by boiling "with acid " and water in the proportion of two pounds of acid to one " hundred pounds of farina." "The roots of the dahlia may be " converted into sugar without being first made into farina " by steeping "the roots in acid and water in the proportion of ten " pounds of acid to one hundred pounds weight of roots, and the " acid will in the course of three days convert the dahlia into " saccharine. The roots are then pressed, to separate the juice, " and the acid is neutralized. That juice is boiled and clarified " in the usual way, and thus sugar is made."

[Printed, 4d. No Drawings. Repertory of Arts, vol. 10 (*third series*), p. 31; London Journal (*Newton's*), vol. 8 (*second series*), p. 31; Register of Arts and Sciences, vol. 5 (*new series*), p. 35; Engineers and Mechanics' Encyclopædia, vol. 1, p. 489.]

A.D. 1829, December 14.—N° 5878.

DEROSNE, CHARLES.—(*A communication.*)—"Certain improve-" ments in extracting sugar from cane juice and other substances " containing sugar, and in refining sugar and syrups."

[No Specification enrolled.]

A.D. 1830, March 6.—N° 5916.

THOMAS, RICHARD GUPPY. — "A new apparatus for granu-" lating sugar." An ordinary sugar pan is set in brick work and heated by a fire or steam below, an inverted vessel or receiver is supported over the pan by two supports resting upon the brickwork at the bottom of the receiver, and fixed within it is a perforated plate. The receiver may be of the same material as the pan, and a pipe leads from the top of the receiver to an exhausting apparatus. In order to use this apparatus the pan is heated by fire or steam, the pan should be fed with liquor, and the exhausting apparatus kept at work, " a constant supply of " atmospheric air will be drawn under the receiver, and being " distributed by the perforated plate will pass into and through " the liquor contained in the receiver till it reaches the dome " head, or air chamber at the upper part of it, whence it will be " drawn out by the exhausting apparatus, while fresh portions of

" air supply its place from below ;" and thus the liquid boils at
any temperature required, " according to the height of liquor in
" the receiver," and also that temperature is maintained " by
" continuing a supply of liquor to the pan equal in quantity to
" the diminution by evaporation and exhaustion, which supply
" keeps the column of liquor in the receiver always at the same
" height."

[Printed, 6d. Drawing. Repertory of Arts, vol. 10 (third series), p. 279;
London Journal (Newton's), vol. 7 (second series), p. 326; Register of Arts
and Sciences, vol. 5 (new series), p. 135.]

A.D. 1830, June 29.—N° 5945.

TURNER, EDWARD, and SHAND, WILLIAM.—" A new method
" of purifying and whitening sugar or other saccharine matter."
This consists of a mode " of employing pressure by means of a
" column of liquid to impel water or other convenient fluid
" rapidly through sugar, in order to displace and remove the
" molasses and any other matter from the crystalline part of the
" sugar " as follows :—A conical mould is formed of dry wood,
made water-tight, and carefully secured by iron hoops. The
surface of the rim of the mould is made very flat and smooth ;
and about an inch below the rim there is attached securely a
thick iron ring, with holes to admit screws. "A circular plate of
" cast iron, made smooth to lie flat on the rim of the mould,"
projects "an inch or two beyond it, being to the same extent as
" the ring." The plate has holes for screws corresponding to
those in the ring, " for the purpose of fixing the plate tight upon
" the rim of the mould." In the centre of the iron plate is an
aperture in which is fixed an upright tube or pipe, " to the upper
" extremity of which is attached a funnel." The sugar allowed
to crystallize in the ordinary manner is purified by pouring water
nearly freezing or other proper fluid into the funnel. The length
of the tube "may be varied in order to increase or diminish the
" pressure, as may be desirable. The pressure of the column of
" liquid may indeed be increased by the application of a forcing
" pump or any other convenient means," if it is desirable. To
diffuse the water uniformly over the sugar in the mould " it is
" advisable to lay over the surface of the sugar a piece of coarse
" cloth."

[Printed, 6d. Drawing. London Journal (Newton's), vol. 9 (second series),
p. 60; Register of Arts and Sciences, vol. 5 (new series), p. 258.]

A.D. 1830, June 29.—N° 5946.

POOLE, Moses.—(*A communication.*)—" Certain improvements
" in apparatus used for certain processes for extracting molasses
" or syrup from sugar." These are, " the application of the
" exhaustion vessel or vessels, the vacuum in which is obtained
" by the condensation of steam or by a Torricellian column "
when used for the above purpose, as follows :—What is called
the separating vessel has a perforated false bottom, on which is
spread a cloth of horsehair, finely woven brass wire, or other
material. The sugar to be operated upon, " damped with water
" or any other liquid," or otherwise, is spread to the depth of
three or four inches upon this cloth. The bottom of the vessel is
concave, with a cock placed for drawing off the liquids. " In place
" of using a pump or pumps for the purpose of exhausting the
" lower part of the separating vessel," an " exhaustion vessel,"
spherical or other shaped of copper, iron, or other material, " one
" of about six cubic feet capacity, is a good size, but it may be
" increased or decreased ad libitum," is connected to the lower
compartment of the separating vessel by a pipe having a stop-
cock placed upon it. " At the top part of the exhaustion
" vessel " is a cock to permit of the air being driven out, and at
the lower part " a cock for drawing off the condensing water."
Steam is to be permitted to flow from a boiler by means of a
connecting pipe having a stop-cock upon it into the exhaustion
vessel, which will drive out the air ; when this is effected the air
cock must be stopped. The exhaustion vessel having become
full of steam is closed, and the steam is condensed in this vessel
by cold water from a reservoir above, distributed in minute jets.
When the steam has been condensed the water is stopped, and
the cock between the exhaustion vessel and separating vessel
opened, the air will press upon and rush through the sugar, and
so " carry down the molasses or syrup into the lower part of the
" separating vessel." " The manner of constructing the appa-
" ratus when a Torricellian column is made use of for the purpose
" of obtaining a vacuum " is as follows : — " The exhaustion
" vessel or vessels," are similar to those above, " but in place of a
" steam pipe a water supply pipe with a stop-cock or valve upon
" it must be used. The condensing water pipe will not be neces-
" sary, neither will the cock placed at the bottom of the exhaustion

" vessel" be necessary, but "the air escape cock or valve and
" the connecting pipe between the exhaustion vessel and the
" separating vessel must be retained. To the bottom of the
" exhaustion vessel a pipe, descending about thirty-three feet
" and communicating with a reservoir below, is fixed, having a
" stop-cock or valve placed upon it."

[Printed, 4d. No Drawings. Repertory of Arts, vol. 16 (*third series*), p. 272; London Journal (*Newton's*), vol. 9 (*second series*), p. 61; Register of Arts and Sciences, vol. 5 (*new series*), p. 258.]

A.D. 1830, July 24.—N° 5962.

GARNETT, ABRAHAM.—"Certain improvements in manufac-
" turing sugar." There are, "the cane liquor being tempered
" and clarified in the usual manner " by "passing through the
" clarifiers and four or five open pans," is "strained and skimmed
" on passing each successive pan " until it reaches the last pan,
" in which it undergoes a considerable degree of evaporation,
" and from which a full charge is given to the teach," over
which "a hemispherical pan or dome of cast iron, rather thick and
"- weighing seven hundred pounds," is let down by means of a
chain worked by a winch or lever, and fits over it by means of
a flange, so as to close it on all sides; "in a few seconds the
" ebullition caused at first by the repulsion of the atmospheric
" air will subside, and a partial vacuum being thus produced, the
" fluid will rapidly evaporate its most aqueous property within
" the closed pan, sending out the surplus steam through the
" safety valve " in the crown of the dome. "At each charging,
" the dome may be raised half an inch or so to admit the liquor
" into the teach and immediately again closed, and then after a
" smart boiling for ten minutes " the liquor will be sufficiently
concentrated. By the skipping pan or ladle the evaporated liquor
is transferred to coolers, and "the sugar after remaining twelve
" hours in the coolers may be placed in oblong boxes with wire
" cloth bottoms," and "in them either clayed or packed in hogs-
" heads." "If the sugar, concentrated" as above, "should
" have to be purified by Inne's patent process, it should at once be
" sent into the reservoir from the teach " by "a connecting gutter,
" and each strike or stratum may be cooled by the application of
" a slight shower of lime water."

[Printed, 8d. Drawing. London Journal (*Newton's*), vol. 3 (*conjoined series*), p. 239; Mechanics' Magazine, vol. 15, p. 92; Register of Arts and Sciences, vol. 5 (*new series*), p. 301.]

A.D. 1830, August 5.— N° 5975.

ROBINSON, MARMADUKE.—" Certain improvements in the
" process of making and purifying sugars." These are, first,
" the method of purifying cane juice by precipitating that part
" of the impurities, which, according to the ordinary method, is
" only got rid of by skimming during the boiling of the liquor."
Two pounds of finings " composed of a saturation of alum and
" lime," diluted in pure water, are added " to every one hundred
" gallons of juice," and well mixed. The juice is then perfectly
neutralized by the addition of small quantities of milk of lime,
at each addition, examining the liquor with the usual test papers
for acids in liquids until no change in the paper takes place;
excess of alkali is detected by a paper for detecting alkalies, and
corrected by the addition of more juice. The liquor neutral is
nearly brought to boiling in a vessel in preference of wood,
heated by coils of copper pipes, through which steam circulates.
After heating as above, the heat is discontinued, and the liquor is
mixed with about three pounds more of the finings, diluted as
before, to every one hundred gallons of juice, stirring the same,
examining in a glass vessel, and futher adding finings until additional finings does not increase the rapidity of the precipitation of
the impurities. The impurities having separated, the clear liquor
is drawn off into the evaporators, either of wood, as above, or as
under the third head.

Second, " the application of finings to the cane juice itself."
It is said of this, that it is "highly useful," though "not
" indispensably necessary."

Third, " the application of high-pressure steam to the boiling
" in vacuo of cane juice, or of other saccharine solutions, for the
" purpose of manufacturing refined or other sugars." No steam
heat is applied outside of the pan.

[Printed, 4d. No Drawings. Repertory of Arts, vol. 11 (*third series*), p. 196 ;
London Journal (*Newton's*), vol. 8 (*second series*), p. 258 ; Mechanics'
Magazine, vol. 15, p. 61 ; Register of Arts and Sciences, vol. 5 (*new series*),
p. 302.]

A.D. 1830, September 29.— N° 6002.

DEROSNE, CHARLES.—(*Partly a communication.*)—" Certain
" improvements in extracting sugar or syrup from cane juice and
" other substances containing sugar, and in refining sugar and
" syrup." These are " discoloring syrups of every description

" by means of charcoal," produced by distilling "bituminous
" schistus alone, or mixed with animal charcoal, or animal char-
" coal alone." The filter does not form the object of the patent.
The charcoal is put in a case in which is a perforated metal
diaphragm, about an inch from the bottom, upon which is a
coarse linen or woollen cloth. The charcoal, " reduced to the size
" of fine gunpowder, is very fit for this operation," moistened
with a little water it is placed in layers upon the cloth, each layer
being gently pressed " till it has come up to the height of fifteen
" or sixteen inches," more or less. On the top of the charcoal
is another metallic perforated diaphragm, upon which is spread
another clear linen cloth, upon which the syrup is poured to the
depth of from four to eight inches. The syrup ought to be made
clear before pouring it on the filter, by a previous filtration,
it should not consist of more than two-thirds of sugar and
one-third water, but may be less. When hot the filtration is
more rapid. A reservoir can furnish several filters at a time. The
last portions of syrup passing coloured, are passed again upon
another bed of charcoal. Any syrup in the charcoal is displaced
by water. Molasses are deprived of their bad taste, and are converted into good kinds of syrups. The syrups used " can be
" obtained directly from the juice of cane, or of beet root, or
" from the saccharine matter produced by the action of sulphuric
" acid upon the farinaceous matters before these juices or liquids
" have been baked for extracting the sugar. The syrup may
" likewise be produced by the solution of all kinds of sugar, and
" of the products of inferior quality which are obtained in sugar
" refining, under the name of 'bastards' and other sugars."
" Instead of using the schistus or animal charcoal of the size of
" gunpowder, it can be reduced to a powder still more fine,
" mixed with sand. In this state, a given quantity of charcoal
" discolors better than powdered less fine; but the filtration is
" slower and more difficult to be regulated. After having tried
" this first method " preference is given " to the other mode, but
" both of them are the object of the Patent."
By a Disclaimer and Memorandum of Alteration inrolled 6th
August, 1836, the title is altered to " A certain improvement or
" certain improvements to be used in the course of the process of
" extracting sugar or syrup from cane juice, and other substances
" containing sugar, and also to be used in the course of the pro-
" cess of refining sugar and syrup, for the purpose, in either case,

" of removing the color from, or whitening or purifying such
" sugars respectively," and " all benefit and advantage" in the
use of bituminous schistus or sand is disclaimed, but retained in
" the use of animal charcoal."

[Printed, 4d. No Drawings. Disclaimer, 4d. Repertory of Arts, vol. 3 (*new
series*), p. 175, vol. 4, p. 77 (note) and 177, vol. 11 (*third series*), p. 19, and
vol. 3 (*enlarged series*), p. 374, for extension; London Journal (*Newton's*),
vol. 7 (*second series*), p. 74, vol. 21 (*conjoined series*), p. 475, for Disclaimer,
and vol. 24 (*conjoined series*), p. 459, for extension; Mechanics' Magazine,
vol. 40, p. 363; Register of Arts and Sciences, vol. 5 (*new series*), p. 233;
Webster's Reports, vol. 1, pp. 152, 154, and 158, and vol. 2, p. 1; Webster's
Patent Law, pp. 11, 47, 67, 87, 114, 115, and 116 (also p. 135, cases 110° and
111, and Supplement, pp. 4 and 19) ; Webster's Letters Patent, pp. 19 and
62 ; Carpmael's Reports on Patent Cases, vol. 1, pp. 664, 689, and 698, and
p. 699 for extension; Gale's Reports, vol. 1, p. 109; Moore's Privy Council
Cases, vol. 4. p. 416; Moody and Robinson's Reports, vol. 1, p. 457 ;
Tyrwhitt's Reports, vol. 5, p. 393; Crompton, Meeson, and Roscoe's Reports, vol. 2, p. 476; Patentees' Manual, p. 14; Billing on Patents, pp. 25,
94, 98, 106, 111, 145, and 176. Extended for six years (see No. 10,389).]

A.D. 1830, October 13.—N° 6009.

ARCHBALD, WILLIAM AUGUSTUS. — "An improvement in
" the preparing or making of certain sugars." This is said to
consist in making loaf, lump, or other fine sugars immediately
from the syrup of the cane juice," instead of first converting the
syrup into some kind of sugar, then dissolving " such sugar again
" into syrup for the purpose of granulating it a second time to
" make refined sugars as is now done." The cane juice is first
prepared as described under the first head of No. 6975, except
that in the second addition of finings in place of adding three
pounds of finings to one hundred gallons of juice, one and a half
pounds are added. The juice thus purified is " converted into
" syrup of any density that may be found convenient, according
" to the object in view." If for keeping or transportation it is
evaporated to about from 36° to 40° Beaume's saccharometer, if
intended for immediate operation the density is about 28°, and
any acidity is corrected by gradual additions of lime water, so
that on testing with test papers the syrup is exactly neutral, any
excess of lime water is corrected by an addition of syrup. The
liquor being thus treated when cold is heated, in preference, by
steam passing through a coil of pipe in a pan. As soon as it
begins to boil, about three pounds more of the finings diluted,
to every thirty gallons of the liquor are thrown in, and after boiling two or three minutes, it is filtered through beds of animal
charcoal "in the proportion of from five to ten pounds of char-
" coal to thirty gallons of syrup, in an ordinary filter, in any way

" most convenient." If it is difficult to filter, it is clarified by mixing well with from five to ten pounds of animal charcoal, stopping the boiling, adding some bullocks blood, stirring, heating to boiling point, observing whether the charcoal precipitates and leaves the liquor clear, and if not, adding more blood and proceeding as above until the liquor is clear, when it is filtered and the syrup proceeded with in the usual way to " produce loaf, " lump, or other fine sugars."

[Printed, 4d. No Drawings. London Journal (*Newton's*), vol. 2 (*conjoined series*), p. 66 ; Register of Arts and Sciences, vol. 6 (*new series*), p. 46.]

A.D. 1830, October 20.—N° 6015.

URE, ANDREW.—" An improvement or improvements in curing " or cleansing raw or coarse sugars." This is " the use and " application of the compound solvent consisting of acid and " alcohol for cleansing sugar," as follows :—" After the sugar has " been well granulated in the boxes called coolers and freed by " drainage or lading from a considerable part of its molasses." The above solvent consisting of strong spirits or alcohol, about forty per cent., over proof, and muriatic acid, " in the proportion " of one gallon of the former to one ounce of the latter," " instead of muriatic acid, sulphuric, nitric, or any other strong " acid may be used in equivalent saturating quantity," is " to be " mixed with it either in the coolers or to be diffused through it " by sprinkling in its transfer into the potting casks or hogsheads, " the proportion being one gallon of the solvent to from two to " four handredweights of sugar ; but other proportions may be " used, and the solvent may be also composed of other propor- " tions of acid and alkohol." After the greater portion of the acidulated alcohol has drained off with the molasses, the sugar is washed with some simple alcohol in successive portions on the top of the sugar in the hogsheads and drained and dried " at a " temperature of from 120° to 160° Fahrenheits," whereby it loses the spirituous flavor.

[Printed, 4d. No Drawings. Repertory of Arts, vol. 11 (*third series*), p. 277 ; London Journal (*Newton's*), vol. 1 (*conjoined series*), p. 413 ; Mechanics' Magazine, vol. 15, p. 348 ; Register of Arts and Sciences, vol. 6 (*new series*), p. 73.]

A.D. 1830, December 23.—N° 6057.

PERTINS, MARIE ELIZABETH ANTOINETTE.—(*A communication.*)—" The fabrication or preparation of a coal fitted for the

D

" refining and purifying of sugar and other matter, and to restore
" the coal which has served for that purpose." This consists in
mixing animal and vegetable substances with soluble salts or an acid,
drying and calcining the mixture, afterwards grinding and washing
the calcined matter to free it from the salts or acid. The substances
which " answer the purpose best " are coal tar, molasses, blood,
peat, tanner's waste or bark, and bones the residue from glue
making. The salts generally used " are muriate of soda or common
" salt and muriate of lime; but all the other alkaline or soluble
" earthy muriates will answer the same purpose." The earthy
substances used " are common clay, sand, and river, seaport, or
" dock mud." Taking " 250 parts of fine sand, 50 parts of com-
" mon clay, 50 parts of common salt or other soluble muriate,
" 100 parts of coal tar, molasses, or blood;" or instead, " 300
" parts of river, seaport, or dock mud, 50 parts of common salt
" or other soluble muriate, one hundred parts of coal tar, molasses,
" or blood." The clay is made into a paste and mixed with the
salt dissolved in a small quantity of water, adding the tar,
molasses, or blood, mixing the whole well together, drying and
calcining the mixture in vessels or retorts, as in making bone
black or charcoal. The resulting matter is ground and washed.
When the other substances above named are used the proportions
and modes of mixing are given. "To restore the property or
" quality of bleaching," &c. to the above charcoal when it " has
" lost or ceased to have that property or quality," " 20 parts of
" common salt or other soluble muriate " are dissolved in water,
and 100 parts of the spent charcoal are mixed with it, and the
mixture is dried and proceeded with as above. In the above mixtures muriatic acid may be used in place of the common salt or
other muriate.

[Printed, 4d. No Drawings. London Journal (*Newton's*), vol. 8 (*second
series*), p. 257; Mechanics' Magazine, vol. 15, p. 476; Register of Arts and
Sciences, vol. 6 (*new series*), p. 133.]

A.D. 1831, January 31.—N° 6068.

BATES, Joshua.—(*A communication.*)—" Certain improvements
" in refining and clarifying sugars." These are, first, "the use of
" a revolving filter whether the same be made with pockets," or
" be brought to its utmost state of simplicity by suppressing the
" pockets and reducing the surface of the filtering cloth to a
" single fold wrapped round the wooden cage or frame." The filter

consists of two circular wooden headings, each covered externally by a metal plate, and of bars of wood passing from one heading to the other. The cage or frame is covered with a cylinder or jacket "of blanketing, or of that kind of woollen cloth " usually employed by sugar refiners as a filter, and in this " cylinder of woollen cloth are made as many pockets " of "the " same material as there are intervals between the bars of the " cage." "These pockets hang between the bars and tend towards " the axis of the cylinder." In order to keep these pockets stretched, in each is placed a piece of wood, wedge-shaped, grooved or furrowed on both sides.

Second, an "arrangement of pipes, cocks, and vessels, whereby " the syrup (when let out of the pan in which it has been mixed " with the animal black or other substances for the purpose of " clarifying it) is received into a vessel placed between the " clarifying pan and the revolving filter and is there kept hot by " means of steam in a jacket surrounding the vessel, and is like- " wise subjected to the elastic pressure of steam in order to " expedite its filtration."

[Printed, 10d. Drawing. London Journal (*Newton's*), vol. 9 (*conjoined series*), p. 224; Register of Arts and Sciences, vol. 6 (*new series*), p. 170.]

A.D. 1831, April 14.—N° 6107.

BRUNTON, THOMAS.—(*A communication.*)—" An improvement " in certain apparatus rendering the same applicable for making " or refining sugar." This consists, first, in the "construction " and combination or application of certain vessels, receptacles, " or chambers," called heating chambers constructed with an " internal bar or bars arranged," so that "if water is heated " therein, such rapid currents will be generated " that when they are applied to the heating of water or generation of steam "for " the purpose of heating pans for the making or refining of " sugar, they will effect that object with greater advantage than " can be produced by any other apparatus now in use for that " purpose." A pan is shown with a water jacket, the improved heating chamber is fixed by keys on to one side of this jacket and open into it. The heating chambers are made of any suitable metal and have a fire underneath, the division plates in these chambers cause the water to circulate and flow into the water in the jacket surrounding the pan and so heat the sugar in the pan.

Second, "in heating pans containing the syrup of sugar by

" means of surrounding such pans with water " heated as above,
such water is "pressed upon by an hydrostatic column," as
follows :—From one end of the water jacket rises a pipe or hollow
column on which is a cistern, the water of which is used for
supplying the difficiency or waste produced by evaporation, and
also to effect a hydrostatic pressure. The pipe or column is
capable of being increased in height at pleasure.

[Printed, 8d.   Drawing.   London Journal (*Newton's*), vol. 9 (*conjoined
series*), p. 216 ; Register of Arts and Sciences, vol. 6 (*new series*), p. 257.]

A.D. 1831, July 27.—N° 6144.

ROBINSON, MARMADUKE.—(*A communication from William
Augustus Archbald.*)—" Certain improvements in the process of
" making and purifying of sugars." These are, first, "boiling
" cane juice or other saccharine solution " by the following means
and "thereby expediting the process of evaporation." A pan
about three feet deep and from four to five feet diameter of copper
or preferably of wood and copper, the bottom of copper and
" copper sides to the height of about fifteen inches from the
" bottom and the remainder of the sides of wood. Through the
bottom of the pan are cut a series of holes in which are fixed
copper cups about six inches in diameter and four inches in depth,
dropping or hanging in the fire, or otherwise. Into this pan at
about three inches from the bottom is introduced a copper flue,
" which passes through the centre of the pan and is open at each
" end corresponding holes being cut through the sides of the pan
" so as to allow of a free circulation of the flame through the flue
" into the chimney. The flue should be about four inches in
" depth and from ten to twelve inches in width." "The liquid
" when the process of boiling is finished, may be discharged by a
" cock or valve, but in discharging the liquid, either a sufficient
" portion should be left to cover the flue or the fire should be
" covered or damped until the vessel is recharged."
     Second, " emptying or transferring the contents of the vessels
" used for evaporating cane juice or other saccharine solution,"
as follows: —An air-tight vessel, round or oval, made of copper,
is placed above the coppers with which it communicates by means
of copper pipes upon each of which is a tap. Another pipe on
which there is a tap connects this vessel with a steam generator
and there is a tap for the discharge of air from it. The steam is
let into the vessel until it issues from the air tap, the air tap and

tap on the pipe of the generator are then closed, the cock on the pipe communicating with that copper to be emptied is opened, the liquid will rise into the air-tight vessel.

[Printed, 6d. Drawing. London Journal (*Newton's*), vol. 10 (*conjoined series*), p. 345 ; Register of Arts and Sciences, vol. 7 (*new series*), p. 35.]

A.D. 1832, December 21.—N° 6353.

GUTTERIDGE, WILLIAM, and STEVENS, GEORGE.—" Appa-
" ratus for the manufacture and refining of sugar and other
" extracts, also applicable to other purposes." This consists,
first, " in an arrangement or combination of apparatus for boiling
" and evaporating syrups on the principle of Howard's vacuum
" pans, as follows :—The boiling pan is placed and fixed in an
outer vessel, which acts as a boiler for generating steam for
obtaining a vacuum and also for communicating heat to the
boiling pan. On the upper part of the boiler is a safety valve.
A steam pipe with a stop cock (to shut as occasion requires),
connects the upper part of the boiling pan with the boiler. There
is a tight cover on the top of the boiling pan. A pipe leads from
the upper part of the boiling pan to the part of the apparatus for
obtaining a vacuum, an intermediate vessel being placed to
receive any syrup or other matter that may happen to boil over.
At the bottom of the boiling pan is a valve opening into a trough
by which the charge is withdrawn. The apparatus for obtaining
a vacuum consists of an outer vessel on the inside of which are
four curved parallel plates bolted together all round thus leaving
three spaces, "the upper and lower spaces being intended for
" water, whilst the middle space is intended for the steam arising
" from the boiling pan." There are the necessary pipes, taps, &c.,
for carrying of the water condensed and otherwise.

Second, by the other arrangements obtaining " spirituous and
" other extracts from solutions and other mixtures of other
" vegetable substances."

Third, " in apparatus for drying loaves of sugar." This vessel
is cylindrical with a hemispherical end and to the front is fitted
by grinding a cover. A steam pipe conducts high pressure steam
inside and nearly around the vessel, a pipe from the inside leads to
a vacuum vessel, as above or " worked from a separate steam
" boiler." The loaves are placed on sliding open shelves or
frames. A tap at the top admits air when it is desired to remove
the cover.

[Printed, 1s. 4d. Drawings. Repertory of Arts, vol. 16 (*third series*), p. 129 London Journal (*Newton's*), vol. 5 (*conjoined series*), p. 8.]

NEWTON, William. — (*A communication.*) — "An improved
" apparatus for boiling, evaporating, and concentrating syrups
" for the production of sugar, and also of saline liquors, or for
" the crystallization of salt, which apparatus may also be employed
" in the process of distillation." This consists, in reference to
" this subject, as follows :—First, in "an apparatus for boiling or
" evaporating syrups or liquids in open pans or vessels," consisting of a steam boiler, working a high pressure engine placed
above it, which engine actuates an injecting air pump, and also
supplies steam into two close vessels each containing a worm or
coiled pipe, through which the air is forced by the engine and
blowing apparatus, or the steam is "conducted by a worm of
" larger dimensions and the air driven through the interior of
" the vessels." The air becomes highly heated in passing through
the close vessel, and finally passes through a series of small pipes
under perforated plates at the bottom of each evaporating pan, of
which two are shown. A second apparatus is described very much
resembling the above in principle, but the steam engine and
blowing apparatus are constructed upon the oscillating principle.

Second, in apparatus for evaporating or boiling in vacuo, that
is in a covered vessel, "heated by a steam bath, to which is added
" an air pump " for the above purposes, as well as the pump, to
obtain a partial vacuum within the pan or boiler, so as to draw off
both the hot air and aqueous vapour.

Third, the application of this apparatus for the purposes of
distillation.

[Printed, 3s. 4d. Drawings. London Journal (*Newton's*), vol. 4 (*conjoined
series*), p. 161; Rolls Chapel Reports, 7th Report, p. 145.]

TERRY, Charles and PARKER, William.—"Improvements
" in making and in refining sugar." These are said to be, first,
" the use of ferrocyanic acid for preventing or diminishing
" fermentation in the process of refining sugar and in the process
" of making sugar."

Second, "the use of sulphuric acid in the making and in the
" refining of sugar for promoting and increasing that effect called
" crystallization and producing larger quantities of sugar."

Three solutions are made as follows :—No. 1 solution.—10 ozs.
of crystallized sulphate of zinc are dissolved in 3 gallons of cold

water and 3 oz. measures of sulphuric acid of sp. gr. 1,845 are added. This is sufficient for one ton of raw sugar.

No. 2 solution.—19 ozs. best prussian blue in powder, 6½ ozs. of unslaked lime in powder, and 13½ pints of distilled water are degested together, say at 120° F., stirring gently till all the blue colour has disappeared when it is cooled and filtered. Of this liquid "ferrocyanate of lime" of sp. gr. 1,020 at 60° F., 10 pints are sufficient for one ton of raw sugar.

No. 3 solution.—10 ozs. of crystallized sulphate of zinc are dissolved in five gallons of cold water, and 5 ozs. measure of sulphuric acid sp. gr. 1,845 added. "This is sufficient for one " ton of green syrups or molasses with or without a mixture of " sugar."

Mode of application.—To the boiling liquor clarified in the usual way solution No. 1 is added and after a few minutes a violent action takes place, when 3 lbs. of powdered chalk are added, and the solution No. 2 stirred in and the whole boiled for five minutes, then the whole filtered, evaporated for crystallization, and completed in the usual manner to produce lumps or loaves in moulds, but it is preferred to add bullocks blood, white of eggs, and animal charcoal before filtration. Green syrups as they run from moulds with any proportion of raw sugar added are submitted to the foregoing process but the solution No. 3 is used in place of No. 1, and 5 lbs. of chalk are used in place of 3.

[Printed, 4d. No Drawings. Repertory of Arts, vol. 1 (*new series*), p. 230; London Journal (*Newton's*), vol. 4 (*conjoined series*), p. 24.]

A.D. 1834, February 27.—N° 6569.

ARCHBALD, WILLIAM AUGUSTUS.—"A certain improvement " in the making of sugars."

[No Specification enrolled.]

A.D. 1835, September 1.—N° 6893.

SAUNDERS, JAMES FERGUSON.—(*A communication.*)—" Im-" provements in clarifying raw cane and other vegetable and " saccharine juices, and in bleaching such raw juices." These are " clarifying and bleaching cane and other juices by precipi-" tation " by means of the materials afterwards named, " when " such process is performed previous to such juices undergoing " the application of heat," as follows:—Any earth "taken suffi-" ciently below the surface to prevent any vegetable substances

" being introduced with it into the juice" is sifted in order to
remove stones and made with water "to about the consistency of
" thick mud, it is to be gradually stirred into the juice" until
streams of clarified juice are observed to "follow the course of
" the stirring instrument or stick, no further earth will be
" required, nor will further stirring be necessary and the quantity
" of earth will generally be found to be about one by measure to
" ten of juice." "The result depending on a porous property or
" affinity, (&c.) which the earth has for the oily, mucilaginous,
" and other impurities;" "matters having similar properties,
" such as for instance, as pulverized pumicestone will have a like
" effect." The mixing operation having been performed and the
precipitate settled the clarified juice is now bleached by "running
" it off into a quantity, (varying a little according to the quan-
" tity of the color contained in the juice) of animal or other
" charcoal having known bleaching properties reduced to fine
" impalpable powder and saturated with water," stirring up the
whole ten or fifteen minutes, adding earth as before "to pre-
" cipitate the whole." "The approximate proportion requisite
" will be $\frac{1}{4}$ lb. of animal charcoal to a gallon of juice; of other
" charcoals $\frac{1}{2}$ lb., using in each case a double quantity in the first
" instance of either charcoal will insure it being sufficient for
" three operations."

[Printed, 4d. No Drawings. Repertory of Arts, vol. 5 (*new series*), p. 223.]

A.D. 1836, May 5.—N° 7082.

PONTIFEX, EDMUND. — (*A communication.*) — "An improve-
" ment in the processes of making and refining sugar." The
patentee states that he is "aware that sugar has been made and
" refined by means of condensing the vapors arising from the closed
" pan in which the saccharine liquid or juice is contained and hence
" producing a vacuum or partial vacuum in the closed pan;" no
claim is made to "such means generally, but only when saccharine
" liquor or juice is the medium for condensing the vapours."
Two sets of apparatus are described for carrying out the above.
The "vacuo pan" is made in the usual way and to this is attached
a worm-shaped condenser with which is connected a receiver for
the water; attached to this receiver is an air pump, a cock draws
the water from the receiver. Above the worm-shaped condenser
is a cistern containing the liquor or juice to be evaporated. A

small top from the bottom of the cistern conveys the liquid into a vessel which distributes the juice over every part of the condenser. At the bottom of the condenser is a pan which collects the juice, and from thence it passes by a pipe into two cisterns which communicate by pipes and a cock with a vessel into which the syrup is drawn up at pleasure by the power of the " vacuo " in the pan.

In the other apparatus " the pan and evaporator are the same " as before described, but a third vessel is introduced to further " economize the fuel." Into this vessel the syrup which has passed over the condenser is conveyed by a pipe, and steam from the generator enters a worm pipe at the bottom and boils the liquid, the vapour from this liquid passes into the worm of " the vacuo pan." While the liquid partly evaporated passes into the cisterns below and is drawn into " the vacuo pan " as before described.

[Printed, 1s. 2d. Drawings. Repertory of Arts, vol. 7 (*new series*), p. 85.]

A.D. 1836, June 18.—N° 7124.

WATSON, WILLIAM.—" Certain improvements in the manu- " facturing of sugars from beetroot and other substances." These are, first, a machine called " a macerator " for pressing the juice from the beetroot. The beet is rasped or in preference " cut " into thin slices by an instrument similar to a turnip cutter and " afterwards submitted " to the above machine. The macerator is an oblong cylindrical or syphon-shaped vessel composed of a series of boxes. Near the top of this vessel and inside of it, is a wheel arrangement, worked by handles, over which is an endless chain on which are fixed (at intervals) a series of iron gratings thirty-three in number, in such a manner as to form a series of compartments as it were all round the vessel. The macerator is nearly filled with water from from a pipe into it near the top, and steam admitted into it by pipes not far from the lower part of the machine raises the temperature of the water to " 70° or 75° " Reaumur," about 130 lbs. of sliced beetroot are filled in between two of the plates or gratings near the top, when the handle is moved so as to bring up the endless chain and present to the filler another empty plate. This filling process is continued until the whole of the plates in the syphon are filled with slices of beetroot. In proportion as the plates are filled and they are moved forward in the syphon formed by the boxes there is discharged by a box and cock at the other side near the top of the syphon " a quantity of liquid corresponding to the volumn of " beetroot put in." The arrangements are complete for discharging the pressed beet, &c. This juice as it runs from the macerator is clarified by heat of steam, lime, and again steam heat, it is also evaporated and whitened by means of animal charcoal and the syrup is evaporated a second time, the modes of doing all which are given but it is said are not claimed.

Second, destroying " the objectionable taste and smell to which sugar obtained from beetroot juice is liable " by employing " cane " juice syrup or raw cane sugar or colonial molasses," by mixing them with such beetroot juice in proportions which are given but which "must be varied according to circumstances."

Third, the composition of what are named " bark finings and " the application thereof to beetroot juice or any other saccharine " solution." " The bark of the West Indian elm (called in the " French islands l'orme pyramidale" while fresh is soaked in a solution of alum, " about 100 lbs. of the bark to 100 lbs. of alum " dissolved in 20 gallons of water and there soaked for twenty-four " hours, the water is then to be drawn off and the solution boiled " to dryness." This preparation of bark imported is compounded with lime as "in Howard's finings" only there should be a slight preponderance of alum. "The proportion of" finings thus composed must be varied according to the quality of the juice under operation.

Fourth, in vacuum pans arranging "the pipes by preference " after the manner following :—There are two pipes, the first is " carried into the pan from the steam boiler" and coiled round the pan at different distances from the sides and is finally carried out of it ; the second pipe is carried from the steam boiler into the pan at certain distances from the other pipe and is then carried out of the pan. "There must be cocks in the usual way for " admitting the steam and letting off the steam or condensed " water." Affixing "to each of the interior coil of pipes and " also to the jacket of the pan cocks communicating with cold " water; as soon as the sugar is boiled the steam is shut off " and before the vacuum is destroyed cold water is allowed " to flow through the apparatus till it is cooled when the vacuum is destroyed.

Fifth, " producing loaf, lump, or fine sugar." The sugar thus prepared should be damped with water or syrup and drained. It is then mixed with syrup made from refined or purified sugar of a strong quality transferred to the moulds and treated in the same manner as other refined sugar at present practised.

[Printed, 2s. Drawings.]

A.D. 1836, November 22.—N° 7231.

GWYNNE, George, and YOUNG, James.—"Improvements " in the manufacture of sugars." These are, first, the application of " the different suitable bodies of which phosphoric, pyro- " phosphoric, and meta-phosphoric acids are component parts," preferring "diphosphate of lime and neutral solution of phos- " phoric acid and soda, in the manufacture of sugars, where the " salts of oxide of lead," triacetate of lead, hydrated (precipitated) oxide of lead, albuminate of the oxide of lead, are employed to precipitate the impurities therefrom. The triacetate of lead is prepared by boiling with stirring a solution of "acetate of prot- " oxide of lead (sugar of lead)" in certain proportions with oxide of lead. The solution of triacetate is drawn from the undissolved oxide of lead. It is advisable that the triacetate solution should be added to the saccharine solution so long as there is a precipitate; an excess of lead salt is observed by the black color produced, when a small quantity of the solution has added to it "a few drops of the hydrosulphate of ammonia " (sulph. amm. of the apothecary)." The mode of preparing diphosphate of lime is given. "And further, when other bodies, " such as sulphate of soda, acidulated ferro-cyanuret of calcium " (prussiate of lime), acidulated oxalic acid, sulphuric acid, bin- " oxalate of potash, bichromate of potash, &c., are used for pre- " cipitating the oxide of lead," "the neutralizing the excess of " such bodies, together with the acid of the salts of lead set free, " by percolation through suitable insoluble materials, such as car- " bonate of lime, carbonate of magnesia, coarse grained charcoal, " &c.," is claimed.

Second, "the application of flexible substances, as hair cloth " and other suitable flexible fabrics between filtering bags, or " forming bags of corded cloth, which produces the necessary " separation of one bag from its neighbour, whereby very exten- " sive filtering surface may be obtained in a small space for the " purpose or purposes pursued in the manufacture of sugar."

[Printed, 1s. 8d. Drawings. Repertory of Arts, vol. 8 (new series), p. 1.]

A.D. 1837, January 24.—N° 7289.

OLIVER, Julius.—(*A communication.*)—"A certain improve-
" ment in the filters employed in sugar refining." This consists
in the use of any " apparatus whereby animal or vegetable char-
" coal may be cleared of the saccharine and other matters which
" had closed its pores, by the passing of hot or cold water through
" it, with or without the aid of a high column of pressure, or
" the use of a forcing or an exhausting pump," by which it can
be restored without the necessity of "calcining or heating it as
" usual." After the use of the hot water fermentation is set up
in the charcoal, when the charcoal is finally washed with cold
water.

[Printed, 8*d*. Drawing. Rolls Chapel Reports, 7th Report, p. 183.]

A.D. 1837, September 30.—N° 7438.

HOARD, Francis.—" Improvements in making sugars." These
are, the application of an apparatus such as is afterwards
described, " in the process of making sugar, and particularly in
" the mode of producing a constant circulation of the liquor
" in such apparatus." There is an oblong vessel divided into
five compartments by partitions. A flue is underneath the whole
of the vessel, and throughout the bottom are a series of tubes
within the flue " (or they may be a series of narrow chambers or
" vessels) in which the cane or other juice circulates from the
" bottom upwards." Two of the compartments farthest from
the fire-place are oblong, and are separated from each other by a
division down the middle of the pan, and each of these compartments communicate with the next compartment by pipes and
cocks, whilst the remaining compartments communicate with each
other by the same means. Around the vessel is a trough, into
which the scum is put, and from whence it flows into a cistern.
The two oblong compartments alternately become the preparing
vessels, and whilst one is being worked off, the other is coming
forward. The liquor is passed from one compartment to the
other.

[Printed, 1*s*. Drawing. Repertory of Arts, vol. 10 (*new series*), p. 93 ;
London Journal (*Newton's*), vol. 15 (*conjoined series*), p. 165.]

A.D. 1837, November 11.—N° 7469.

CROSLEY, Henry.—(*A communication.*)—"Improved means to
" be employed in manufacturing beetroot and other vegetable

" substances for the purpose of obtaining saccharine matter
" therefrom." This consists in cutting, drying, bruising, and
breaking up vegetable substances, preparatory to obtaining a
saccharine extract therefrom, and treating vegetable substances
" when so prepared, with milk of lime, and thus obtaining there-
" from a clammy or glutinous mass," without the use of sulphuric acid, and treating the same in a series of vessels, and
" obtaining therefrom the saccharine extract required in a con-
" centrated state, by the simple process of percolation," as
follows :—The process is, by preference, employed upon beetroot,
although it may "with suitable modifications be equally well
" applied to and for extracting the saccharine matter contained
" in other vegetable substances, such as the sugar cane, the stalks
" of Indian corn, potatoes, and other roots." The roots cleansed
are " cut into pieces somewhat of the shape of a parallelopiped,"
and put into open worked boxes or baskets, and these boxes are
placed upon open worked drawers in drying chambers, into the
bottom of which air is passed, the air and vapour passes off from
the top into the general chimney. The temperature ranges from
54° to 188° F. The cut pieces of root having been dried are
reduced by grinding, &c. as coarse as malt, and may be mashed
in an ordinary brewer's mash ton with a milk of lime of about
4¼ lbs. of lime to 5 gallons of warm water, to about 100 lbs. of
ground roots. This clammy or glutinous mass is put into a
series of wooden or other tubs with false bottoms, and water
from 166° to 212° placed on the first vat filters through the series
of vats. The extract is filtered through charcoal, and evaporated
in the ordinary manner.

[Printed, 10d. Drawing.]

A.D. 1838, February 24.—N° 7573.

STOLLE, EDWARD.—" Improvements in making sugar from
" sugar cane, and in refining sugar." These are, in the employment of "a new chemical agent for the discolouring of saccharine
" matters (sugar), for which purpose animal charcoal has been
" generally employed," as follows : — One or two thousandth
parts of lime are employed, and when boiling, the scum is taken
away, and "twelve pounds of liquid sulphurous acid (marking
" no more than four degrees by Beaumé's areometer) are slowly
" poured into the juice with care" into 1,000 parts of juice,

g

after which the juice is evaporated to about twenty or twenty-two degrees, passed through a filter of flannel, &c., and concentrated to crystallization. For refining sugar of a very bad quality using concentrated alcohol or spirit charged with about two per cent. of sulphurous acid and mixed with sugar, " so that a small portion " only of the liquid swims upon the sugar;" " it is then stirred " several times, and after about two hours the liquid is drawn " off, after which the sugar is washed in pure alcohol." The alcohol in the molasses washed out is separated from them by distillation, "and can be employed again for the same purpose."

[Printed, 4d. No Drawings. Repertory of Arts, vol. 10 (*new series*), p. 233 ; London Journal (*Newton's*) vol. 13 (*conjoined series*), p. 271.]

A.D. 1838, March 8.—N° 7587.

LAWRENCE, Morton William.—" Certain improvements in " the process of concentrating certain vegetable juices and saccha- " rine solutions." These are, first, " evaporating in an open " vessel the juices and solutions " to " a point of concentration " as near that which " it is desired that they should ultimately " attain as they can be safely made to attain in such open vessel, " and then completing the process of concentration by further " evaporating such juices and solutions in vacuo either with or " without the application of heat," and whether "carried on in " any close vessel of the kinds now in use," or of the kind described under the second head.

Second, substituting for the usual mode of fastening the cover of close vessels for the above purpose, which was "by bolts and " cement, or other fixed and permanent fastenings an air-tight " junction" effected somewhat in the following manner :—When no heat is to be applied during evaporation in vacuo, the lower part of the vessel is enclosed in a casing of wood containing saw-dust, charcoal, &c. If heat is to be applied, another material must be used. The upper part or cover of the vessel should drop into the lower part, and " rest when down upon the interior sur- " face of the bottom of such lower part," and "when let down " upon the bottom it may enclose nearly the whole thereof." It is made with a flanch with a packing of sail or other cloth, and it is rivetted or otherwise. There may be a flanch on the lower part from three to four inches inches broad, "provided with rims at " its external and internal circumference, so as to form a channel " between the rims of about an inch and a half in depth, and

" of sufficient width to receive the edge of the cover intended
" to drop into it." The cover has " a flanch " about the width of
two inches, and there is a packing of cloth between, which is
moistened by the syrup.

[Printed, 1s. Drawing. Repertory of Arts, vol. 13 (*new series*), p. 109.]

A.D. 1838, March 26.—N° 7601.

OLIVER, JULIUS.—(*A communication.*)—" A certain improve-
" ment in the filters employed in sugar refining." This consists
in the use of any form of apparatus whereby animal or vegetable
charcoal may be cleared of the saccharine and other matters which
had closed its pores by the passing of hot or cold water through it
with or without the aid of a high column of pressure, or the use
of a forcing or an exhausting pump," without "the necessity of
" again heating it in close vessels as is usually the case, but
" simply by the passing either of hot or cold water, as above
" indicated, through it, and fermenting it after the use of hot
" water," finally washing it with cold water.

[Printed, 8d. Drawing.]

A.D. 1838, September 6.—N° 7796.

BERRY, MILES.—(*A communication.*)—" Certain improvements
" applicable to certain parts of the process generally used for the
" manufacture and refining of sugars."

[No Specification enrolled.]

A.D. 1839, June 17.—N° 8107.

LOOS, EDWARD.—" Improvements in extracting the saccharine
" matter from sugar canes and other substances of a saccharine
" nature; which improvements are also applicable in extracting
" colouring matters from wood and other matters used in dyeing."

[No Specification enrolled.]

A.D. 1839, July 6.—N° 8147.

FAIRRIE, JOHN. — " Improvements in making and refining
" sugar." These are, first, " dispensing with the high degree of
" temperature heretofore resorted to in the process of mixing and
" filtration " of sugar. " About forty-five parts of sugar and
" fifty-five parts of water " (cold), " have been found convenient

" in practice," the sp. gr. of the solution being from 1199 to 1200 ; in preference, the sugar is dissolved "in lime water, or pure " or spring water," alone; " this process is equally applicable " whether bog or charcoal filters or other filters be employed."

Second, concentrating sugar, by means of what is called an " evaporator." This "consists of a copper pipe of about twelve " inches diameter and twenty-six feet length, containing nine " copper steam tubes, each twenty-four feet long and one inch " and a quarter diameter, which is arranged at nearly equal " distances from each other in the lower portion of the interior of " the larger pipe." These tubes are arranged for steam to pass through them ; the evaporator closed in is connected to the upper part of a common vacuum pan. "The clarified sugar liquor is " passed through the evoporator in a continuous stream," and enters the vacuum pan not quite at the crystallizing point. The evaporator may be made open at top and placed over an open fire, or high pressure steam may be used without a vacuum.

❧ Third, "applying heating apparatus after the sugar has left the " vacuum pan." This is an open copper vessel or trough about 36 feet long, 1 foot wide, and 1 foot deep, heated at the bottom by steam into a double bottom, and by three steam pipes inside there are two valves for drawing off the sugar, &c.

[Printed, 4d. No Drawings. Repertory of Arts, vol. 14 (new series), p. 43; Inventors' Advocate, vol. 2, p. 131.]

A.D. 1839, November 7.—N° 8258.

CROSLEY, HENRY.—" An improved battery or arrangement of " apparatus for the manufacture of sugar." This is said to consist, first, in "the application of steam to heat the liquids in the various " vessels of the battery by the introduction of coils or lengths " of pipe."

Second, " the crystallizing vessel."

Third, "the general arrangement of the battery or apparatus."

There is a steam boiler from which is a pipe which conveys steam into a series of pipes at the bottom of an oblong square cistern, into which the cane juice is pumped "previous to temper- " ing it with lime," the condensed steam as hot water flows into a cistern in which there is a pump for feeding the steam boiler. The heated juice is discharged by a tap on a pipe into a clarifying pan heated by means of a coil of pipe at the bottom. The clarified sugar or cane juice is discharged, by means of a shifting gutter, into either of two pans, in each of which a coil of pipe is fixed in the bottom, and is "finally cleansed or cleared and partially "evaporated into syrup," preparatory to "discharging it into "the striking teache or pan," which also has a coil of pipe at the bottom and sometimes around the side; "in this pan or teache "the syrup is finally evaporated to the consistence for crystalliz- "ing," when it is discharged by a gutter into a crystallizing vessel of an oblong shape with "a rod or pricker, with stays above "it, passing along the apex or bottom of the vessel," which, when the crystallization is completely effected, is withdrawn and forms a hole for the molasses to drain. The rod is introduced from time to time. This crystallizing vessel has a space around it for the circulation of steam around its sides.

[Printed, 8d. Drawing. Inventors' Advocate, vol. 2, p. 227.]

A.D. 1840, March 31.—N° 8461.

BANCROFT, Peter, and MAC INNES, John.—"An im- "proved method of renovating or restoring animal charcoal after "it has been used in certain processes or manufactures to which "charcoal is now generally applied, and thereby recovering the "properties of such, and rendering it again fit for similar uses." This consists "in the application, use, or employment of an "alkali, in order to dissolve and carry off the coloring matter and "other impurities left or deposited in the animal charcoal after "having been used in decoloring syrups and other fluids." The spent charcoal is washed to remove any syrup, and saturated with a solution of caustic potash or soda of sp. gr., "about 1·06, more "or less," and allowed to remain for a few hours. The caustic solution "charged with the coloring matter, &c. is to be run off "by a plug or tap placed at the bottom of the vessel or in any "other way;" the charcoal is then washed till free from alkali ascertained by the use of turmeric paper or any other known test. In place of caustic potash or soda, the "carbonates or liquid am- "monia will effect the same purpose, though not in so rapid or "perfect a manner." Animal charcoal in a granular state may be thus purified without removing it from the filter, "but if fine "pulverized animal charcoal is to be similarly operated upon, it "must be washed by subsidence and decantation."

[Printed, 4d. No Drawings. London Journal (*Newton's*), vol. 20 (*conjoined series*), p. 355. Mechanics' Magazine, vol. 33, p. 381. Inventors' Advocate, vol. 3, p. 213.]

A.D. 1840, November 2.—N° 8675.

SCHRODER, HERMAN. — (*Partly a communication.*) — " Im-
" provements in filters." These are, employing frames and
bags; " constructing and applying filter bags ;" also, " construct-
" ing and applying filter bags by drawing each bag within itself ;"
likewise, " the mode of combining bag filters " as follows :—
First, a quandrangular box with a cover has a perforated false
bottom ; a series of bags which may be plain, but in preference,
are plaited or gathered and the edges bound, the mouth or upper
part the same, and also bound with tape or web so as " to obtain
" as extensive a surface as possible." A frame is introduced into
the mouth of each bag, and these frames (with their bags) are
wedged together in the above box, the bottom of each bag resting
on the false bottom. The liquor flows into the bags from a vat
above, and a cock is in the box below the false bottom for
running off the liquor as it is filtered.

Second, an " outer bag, which, when full of liquor, should lie
" closely against every part of the interior surface of the box or
" chamber," a series of smaller bags have their lower parts affixed
to the bottom of the outer bag, they are tied up to two loops
fixed to the upper part of the interior of the bag. The mouth of
the outer bag is drawn together and tied or secured round the
nozzle of the pipe supplying the liquor to be filtered. In preference, " the bags should be plaited in place of plain," as also
" the filters hereafter explained."

Third, " making bag filters whereby each bag is drawn into
" itself one or more times." The box or chamber is same as
above, and each bag has its frame in its mouth, but the bags are
much longer and each bag is drawn up into itself.

Fourth, an outer bag in which are a series of inner bags closed
on all sides, they are kept upright in the inner bag by strings, and
have each an outlet at the bottom of the inner bag, just above the
false bottom.

[Printed, 1s. Drawings. Mechanics' Magazine, vol. 34, p. 383; Inventors
Advocate, vol. 4, p. 293.

A.D. 1840, December 2.—N° 8731.

ROBINSON, JAMES.—(*A communication.*)—" A sugar cane mill
" of a new construction, and certain improvements applicable to

" sugar cane mills generally, and certain improvements in appa-
" ratus for making sugar." These are, first, " constructing sugar
" cane mills whereby the sugar canes in passing shall be pressed at
" least three times, and the roller tied with malleable iron straps."

In effecting the above in the first arrangement there are four
rollers, a large roller above and three small rollers below; in
another arrangement are three pairs of rollers, one after the other.

Second, "the application of hot water or steam to sugar canes
" as they are passing through sugar cane mills," by means of
" a pipe perforated with numerous small holes."

Third, "feeding sugar canes into sugar mills of any construc-
" tion." The canes are first put on an inclined plane or table,
raked down on to an endless band, which is kept constantly
moving towards the sugar cane mill at a proper rate or speed by
means of pulleys of different sizes. There are other arrangements
modifications of this.

Fourth, "constructing driving wheels of sugar cane mills,"
capable of "giving way on an extra strain being brought on the
" mill, which might otherwise destroy some part of the machi-
" nery." A wheel has a groove in the periphery, and also several
set screws passing through the outer ring. On the periphery of
the wheel is applied a toothed ring, which has also a groove
corresponding with the groove in the periphery of the wheel.
Within these grooves friction brasses are placed, and held by the
set screws. By excessive pressure the toothed ring will slide
round on the friction brasses and thus prevent any damage
being done.

Fifth, in place of making the necks or axes of sugar rollers of
separate parts the necks or axes are cast with the roller itself.
They are also cast hollow for steam.

Sixth, "forming a rim around a set or series of sugar pans by
" casting part of such rim on each pan," so that "when a series
" of pans are attached by screws " "they should together produce
" the requisite rim and allow of the pans boiling over from one to
" the other, by the rim at the junction of the two pans."

Seventh, setting sugar pans with flat bottoms with shields of
brickwork or masonary under them, by which the action of the
fire will be "materially less, and the liquor over these places will
" be more quiet than elsewhere, and the scum will collect over
" such shielded parts of the pans, and allow of its being readily
" removed."

Eighth, "the application of iron pans tinned or coated with tin
" or alloys thereof."

[Printed, 2s. 2d. Drawings. Mechanics' Magazine, vol. 34, p. 446, and
vol. 35, pp. 257, 362; Inventors' Advocate, vol. 4, p. 372, and vol. 5,
p. 313.]

A.D. 1841, May 6.—N° 8952.

MORLEY, PHILEMON AUGUSTIN.—" Certain improvements in
" the manufacture of sugar moulds, dish covers, and other articles
" of similar manufacture." These are, in reference to this subject,
in making of sugar moulds, first, of "apparatus for holding the
" corner of the plate during the operation of cutting, and dis-
" pensing with the hole usually made therein." A strong iron pin
passes through a bench and is fixed therein by a nut and screw,
upon this pin as a centre turns a flat piece of iron; this iron is
secured to the pin by the head of the latter being rivetted, but so
loosely as to allow the plate to turn about freely; to this piece of
iron is attached the bridge piece, through which passes the screw
turned by a handle. The handle turned, the bottom of the screw
is raised from the plate, the sheet of iron introduced under the
screw and the handle reversed, the iron plate is held securely. On
introducing the edge of the plate between rototary shears its edge
is cut into the form required. The edges of the plate are next
turned in reverse directions. It is said that " the apparatus used
" for this purpose is of the common kind, and therefore needs no
" explanation."

Second, a bending machine; a block formed into a conical figure
the size of the mould to be formed, is mounted on an axis, and put
in motion by a winch, the edge of the plate is secured to the cone,
the winch is turned by one hand whilst with the other the metal
plate is pressed against the surface of the cone and, passing under
a gauge bar or plate affixed a short distance from the plate, the plate
is wrapped round it, and by a "grooving stake" the metal cone is
more contracted, the hooked edges are joined and punched together.

Third, the wiring operation; any excess of metal is cut off, and
the edge is brought between two rollers so as to form a U groove
into which a wire is deposited, and the edge is next brought under
a creasing machine, consisting of an arrangement of two rollers
which are made to approach or recede from each other by means of
a treadle, and finished.

[Printed, 1s. Drawings. Mechanics' Magazine, vol. 35, p. 399; Inventors'
Advocate, vol. 5, p. 310.]

A.D. 1841, November 23.—N° 9162.

MANWARING, WILLIAM. — "Certain improvements in the " manufacture of sugar." These are, first, "the sole and exclu- " sive use of every apparatus and combination of apparatus " whereby sugar may be "made to cleanse itself." The apparatus described under this head consists of a pan or pans " surmounted " with a rim or bevelled ledging, projecting above and over the " top and sides of the pan or pans, forming an expanded space " into which the froth, scum, and feculent matter that arises " from the liquid or solution ascends." From thence it flows into gutters, where it is exposed to the air so as to bring it into a liquid state, and from thence it flows "through strainers or any well " known filtering medium," and again through other strainers or vessels containing a filtering medium. "These gutters or vessels " for containing the filtering medium are so constructed as to be " easily removed " and "replaced by other and clean ones, &c."

Second, charging the juice up into the evaporator or evaporators, or to any other required vessel, as follows :—The liquid from the aforesaid filters or strainers flows into one or more receiving vessels until they are " charged to a proper height, a stop-cock is opened " in a pipe attached to the upper part of this vessel, and com- " municating with a high pressure steam boiler," the steam forces the liquid up through the pipe into the evaporator or evaporators ; when the receiver is discharged " the steam cock is closed, and the " cock from the strainers or filters opened, and the operation " goes on as before."

Third, the application of atmospheric air in the final evaporation, by means of a series of perforated pipes " fixed in the pan or pans " near to and immediately above the coil of steam pipes," where steam is used as the agent of heat or near to the bottom of the pan or pans where steam is employed in a jacket or where fire is employed. The air is injected by means of " pumps or any well " known blowing contrivance."

[Printed, 10d. Drawing.]

A.D. 1842, November 25.—N° 9522.

TARGET, FELIX NAPOLEON.—" A new method of refining or " manufacturing sugar."

[No Specification enrolled.]

A.D. 1842, December 3.—N° 9533.

POUCHANT, Don PEDRO.—"A certain improvement or im-
" provements in the construction of machinery for manufacturing
" sugar."

 [No Specification enrolled.]

A.D. 1842, December 28.—N° 9574.

CROSLEY, HENRY, and STEVENS, GEORGE.—"Certain im-
" provements in the manufacture of sugar and the products of
" sugar." These are, first, "expressing or squeezing out the
" saccharine juice from sugar canes or rattoons in such lengths
" as they are ordinarily brought from the fields, or cut into
" shorter lengths, if thought advisable, by subjecting them, in
" heaps of more or less bulk, to a strong insistent pressure in
" one given line of direction, whether such pressure is produced
" by means of a hydraulic press," which is preferred, or "by any
" other equivalent mechanical means."

Second, "laying of the canes or rattoons, or pieces thereof,
" (lengthwise or endwise when of suitable length for the purpose),
" according as the said pressure is applied vertically or laterally,
" whereby they are split or riven in the direction of the fibre only,
" and whether the canes or rattoons are all laid one way, or laid
" crosswise, or at any angle with respect to one another, as long
" as the insistent pressure is applied in such a line of direction as
" to split and rive the canes or rattoons in the direction of the
" fibre."

Third, employing for "filtering, clarifying, and purifying sugar-
" cane juice, molasses, or syrup, or other residual products of
" sugar; of anthracite coal, coke of bituminous or other coal,
" burnt megass ashes, cocoa-nut charcoal, and clean coarse sand,
" whether the same are used" in the proportions named in this
Specification or "in any other available proportion, and whether
" combined with the method or system of pressing the canes
" before described, or combined with the ordinary sugar mills, or
" any other known means of compression." The filtering vessel
has a false bottom on which is a perforated sheet of metal or wire
gauze, and over this a cloth of some fibrous material. A small
air pipe is passed from the top downwards to below the false
bottom. A layer of filtering materials of two, three, or more feet

is placed, and again a perforated plate, leaving sufficient space for a head of liquer. An exhaustion may be applied under the filter.

[Printed, 10d. Drawing.]

A.D. 1843, January 11.—N° 9584.

RITTER, WILLIAM.—(A communication.)—"Improvements in "crystallizing and purifying sugar." These are, first, in "puri- "fying and crystallizing vessels." The purifying boxes are made of wood and are larger at the top than at the bottom. They are placed one within the other; each has a zinc lining at and near to the bottom; there is a lip or gutter formed in the zinc lining to carry the syrup to a funnel, which dischages it into a gutter or channel. There are iron wire supports fixed near the bottom of the purifying box, over which the metallic web, wire gauze, or other sieve is stretched, and on this the sugar is placed. The syrups used for washing the sugar are placed over the smooth surface of sugar and percolate through it. The crystallizing cases are arranged one within the other; their bottoms are "slightly "inclined from back to front, say, half an inch." The bottom of the front is pierced with seven holes, one inch diameter, to allow the syrup to run off. These holes are stopped by seven stoppers, the ends of which project beyond the back of the case. The plugs are "turned by a key to loosen them from the crystallized "sugar and allow of removal." The front of the case has a lip to collect the syrup and causes it to drop into a channel without flowing over the bottom of the case.

Second. The process of "continued purification" of the sugar by repeated applications or coverings of sugar syrup of rather superior quality, in each operation to the sugar under treatment, until finally pure sugar is used in the last washing.

Third. Collecting "crystals of sugar from exhausted syrups "by boiling or mixing them with impure sugar," and also "by "boiling very weak syrups at a very low temperature and allowing "them to stand in large reservoirs until crystallization has taken "place, in combination with the subsequent use" of the continued purification as above.

Fourth. "The concentration of solution of pure sugar when "heated to about 230° F., by the addition of a proper quantity of "pure sugar instead of evaporating away the water to produce

" the said concentration for removing into the water or cooler, as
" the case may be, and being then conveyed into the conical
" moulds in the usual manner."

[Printed, 1s. 10d. Drawings. London Journal (*Newton's*), vol. 24 (conjoined series), p. 112.]

A.D. 1843, October 5.—Nº 9898.

HARDMAN, LAWRENCE.—" Certain improvements in machinery
" or apparatus to be employed in the manufacture of sugar."
These are said to be, " the application of the principle of centrifugal
" force for the separation or dispersion of molasses and other
" soluble impurities and colouring matter from sugar, and as a
" means of syruping or watering sugar from which its molasses
" and impurities have been in part removed, in order further to
" cleanse and purify the same." It is stated " that a great
" variety of modifications and arrangements of mechanism may
" be readily adapted to the same purpose." There is described, a
revolving case or drum divided by fine sheet wire gauze or cloth
so as to form two circular chambers, and the whole is enclosed in
two thicknesses of wire gauze or cloth. There are in the upper
plate doors, valves, or openings for the admission of the sugar and
air, &c. The mode of operating is as follows :—The raw sugar
(combined with molasses, &c.) is placed in the outer compartment,
and a high velocity (say, 800 or 1000 revolutions per minute)
given by means of the driving shaft (which is hollow and perforated with holes), when the molasses and other soluble matters
will (by centrifugal force) be immediately dispersed through the
gauze into the outer casing, and pass off by the pipes below. The
sugar thus dried may be removed by discharge valves. " When
" the molasses is sufficiently dispersed from the sugar, a quantity
" of syrup may be introduced by the hollow shaft, and passing
" through the perforations and gauze " by the revolution of the
machine, passes through the sugar, carrying off a portion of
impurity which may adhere to it.

[Printed, 10d. Drawing. London Journal (*Newton's*), vol. 24 (conjoined
series), p. 153.]

A.D. 1844, March 28.—Nº 10,125.

COOPER, JOSEPH.—" Certain improvements in the purification
" and clarification of sugar, which improvements are also appli-
" cable to the purifying and clarifying other articles of commerce."

These are, "employing apparatus suitably arranged for allowing
" the crystallizing process to be performed therein, and then to
" admit of the vessel being opened to a vacuum, or partial
" vacuum, so as to bring the pressure of the atmosphere to
" operate on the mass and displace the fluid matters therefrom,"
The crystallizing vessel has a curved surface at the bottom, which
is perforated with small holes, and "on the interior is placed a
" tube" "having a long split or opening, which being turned
" upwards (the interior made vacuous) brings the pressure of the
" atmosphere to act on the sugar in the crystallizing vessel, and
" thus facilitates the removal of the fluid matter from the crystals
" of sugar." This tube turns within the perforated curved bottom
as a plug in a cock, and one end of the tube is connected by
means of a branch pipe with a proper vessel to receive the syrup,
while the other end or branch communicates with an air pump or
other suitable apparatus for exhausting the air from the receiver.
Other arrangement of apparatus may be employed.

In a Disclaimer enrolled the 28th September 1844, it is stated,
that the invention intended to have been described under the following part of the title, namely, "which improvements are also
" applicable to the purifying and clarifying other articles of
" commerce," is not of such practical utility as would make it
" desirable to retain it," and it is disclaimed.

[Printed, 8d.   Drawing.   Repertory of Arts, vol. 4 (*enlarged series*),
p. 342.]

A.D. 1844, April 30.—N° 10,171.

CONSTABLE, JOHN.—(*A communication.*)—"Certain improve-
" ments in the manufacture of sugar." These are, first, "cutting
" the cane into thin slices in place of bruising and pressing the
" cane in a mill."

Second, "neutralizing the acids and fixing or solidifying the
" other extraneous matters, which would otherwise be liable to
" act chemically on and injure the juice."

Third, "extracting the saccharine matter from the cane by
" infusion or immersing the slices of cane in water."

Fourth, "the general arrangement of the apparatus and mode
" of conducting the manufacture" as follows :—

The cane is cut into slices from one-thirtieth to one-tenth of an
inch in thickness; the machine should be without vibration "so as
" to prevent the acids of the plant during the cutting operation

" from mingling with the saccharine matter." A machine is minutely described, the cutting part of which is a series of discs or cutting blades on a shaft, kept at equal distances by brass collars one-tenth of an inch thick. Other modes of making such a machine are described more suitable for unskilled workmen. The machine, it is said, admits of many other modifications, and " in fact, cutting apparatus of totally different constructions may " be employed as " on the principle of an ordinary chaff-cutting " machine," and " on the principle of the guillotine," both of which are described. The slices of cane when cut by either of these machines are transferred into an open wire basket, which is attached to a carriage travelling on a rail above a series of vessels containing water at a temperature of 176° F., in which there is a quantity of hydrate of lime to neutralize the acids in the cane. By immersing this basket in this solution in a particular way, solutions are obtained and run from these vessels, are filtered and evaporated in flat vessels, and by passing over or through cones, &c. heated by gas, and through filters heated by gas, passing over columns similarly heated. After the last evaporation the juice is run into moulds. The ligneous matter after removal of the saccharine matter, dried in the sun or by pressure, may be distilled in a furnace similar to those for smelting iron, but smaller; the gases pass into a small purifying apparatus used as a stove for drying the sugar placed over it, and the gas passes through pipes to be used as above along with air.

[Printed, 6s. 4d. Drawings.]

A.D. 1844, June 6.—N° 10,221.

HIGHAM, WILLIAM, and BELLHOUSE, DAVID. — " Im-" proved construction of boilers for evaporating saline and other " solutions for the purposes of crystallization." This consists as follows :—An oblong boiler is supported by brickwork, at one end is the fire place, and the flue runs through the boiler, the steam generated in the boiler rises into a chamber placed above it, and when it is at too great a pressure it is allowed to blow off at two escape valves in the top ; the evaporating pan is placed immediately above the boiler, and is divided longitudinally by the steam chamber. The boiler is filled with water by a pipe into that part of the boiler over the fire-place. There are two cocks near the top for ascertaining the height of the water, and a cock below for emptying the same when required. " In order to prevent the

" boiler from colapsing when but a small quantity of steam is in
" the chamber," there is a self-acting valve in connection with
the escape valve for " admitting air into that chamber, and by this
" means a pressure equal to that of the atmosphere is always
" preserved therein." " Instead of passing the flue directly
" through the boiler to the chimney," sometimes returning it
under the same, and thus obtaining " a greater heat and draught."
" The boiler may also be varied."

[Printed, 6d. Drawing. London Journal (*Newton's*), vol. 26 (*conjoined series*), p. 84.]

A.D. 1844, October 10.—N° 10,345.

ROBINSON, HENRY OLIVER. — " Certain improvements in
" steam machinery and apparatus for the manufacture and refining
" of sugar." These are, first, " the combination of a steam engine
" of any kind and a sugar-cane mill upon one base-plate or
" base-frame of iron or timber."

Second, " the arrangement of the same so that the lower extre-
" mities of their fixed parts shall come into the same plane, and
" be capable of being affixed to a frame or plate of any kind."

Third, " the construction of a sugar-cane mill with four rolls
" or pressing cylinders placed in pairs, and the upper rolls
" advanced forward with respect to the lower ones."

Fourth, " the evaporation or partial evaporation of cane juice in
" a steam close vessel, under a pressure of the steam emanating
" from the liquor underneath undergoing evaporation, and thus
" obtaining steam for useful purposes, economizing fuel, and
" dispensing with water."

Fifth, " close filtering vessels operated upon by a pressure of
" steam urging onwards the liquor or syrup and keeping it hot
" during the process of filtration." The steam is led by a pipe
from the above close vessel, and it may be from a steam boiler
into the " surface filtrator," a steam-tight vertical cylinder with a
spherical bottom in which are alternate layers of " textile fabric
" or filtering cloth and wicker or pierced metallic plates ;" from
this the syrup flows into a similar vessel with animal or bone
charcoal.

Sixth, the employment of steam for the above purpose.

Seventh, " the form of vacuum evaporating vessel " with
" straight internal tubes and compartments at each end." This,
in preference, is of a cylindrical form placed horizpntally ; at each

end is a space formed by plates of metal in which are the above tubes, and steam is in the tubes and compartments for evaporating the syrup in the body of the pan.

Eighth, the construction of a vacuum pan or vessel of any shape " of cast-iron, or cast iron coated with or alloyed with any " other metal."

[Printed, 10d. Drawing. Practical Mechanics' Journal, vol. 4, pp. 181, 195; Engineers' and Architects' Journal, vol. 8, p. 168.]

A.D. 1844, October 17.—Nᵒ 10,351.

BORRIE, PETER.—" Certain improvements in the machinery or " apparatus for the manufacture of sugar." These are, first, " the " construction and arrangement of a machine for cutting the sugar " canes into proper lengths when the juice is to be expressed by " a hydraulic press." There are fixed blades and moveable blades working on the centres, so that they pass close to the edges of the fixed blades in the manner of shears.

Second, " the use of a friction wheel and self-acting feeding " apparatus" in " the construction of horizontal cane mills." In this friction wheel the tightening is self-acting, and in the feeding apparatus there is a toothed or fluted roller in connection with a disengaging clutch and lever, to prevent too heavy a feed going into the mill.

Third, " the use of hydraulic presses to express the juice from " the cane, of the peculiar construction shown in the drawings." The receiving box is made of two thicknesses of malleable iron plates with iron bars between them ; the inside plates are pierced with holes ; the top plate is in two pieces. Between the layers of cane are double perforated plates or small perforated malleable iron tubes mixed with the canes, which convey the juice into the spaces between the double sides of the box. There are other arrangements in these presses. Steam is sometimes introduced into the receiving box amongst the canes.

Fourth, the use of the hydraulic pumps. Four pumps are placed in a cistern ; there is a rocking shaft for working the pumps by means of double-ended levers fixed upon them. Two of the pumps are larger in diameter than the others, and other arrangements.

Fifth, the general arrangement and combination of a steam boiler which enables the flame and heated air to traverse in various ways about the boiler.

Sixth, apparatus for raising the juice or liquor from a low to a higher level. It is stated that "when a pump is used for this " purpose the churning motion of the piston or bucket of the " pump agitates the juice or liquor so much as to cause it to fer- " ment;" and to obviate this evil apparatus is used where the juice is raised through a pipe by means of an exhaust and also by the pressure of steam upon the surface of the juice.

Seventh, apparatus for raising the temperature of steam. The steam under pressure comes along a pipe from the boiler, enters an expansion vessel, and from thence passes into a series of vertical heated pipes by which the temperature is raised.

Eighth, treating defecating vessels or clarifiers by surcharged steam. There are a series of vessels having a double bottom between which the steam enters.

Ninth, the filters, consisting of a cylindrical vessel, the upper part or head being larger than the lower part, a filter box is in the middle. Heating the filter by means of a casing of steam round the filter; also employing a partial vacuum at the lower side of the filtering medium.

Tenth, the application of surcharged steam to heat the different vessels of the system. There are four pans with double bottoms, between which surcharged steam is introduced to heat the juice in the pans.

Eleventh, the use of surcharged steam to heat the syrup in a vacuum pan. The vacuum pan may have a double bottom for steam and also sometimes small tubes are in the bottom of the pan itself communicating with the double bottom.

Twelfth, in air pumps a pump producing double action, and an air pump of a rotatory description.

Thirteenth, in sugar moulds having them fitted with flanges so as the lower sides fit by means of a joint of canvas, &c. into a box in which a partial vacuum can be formed.

[Printed, 4s. 6d. Drawings. Artizan, vol. 5, p. 218.]

A.D. 1844, November 9.—N° 10,389.

DEROSNE, CHARLES.—"Certain improvements in extracting " sugar or syrups from cane juice and other substances containing " sugar, and in refining sugar and syrups."

This was a prolongation for six years of the grant of Letters Patent to Charles Derosne, 29th September, A.D. 1830, No. 6002.

For the abridgment of the Specification see page 46.

A.D. 1844, December 5.—N° 10,418.

RONALD, JOHN.—" An apparatus for boiling sugar-cane juice
" and other liquids." A vessel is described heated by a continuous length of eight strong pipes along the bottom, and the
invention is said to be, first, admitting the steam into the pipes at
two or more different places.

Second, " creating a much greater degree of boiling " in one
part of the chest or vessel by having in that part piping of greater
diameter than in another. This also causes a rolling motion in
the liquid, and this rolling motion, it is said, may also be caused
by additional pipes, or leaving out pipes, in a part of the vessel,
or the rolling may be caused " by any other plan."

Third, " driving a revolving fan or agitator by means of the
" rolling motion of boiling liquid in manner like an under-shot
" water wheel; " also, " half cylinders of copper or other metal
" or wood " may be introduced, " which the rolling motion of the
" liquid may rush upon and over them," or any other contrivance whereby evaporation is aided or increased.

Fourth, " joining the lengths of pipes by bends with external
" and internal screws."

Fifth, " the plan of forked supports for the piping fixed
" through the bottom of the chest, each support being about
" one-sixteenth of an inch shorter than the one preceding it."

Sixth, " affixing all the entrance and exit pipes at one end or
" side of the chest," so that the expansion or contraction of the
piping will not injure the chest or joints, and " will render
" unnecessary the usual and troublesome plan of stuffing
" boxes."

Seventh, a " self-acting expansion valve for keeping the pipes
" clear of the water formed by condesation."

[Printed, 10d. Drawing. Engineers and Architects' Journal, vol. 8, p. 226;
Mechanics' Magazine, vol. 43, pp. 49, 78.]

A.D. 1845, January 16.—N° 10,474.

GADESDEN, AUGUSTUS WILLIAM.—"Improvements in the
" manufacture of sugar." These are, " effecting the evaporation
" of syrups at a low temperature" by revolving in the syrups
which are heated, cylinders which are partially immersed,
" although a cylindrical vessel is best," a conical or other figured

surface might be used. The heated vapours may be removed by having coverings over the apparatus, "with a pipe leading to a " chimney or other outlet, care being taken to maintain such a " draft as will cause the heated vapours to be quickly removed " from such evaporating vessels."

[Printed, 6d. Drawing. Repertory of Arts, vol. 6 (*enlarged series*), p. 230; London Journal (*Newton's*), vol. 27 (*conjoined series*), p. 183; Engineers and Architects' Journal, vol. 8, p. 357.]

A.D. 1845, January 31.—N° 10,505.

JOHNSTON, JAMES. — " New and improved processes and " machinery for making and refining sugar." This consists as follows :—First, "an apparatus by which vacuum pans may be " emptied without the vacuum being destroyed," by attaching " a pipe to the bottom of a vacuum pan." This pipe is named " the barometer pipe," and is "of such length that the weight of " the saccharine fluid within it shall counterbalance or rather " exceed the weight of the atmosphere, so that when a stop-cock - " at the bottom of the pipe is opened the saccharine fluid within " the pan will of its own weight rush out by the pipe till the pan " is emptied, but the pipe will remain filled with saccharine fluid " in the same manner as a common barometer tube remains full " of mercury." The pipe will be upwards of twenty-five feet long, but the length of the pipes depends on the sp. gr. of the fluids, "the heavier the fluids the shorter the pipe may be made." A pump may be used without destroying the vacuum, but "the " barometer pipe when it can be conveniently used " is preferred.

Second, "the evaporating of saccharine fluids by distributing " them " in a thin strata obtained by causing the fluids to "run " by their own weight from the high to the low end of heated " inclined surfaces, whether curved or plane."

Third, "exposing to the influence of a vacuum saccharine " fluids in a state of minute division or thin strata (as above), for " the purpose of bringing a large amount of the surface of the " fluid in contact with the vacuum."

Fourth, "instead of bags for removing mechanically suspended " colouring matter," passing " saccharine fluids through beds of . " mineral substances unprepared, such as sand or earths, or " through prepared minerals, such as potter's refuse, unglazed " earthenware or pounded bricks."

By a Disclaimer, enrolled January 17, A.D. 1848, the second, third, and fourth parts of the invention are disclaimed, as also the words " *new and* " and likewise the words " *processes and* " from the title.

[Printed, 4d. No Drawings. Repertory of Arts, vol. 6 (*enlarged series*), p. 184.]

A.D. 1845, February 4.—N° 10,510.

BROWNE, Henry Nibbs.—" Improvements in the manufac-
" ture of sugar."

[No Specification enrolled.]

A.D. 1845, March 13.—N° 10,557.

PURBRICK, Robert Barr.—" Certain improvements in certain
" apparatus used in the manufacture of sugar, which apparatus
" is commonly called sugar pans or coppers." These are in the
form of the pans or coppers, " being rectangular in their horizon-
" tal plan (instead of circular as in the common sugar pans or
" coppers heretofore commonly used) and their bottoms being
" portions of cylindrical surfaces concave and convex," concave
in the middle part of the bottom, in order that the last "portion
" of syrup can be ladelled out, so as to empty the pan or copper
" completely " " (instead of being portions of spherical concave
" surfaces, as in the said common sugar pans or coppers), and with
" flat vertical surfaces at those parts which will be adjacent, when
" a number of such pans or coppers are set in a row or series
" suitably for joining one to another at such flat vertical sur-
" faces," without "leaving any interval between the several pans
" or coppers." Also, in " combining a number of such pans or
" coppers together in a series, and setting them in brickwork,"
whereby " such series of pans or coppers so combined will become
" as it were one long vessel with parallel sides and an undulating
" bottom, and having partitions accross its width at the highest
" parts of the undulations of the bottom, so as to divide its
" interior capacity into compartments, and with the whole of
" the said undulating bottom covering over and forming the
" upper part of the flue or passage from the fire-places to the
" chimney."

[Printed, 1s. Drawing. Repertory of Arts, vol. 7 (*enlarged series*), p. 157; London Journal (*Newton's*), vol. 27 (*conjoined series*), p. 172.]

A.D. 1845, March 17.—N° 10,561.

CHAMPION, CONSTANT.—(*A communication.*)—"Improvements
" in burning animal charcoal." These are, first, the use of a
series of tubes, the drawing shows thirteen, " made of fire-clay or
" or other analogous material capable of sustaining a high degree
" of heat, such pipes or tubes being of small diameter placed
" vertically in a chamber or oven " and closed at their lower
extremities by valves.

Second, " heating the pipes or tubes on all sides equally by
" surrounding them with flame from one fire."

Third, " the arrangement, construction, and mode of working
" the valves for discharging the charcoal;" the valves are opened
and closed by means of rods and cranks or handles attached to
each rod; each rod is connected to the valves closing or opening
four or five pipes.

Fourth, " allowing the charcoal in a red-hot state to fall into
" the open air, where it is cooled without water by being kept
" stirred whilst cooling, so as to prevent it from whitening."

The charcoal to be revivified, after being washed, is placed on
the drying floor furthest from the furnace, and is gradually moved
towards the furnace. The drying floor is made by a horizontal
flue running from the fire heating the pipes, to the chimney. The
pipes are supported at the top and bottom in sockets in iron
plates; large fire-clay tiles close the chamber or oven which contains the tubes. The pipes are kept well filled with charcoal by
keeping it piled " up in a large heap above the pipes." Bones
are burned in such an apparatus with fewer pipes but of larger
diameter, and with funnel-mouthed pipes adapted at the upper
ends to conduct off the gases to be condensed or otherwise.

[Printed, 8*d*. Drawing. London Journal (*Newton's*), vol. 31 (*conjoined
series*), p. 127.]

A.D. 1845, March 18.—N° 10,570.

MOLINEUX, FRANCIS.—" Improvements in apparatus for cut-
" ting and dividing sugar." These consist of a frame carrying a
circular saw on an axis which is supported in bearings, and
motion is given to the axis by means of a girt or band passing
over a pulley on the axis of the circular saw, and over a pulley
on the axis of a fly wheel " worked by hand, or it may be driven
" by a steam engine or other suitable power." The loaf of sugar

F

is placed on to a frame which is guided by rails, and so brought
in contact with the circular saw, " so as to cut a groove in the
" direction of its length.   I then reverse the position of the loaf
" of sugar so as to bring the part of its surface directly opposite
" to that which was first cut," and having made two cuts, a wedge
separates the two parts or half loaves ready to be cut into pieces
by placing them in a carriage, upon which there is a stop for the
end of the loaf to rest against, the carriage runs on rails slightly
inclined towards the circular saw, so that the loaf by its weight is
borne against the circular saw, when a portion equal in thickness
to the distance between the stop and the side of the saw will be
cut off.   The carriage will " then be drawn back, and the loaf of
" sugar again moved, so as to present a fresh part to be cut off,
" and so on till the part of the loaf is cut up."   These pieces are
further cut into lumps by means of plates having knife edges
carried on a roller on the axis of the fly wheel, and plates with
knife edges on another drum.   The lumps are conducted by an
open wire trough into a receiver, and the dust, &c., falls into a
box below.   A roller with a series of serrated surfaces may be
used instead of the plates with knife edges.   A machine is
described resembling the above, but having a series of circular
saws placed at a suitable distance apart in the shaft.   The motion
is given by means of a treadle.   A machine is also described in
which the loaf of sugar is held in suitable chucks, in such manner
that the loaf may be caused to revolve, and while revolving it is
subjected to a scoring action from instruments, acting by their
weights, and guided by the framing of the machine.   A machine
is likewise described for dividing loaf sugar into small particles
for confectioners use, consisting of a fly wheel driven by hand or
other power, having an axis carrying a bevelled tooth wheel,
which takes into and drives a pinion on a spindle, on which is
affixed a conical disc serrated on its upper surface, and the loaf of
sugar in a frame rests its own weight or otherwise upon the
serrated disc, which cuts particles from its lower end as it revolves
on turning the fly wheel.

[Printed, 1s. 6d. Drawings.]

A.D. 1845, June 26.—No 10,740.

ST. CLAIR, Bower. — " Improvements in the manufacture of
" sugar."   These are, apparatus in which the heat is applied in

" a new manner, namely, upon the "upper surface of the saccha-
" rine liquids after they have been filtered and prepared to a
" convenient degree of concentration." The pans or cauldrons
are placed in a row, and the heat from a furnace passes up through
a casing over the top of each of the cauldrons, and finally into a
chimney; "the vapour arising from the liquids "in the cauldrons
" is directed out from the sides of the furnace by a horizontal
" plate." The sides of the vessel or vessels next the furnace are
preserved from the direct action of the heat from the furnace by
means of washers or other means, &c.

[Printed, 1s. Drawings.]

A.D. 1845, July 3.—N° 10,755.

SYMES, WILLIAM.—" Certain apparatus for dividing lump
" sugar." This consists of a "frame or stand to be fixed by
" bolts or screws on to the top of the counter, table, or shelf;"
the lower blade is fixed to the frame or stand, the upper blade
" is attached to a moveable lever over the lower cutter, the fulcrum
" or joint on which this lever works is at one of its extremities,
" and at the other extremity the power is applied by which the
" blade or cutter attached to the lever is moved to and from the
" the blade or cutter fixed to the stand or frame." The blades
may be made either wholly or partially serrated or otherwise.
" The power may be applied to the apparatus at the end of the
" lever, either by the hand or any known machine or engine."

[Printed, 6d. Drawing. London Journal (*Newton's*), vol. 28 (*conjoined series*), p. 96.]

A.D. 1845, August 14.—N° 10,817.

SALEMBIER, HYPOLITE LOUIS FRANÇOIS.—" Improvements
" in the manufacture and refining of sugar."

[No Specification enrolled.]

A.D. 1845, August 21.—N° 10,818.

PEARSE, HENRY, and CHILD, WILLIAM DIMSDALE.— (*A communication.*)—" Improvements in the manufacture of sugar."
These are, first, "the mode of heating vacuum pans by the hot air
" passing off" from "the flues of the clarifyers or of the pans
" or teaches," being conveyed into a chamber in which the
vacuum pan is set.

- Second, arranging filters by " so combining two or more filters
" when manufacturing sugar " that " the filtered product of one
" may flow into the next in succession, and below and from the
" last or lowest into one of the teaches or pans." The filtering
medium being composed of horse hair, wool, or cotton.

Third, " the mode of setting teaches or pans," by dispensing
with the arches of masonry and using metal, by which they are
more thrown open to the heat in the flues. The flue leads
" from the fire-place under the pans or teaches to the clarifyers,
" and then to the chamber under the vacuum pan, and from
" thence to the chimney," there being dampers, &c., to control
and direct the heat.

[Printed, 10d. Drawing. London Journal (*Newton's*), vol. 28 (*conjoined series*), p. 26.]

A.D. 1845, November 18.—N° 10,950.

WRIGHT, RICHARD.—" Improvements in refining sugar."

[No Specification enrolled.]

A.D. 1845, November 20.—N° 10,960.

GYE, FREDERICK. — (*A communication.*) — " Improvements in
" moulding sugar." These are, " the shaping of sugar into suit-
" able forms for use by compressing the same in moulds," as
follows. The refined sugar taken from the moulds when nearly
dry is crushed by the workman simply pressing with a board, and
slightly moistened by minute streams of water, and put into the
moulds; " a quadrangular frame of moulds, suitably arranged for
" compressing and moulding sugar into cakes," and worked by a
screw press, is shown; but moulds for making other shapes and
other arrangements of presses may be employed. The sugar is
afterwards dried " in a stove such as is now used for drying
" sugar."

[Printed, 10d. Drawing. London Journal (*Newton's*), vol. 29 (*conjoined series*), p. 160.]

A.D. 1846, February 25.—N° 11,111.

BRITTEN, JOHN.—" Certain improvements in the method of
" applying heat for the purposes of heating, cooking, and evapo-
" rating, and in the apparatus connected therewith." These are,
in reference ¦to this subject, in the manufacture of sugar, " con-
" densing the steam generated in the ordinary vacuum pan, by

" using it to concentrate the weaker solution of sugar before it
" enters the vacuum pan, which weaker solution must be kept at
" a sufficiently low temperature by any ordinary method of ex-
" posing a large and continually changing surface to the air."
In preference, a number of tubes are placed perpendicularly in a
vessel, and these are surrounded and inclosed by one large cylinder, or a number of segments of smaller cylinders, "this interval
" part of the apparatus" is called "the heater." The steam from
the vacuum pan is drawn through the heater, which is surrounded
by the weaker solution of sugar above named.

[Printed, 6d. Drawing.]

A.D. 1846, April 3.—N° 11,158.

CROSLEY, HENRY. — "Certain improvements in the manu-
" facture of sugar, and in the machinery and apparatus employed
" therein." These are, first, the employment in sugar-cane crushing and grinding mills of "rollers (whether three or any greater
" or lesser number be employed)," which "shall either be of
" different diameters and velocities, or of the same diameters and
" different velocities."

Second, the employment in sugar-cane-crushing and grinding
mills "of rollers (two, three, or more) with rounded teeth working
" into each other."

Third, "a machine for pressing or crushing canes," which is
named "a circular press." A feed table delivers the canes between
two rollers, whence they pass on to a curved bed, up which they
are carried by the action of one of these rollers to the discharge
point. There are arrangements for heating the apparatus with
steam.

Fourth, "a machine for crushing and grinding canes, in which
" a combination of rotary and reciprocating movements is em-
" ployed." There is a horizontal table, perforated with numerous
holes, which is moved to and fro on three rollers underneath, by
means of a crank and rod, or otherwise. There are three rollers
mounted in loose bearings a little way above the table. The centre
roller is about twice the diameter of the two side rollers, which are
of equal diameters, and fixed at such a height that their peripheries come much closer to the face of the table than the periphery
of the centre roller. An inclined plane is on each side of the
centre roller, by which the canes are introduced from both sides

between the rollers. The apparatus is constructed so that when a feed of canes is entering at one side of the main roller the feed from the other side may be stopped till the first has been cleared, &c.

Fifth, "the application of steam to the extraction of the sac-
" charine matter in megass or begass," in "close vessels of any
" suitable form." The steam pipes, perforated, are arranged near the bottom of the close vessel, or radiate from, and communicate with, a central vertical pipe, or in any other convenient way; the juice flows from the pan below.

Sixth, "extracting the juice from sugar canes without crushing
" or grinding, and by the agency of steam and water, or steam
" alone, or fire alone." The canes are cut or sliced into small pieces, as directed in No. 7469, using lime water in place of milk of lime, macerating and digesting the canes in vessels with or without jackets, and a coil of pipes carried round the inside of the the vessel, and in a series of vessels communicating with each other.

Seventh, "drying the megass for fuel." The megass is placed on perforated plates, which cover the flues, running through a drying house; and there are arrangements for admitting air, carrying off moisture, &c. Instead of employing a furnace and flues, steam pipes are preferred.

Eighth, a sponge filter, as adapted and applied to the filtration of cane juice. At the bottom of a cylindrical vessel is a perforated false bottom; above this is a perforated diaphragm; a mass of sponge is inserted between the diaphragm, and may be compressed by turning a wrench to any extent desired. From time to time the sponge may be removed and washed.

Ninth, "the employment of corrugated pans for the evapora-
" tion of sugar-cane juice." Three evaporating corrugated pans are on the same level, and the concentrating pan or teach, also corrugated, is on a lower level, and immediately adjoining the furnace; and "the application of ærating and cooling wheels" of the following construction to the processes of concentration and crystallisation. The wheel mounted over the teach carries a number of perforated buckets.

Tenth, employing pans "for evaporating and cleansing the
" clarified cane-juice, and concentrating the cleansed liquor or
" syrup," as follows. The pan is formed of a top and bottom part firmly bolted together, a second bottom supports a series of

flat vertical chambers, open at the bottom, and terminating below the liquor in the pan. There is a pipe for carrying steam into these chambers, and also a pipe for carrying off any water condensed in these chambers, besides a pipe for carrying off any air, &c.

Eleventh, "applying animal charcoal to the decoloration of the " syrup obtained from the sugar-cane," by passing the syrup " through a series of ordinary sized vessels filled with the charcoal " (say six, eight, or ten), and communicating with each other by " pipes and cocks," in place of passing it through one great bed.

[Printed, 3s. 2d. Drawings.]

A.D. 1846, July 6.—N° 11,280.

WRIGHT, RICHARD.—" Improvements in refining sugar." These are, first, " the process of refining sugar by submitting it to the " action of high pressure steam and pressure," as follows. " A " sheet iron cylinder of about 30 inches diameter and 25 feet in " length " is fixed and fitted with a large funnel. " On different " sides of the cylinder, about 5 feet distant from each other, are " to be inserted small steam pipes (half inch bore), connected " with the main pipe from a steam boiler, having the ends inside " the cylinder turned downwards." " A large box for receiving " the sugar is to be placed on the floor beneath the lower end of " the cylinder." The sugar, roughly broken or ground, is passed through the cylinder while " high pressure steam, at from 40 to " 50 lbs. the square inch, is to be turned on from the boiler." The sugar falling into the box below, warm and slightly moist, is made " into cakes of about 18 inches square and 3 or 4 inches " thick ;" wrapped in coarse linen cloth, and submitted to severe pressure " for three or four hours, it may be taken out and passed " through a mill, when it will be in good marketable condition, " and of a fine color."

Second, " the use of sulphuric acid in cleansing solutions of " sugar, and in promoting their crystallisation." The drainage from the above pressing of the sugar is dissolved in boiling water until 20° on " Beaumé's saccharometer, and to every 100 gallons " is to be added from 1 lb. to 1½ lbs. of sulphuric acid of sp. gr. " 1845°, previously mixed with one gallon of water. After allowing " the mixture to boil for five minutes, the acid is to be saturated " with carbonate of lime or chalk, and the solution of sugar is to " be boiled with animal charcoal in the usual way, or filtered."

Third, "arranging the filtering apparatus." At the top there
are four cisterns with steam pipes (to keep the syrup warm) for
receiving the syrup; each of these has a cock near the bottom,
Below these are four filters, and again below these filters are
four cisterns, to receive the filtered syrup; in these are steam
pipes. The syrup can be pumped up from the cisterns and
passed through all the four filters, where it is evaporated and
crystallised, &c.

[Printed, 6d. Drawing. Repertory of Arts, vol. 9 (enlarged series), p. 352;
London Journal (Newton's), vol. 30 (conjoined series), p. 251; Patent
Journal, vol. 2, p. 555.]

A.D. 1846, July 23.—N° 11,312.

NEWTON, ALFRED VINCENT.—(A communication.)—"Certain
" improvements in the manufacture of sugar." These are said
to be, "the construction of filtering apparatus," "the application
" of the partition and channel to the boiler," "rotating paddles
" or leaves for skimming" the liquor, "the application of a lift
" pump," and "the general arrangement and construction" of
the apparatus. There is a long open boiler set in brickwork, at
one end of which is a furnace, from which is a flue extending its
whole length into a shaft beyond it. This flue is between two
tanks provided with steam jackets, and each are connected with
the boiler by means of a pipe. Beyond these pans is an arrangement of an endless web of wire cloth, which passes over two rollers
over a water tank. The liquor of the crushed cane is run down a
trough or channel from the mill on to the endless wire cloth,
and the liquor percolates through the trough into a tank below,
" while the feculences or refuse matters will be carried forward
" and fall into the water tank." The endless web is washed of
its impurities in the water in the trough. The liquor flows from
this tank into the two pans with the steam jackets, and from
thence into the long open boiler. In this boiler, at about the
middle of its length, is a partition which divides it into two, and
attached to this partition is a channel for the purpose of carrying
off the scum which is thrown on to the surface of the liquor in
the first division of the boiler by the action of the heat in the flue
beneath." At one part of the boiler nearer the fire is a paddle-
wheel, which projects the scum on to an inclined plane provided
to receive it. There is a large tank intended as a receptacle for
all the matters swept out of the second division of the boiler, and

in which, after settling some time, there is a small lift pump to pump up the clear liquor into the first division of the boiler.

[Printed, 10*d.* Drawing. London Journal (*Newton's*), vol. 30 (*conjoined series*), p. 85.]

, A.D. 1846, August 11.—N° 11,331.

JENNINGS, HENRY CONSTANTINE.—"A new method or ap-
" paratus or machine for the better or more economic evaporation
" of fluids or liquids containing crystalline and other matters to be
" concentrated or crystallized." This consists first, in obtaining
" great evaporating surface by the elevation of a column of
" liquid containing sugar," &c. " to be evaporated or crystallized,
" and causing the same to pass or circulate over a certain number
" of double cylinders heated within by steam or hot air. The
" said elevation and circulation of the liquid is obtained by the
" injection of atmospheric air, heated or cold, by means of a
" pneumatic forcing pump of proportionate power or capacity to
" be worked by manual labour, animal or steam power," as
required.

Second, "the application of electricity to facilitate evaporation,
" in the concentration and crystallization of liquids." In the
lower pan containing the liquid to be circulated as above, may be
placed " a galvanic pile or battery of three elements composed of
" copper and zinc ; the first element is twenty inches diameter,
" each metal being half an inch apart, the second is sixteen
" inches, et cetera, the third is twelve inches, et cetera, and
" would be supported upon a wood frame to partially insulate it
" from the copper or metal bottom of the pan." "The copper
part of each element should be about six inches high or deep, or
one-sixteenth of an inch thick ; the zinc circles should be the
same height. Different sizes of batteries may be employed.

[Printed, 1*s.* Drawings. London Journal (*Newton's*), vol. 30 (*conjoined series*), p. 153. Repertory of Arts, vol. 9 (*enlarged series*), p. 336.]

A.D. 1846, September 10.—N° 11,370.

RICHARDSON, CHARLES.—(*A communication.*)—" Certain im-
" provements in making and refining of sugar, and in the
" machinery and apparatus employed therein." There are, first,
the employing in clarifying saccharine extracts or solutions of
three several compounds. No. 1 compound is formed by taking

about 100 lbs. temper·lime in powder, 5 lbs. pure clay, as free as possible from iron, or 5 lbs. of Spanish whiting, 20 lbs. " Howard's " finings," dry, 30 lbs. of powdered animal charcoal, 5 lbs. sulphate of zinc, although this last may be omitted, mixing the whole intimately together, and keeping it in a close vessel till wanted for use. No. 2 compound, 100 lbs. of dry finings, prepared in the same way as " Howard's " (except that it is preferred to substitute sulphate of alumina, or a mixture of alum and sulphate of alumina, for alum), are added to about 200 lbs. of powdered animal charcoal, about 5 lbs. of clay, as above, or of Spanish whiting, and 10 lbs. of sulphate of zinc. The whole is well mixed together in a wet state; the water is removed by a filter. Or mix only the finings and clay solution together, filter, dry, and compress, and " add the powdered animal charcoal and " sulphate of zinc just previous to the compound being employed." No. 3 compound, take 100 gallons of bullock's blood, 200 lbs. of powdered animal charcoal, thicken with dry sawdust (white preferred), dry, and then mix with it about 40 lbs. of " Howard's " finings," and dry.

Second, passing saccharine extracts or solutions from the clarifiers, and before boiling, through filters of cloth or fine wire gauze, &c.

Third, the peculiar combination of filters and receivers, and the peculiar mode of working them. There are three filters and two receivers below; each receiver is provided with a pump, so that its contents may be pumped up again into any one of the filters, as required. When the liquors run dark they are run into another receiver, in which they may be heated, or they are filtered hot by any method.

Fourth, employing " for the evaporation of saccharine solutions " of a fire pan or pans combined with a number of pipes."

Fifth, "the steam evaporating apparatus, called the steam " ladder evaporator." This is of an inverted pyramidal form, and is composed of two hollow side frame pipes, about twelve feet high, separated from each other by a series of horizontal pipes; steam is passed through these pipes; the juice is allowed to trickle down in a shower upon the topmost of the horizontal pipes, and is caught in a receiver at the bottom. The steam is under pressure.

Sixth, " applying the heat by means of additional fires equally " to the whole or any number of the ordinary coppers, for the

" evaporation of the purified cane juice or other saccharine
" extract."

Seventh, " the evaporating the juice or other saccharine solu-
" tions in vacuo by the joint agency of fire and steam, or hot air
" and fire, or hot air alone, or hot air and steam, and by the
" employment of a vacuum pan of a cylindrical or conical form,
" with the addition of a manhole in the bottom of the pan ; a
" vessel for drawing a sample for testing the density of the
" liquors," and a cock in or near the bottom of the pan for
drawing off the liquor.

Eighth, evaporating saccharine solutions in a hot air chamber,
fitted up with wooden or metallic bars or racks; the roof is a
metallic shallow vessel, the bottom of which is perforated with holes.
This room is heated to a high temperature, and " the liquor,
" being placed in the vessel at the top, falls like rain on the bars
" or racks," is evaporated, the vapour escaping by a hole at the top.

Ninth, " making concrete juices or other saccharine matter "
by evaporating them "upon a heated surface."

Tenth, cooling and crystallising of the syrup by means of one
cooler only.   Each strike or boiling is put into one and the same
cooler till the cooler is full; each strike, as it is thrown in, is to
be well mixed, and when full the whole is well stirred, and left to
cool, when it is casked.

Eleventh, reducing saccharine extracts and solutions, and also
molasses, to a concrete or solid state, "by continuing the evapo-
" rating or boiling process as much beyond what is called the
" 'sugar point' as may be necessary" to ensure no escape of
liquid; this is best effected by means of a vacuum pan, though
any method not carbonising the sugar may be employed.

Twelfth, the concentrated juice, purified by means described
under the first head, is heated to the temperature at which refined
sugar is ordinarily put into moulds, and put into moulds of any
convenient form, avoiding remelting and reboiling.

Thirteenth, applying "to the concentrated material aforesaid, or
" after it has been imported, the well-known pneumatic process."

Fourteenth, to convert concentrated juice or syrup, not treated
by the compounds under the first head, into loaf or other fine
sugar; it is then clarified and, if needful, whitened with charcoal
or any other agent, or it may be clarified as refiners treat raw
sugar, or, what " is preferable, by the application of the said com-
" pounds " in certain proportions.   To the sugar obtained the

pneumatic process is applied, and if desired the sugar so produced is converted into stamped loaves.

Fifteenth, separating "·the crystals from the syrup after the " concentrated material has been preserved some time," " how- " ever it may be carried into effect."

Sixteenth, "employing saccharine matters in a concrete state " to manufacture sugar. The concrete molasses, the result of the " first drawings of raw cane or other sugar, may also be employed " to make sugar by melting such concrete molasses, and, if need- " ful, clarifying in the usual manner; or " it may be converted " into ordinary molasses by reducing it to the necessary density." " The employment of such concrete molasses for making sugar " or ordinary molasses."

Seventeenth, combining or mixing "with raw sugar, or the " liquor produced therefrom, cane juice or molasses," or "both " together." "Mixing with the 'syrups' produced in the refining " raw sugar either cane juice or molasses, or both together." The proportions in which they are mixed are given, but these pro- portions may be modified. "The application of the pneumatic " process to sugar produced from cane juice alone, or from one " or other of the aforesaid mixtures, after such mixtures have " been whitened with animal charcoal, or such like substance as " may be found a substitute for animal charcoal."

Eighteenth, in making sugar in the colonies the juice is brought to the sugar point in the ordinary way, and when sufficiently cold it is put into large tight vessels and there pre- served with the molasses as long as convenient. When required the molasses are separated.

Nineteenth, "mixing raw sugar or molasses and sugar, or con- " centrated or partly concentrated cane juice, or the whole together, " with raw cane juice." "The clarifying cane juice alone, or " mixed as aforesaid," either with the compounds described under the first head, or with blood.

Twentieth, "the employment of large vessels in crystallising " sugar," in preference of wood; they should not contain less than 300 lbs., but it is preferred that they should contain 4 or 5 tons, or more.

Twenty-first, "liquoring of sugar with colonial molasses, or " molasses mixed with sugar," the " same having been previously " clarified and whitened for the purpose."

[Printed, 1s. 4d. Drawings. Patent Journal, vol. 2, p. 718.]

A.D. 1846, September 17.—N° 11,376.

BOWMAN, JAMES WILLIAM.—"Improvements in re-burning
" animal charcoal." These are, applying revolving retorts for
the above purpose. Two fixed end plates form the bearings, in
which the cylindrical retort turns. The retort has a pulley formed
at each end to receive an endless chain, "or other means may be
" resorted to for giving motion to the retort." On the interior
of the retort are a series of projections, which as the retort revolves
cause the animal charcoal to be deflected towards the centre. At
each end of the retort is a door or cover, readily removed, and
when so, suspended by a chain which passes over pullies and has
a counterbalance weight attached to it, there is an opening with a
cover for observing the progress of the process, and a tube is in
the back cover, "through which any vapours driven off from the
" animal charcoal may pass away to a suitable condenser or
" otherwise." On the axis are fixed two pulleys which give
motion to the endless chains which move the retorts."

[Printed, 1s. 2d. Drawings. Repertory of Arts, vol. 9 (*enlarged series*),
p. 275; Patent Journal, vol. 2, p. 714.]

A.D. 1847, May 27.—N° 11,722.

JOHNSTON, JAMES. —" Certain improvements in the manu-
" facture of sugar." These are, first, " the medicating of
" saccharine fluids." A self-acting apparatus for regulating the
flow of saccharine fluids on to an inclined plane is described; it
consists of a small fixed cistern at the top; below it is a moveable
cistern suspended from a spring balance by a framing. In the
bottom of the fixed cistern there is a valve seat, and suspended
from the framing there is a valve, which works into the valve
seat which is in the bottom of the fixed cistern. In one side of
the suspended moveable cistern there is a narrow perpendicular
slit. The fixed cistern " is always kept full of the liquid to be
" boiled. It should communicate with one of the ordinary
" cisterns in a sugar house." Above the inclined plane is fixed
a small cistern provided with lime water and a pipe and tap. If
sufficient lime is not added " the foaming from fermentation
" would not have entirely ceased."

Second, " the evaporation of saccharine fluids, or bringing
" them to the crystalline point or proof." " Instead of warming
" a large quantity of any saccharine fluid in an open caldron and

" then transferring it to a vacuum pan," as is done in No. 7587,
one of the two following apparatus is used, " either of which will
" raise the saccharine fluid to the necessary degree of temperature
" whilst it is being fed into a vacuum pan." In one the inclined
plane on which the fluid is heated and apparatus described above
is used. In the dome of the vacuum pan is " a colander," through
which the fluid passes, which allows the liquid to pass in a hot
shower into the vacuum pan, which is of the nature described in
No. 10,505. The second apparatus " is to be used for heating or
" parboiling saccharine solutions that have been previously
" clarified." It is a steam casing, containing a number of pipes
through which the saccharine fluid is passed into the vacuum
pan.

Third, the potting or moulding of sugar. In common sugar
moulds making them of cast iron; for coarse kinds of sugar,
such as bastards and pieces, they are not coated; for fine goods,
as loaf moulds, they are coated with alloys of tin or zinc or
enamel; and in economizing space moulds are made six-sided
or otherwise and of any suitable material.

[Printed, 1s. Drawings. Repertory of Arts, vol. 10 (*enlarged series*), p. 356;
   Patent Journal, vol. 4, p. 54.]

A.D. 1847, July 12.—N° 11,790.

SIEVIER, ROBERT WILLIAM.—" An improved material or
" materials for purifying or decolorizing bodies, which material or
" materials may also be employed as manure and pigments, and
" for other like purposes." This consists, first, in a filtering
material for filtering various substances, among others which are
named, are oils, manures, or as pigments, solutions of sugar, and
cane juice, for " purifying or decolorizing the same;" by the carbonization of sugar by the action of sulphuric acid, in preference,
using coarse unrefined sugar; on mixing the sugar and acid
together in an earthen ware or leaden vessel, and allowing them
to stand a hard or crisp mass results, which is thoroughly washed
with water until the washings are nearly neutral to litmus paper.
When the above material is spent, it is washed and may again be
used, but it " will not act so efficiently as it did in the first
instance unless it be prepared according as follows :—

Second, preparing porous or fibrous matters or substances, as
" pieces of sponge, pumice stone, cloth, tow, or other similar
" materials," " wood charcoal and coke," " animal charcoal and

" carbonized sugar," by impregnating them with sulphurous acid, either by submitting them when moist to gaseous sulphurous acid, or by steeping them in a solution of the acid. When these substances are spent they are renovated by washing and again impregnating them with sulphurous acid.

Third, applying for purifying or decolorizing bodies, and also for the other above purposes, alumina, oxalate of alumina, and metallic salts, as nitrate, sulphate, and chlorides of iron, yellow prussiate of potash, oxalic acid, alum, carbonate of ammonia, acetates of lead, sulphurous acid, sulphuretted hydrogen, lime water, or chalk, crystallized nitrate of bismuth, soda. But by two disclaimers dated respectively June 21, A.D. 1849, and March 1, A.D. 1850, the words of the title " which material or materials may " also be employed as manure and pigments, and for other like " purposes " are disclaimed, and in the specification the words oils, manures, or pigments, and all the matter contained under the third head are disclaimed.

[Printed, 1s. 2d. Drawing. London Journal (*Newton's*), vol. 32 (*conjoined series*), p. 434: Mechanics' Magazine, vol. 51, p. 22; Patent Journal, vol. 4, p. 221.]

A.D. 1847, September 9.—N° 11,863.

STEINER, FREDERICK.—(*A communication.*)—" Improvements " in the manufacture of sugar." These are, first, " the combina- " tion of an inclined trough within a vacuum pan, and applied " for evaporating and concentrating saccharine fluids in the " manufacture of sugar." The syrup drops into a spiral-shaped trough in the vacuum pan; this trough forms several convolu- tions or coils before it reaches the lower part of the vacuum pan.

Second, " the combination of apparatus for condensing the " steam and obtaining vacuum, adapted and applied to vacuum " pans employed in the manufacture of sugar." The condenser for forming the vacuum in the pan is made of an upper and lower cylinder; the upper one is provided with a series of pipes through which the vapour passes as it is brought from the vacuum pan the spaces between these pipes have cold water passing through them. The lower part of the condenser is a conical shaped pipe passing through a series of perforated plates, through which the vapour passing it will " be condensed and almost form a perfect " vacuum in the pan."

Third, " the combination of apparatus " called " a self-adjusting " vacuum regulator, for maintaining " the " partially formed

" vacuum in the condenser at the same degree of rarification."
This is a pillar forming an air chamber; the upper end is closed
by a lid, and the lower end by a plate, which rests upon a stand
fixed to the lower floor of the building. The upper end is in
connection with the condenser by a small pipe adjusted to the lid,
a tube passes through an opening in and is fastened to the lid;
this tube is open at both ends, and reaches nearly to the bottom
of the pillar, and it is supplied with the requisite quantity of
mercury upon which rests a float connected to a lever the
fulcrum of which is fixed on the lid.

Fourth, "supplying steam of a uniform temperature" to an
inclined trough within a vacuum pan in the manufacture of sugar.
This is effected by means of a throttle valve, which is opened and
shut by the action of the steam upon the quicksilver contained in
a vessel below.

[Printed, 1s. 4d. Drawings. Patent Journal, vol. 4, p. 400.]

A.D. 1847, September 30.—No. 11,881.

JAY, CHARLES. — "Certain improvements in apparatus for
" evaporating and concentrating saccharine and saline solutions,
" and which may also be applicable to the evaporation and con-
" centration of vegetable and other extracts." These are, in
reference to this subject, "a complete apparatus heated by steam
" with clarifiers and filters, and a steam blast adapted to the
" evaporating vessel, the whole apparatus being calculated for the
" manufacture of fine sugars." The cane juice from the mill is run
into the first clarifier situated above, here it is "treated with clari-
" fying ingredients" and then boiled up for a few minutes, after
which it is run through filters into a cistern below, where it may
be mixed with some finely powdered bone black, say, "6 lbs. for
" every 100 gallons of juice." The mixture is then run into two
boilers for the purpose of partially concentrating the juice and to
generate steam for heating the syrup in the evaporating vessel
placed on a lower level. When the juice has been boiled long
enough with the bone black it is run into an open pan and treated
with a little blood or albuminous matters, and boiled and passed
through two filters below containing bone black into a cistern
below. It is finally concentrated in the evaporating vessel named
above, to which it is "transferred in proper charges either by
" means of a montjus" or through a pipe by rarifying the atmosphere in the vessel, to which is connected by pipes a steam blast

from a boiler and dry air from a heating apparatus. The syrup is finally run off into a crystallizer or cooler.

There is shown " a vacuum pan and cane juice boiler for the " manufacture of raw sugar, in which the current of dry heated air " which in the foregoing apparatus is made to percolate through " the saccharine solution is dispensed with."

[Printed, 1*s.* 4*d.* Drawings.]

A.D. 1847, October 21.—No 11,920.

PLAYFAIR, Patrick, and HILL, Laurence, junior. — " Im- " provements in the manufacture of sugar." These are, " sepa- " rating liquids from the crystals of sugar by causing such " matters to be put in motion by centrifugal force," as follows :— There is an axis to which quick motion is given by any suitable power; on this is fixed a vessel, the periphery of which is per- forated with fine holes; this vessel is surrounded with another vessel to receive the fluid percolating through the inner vessel, and from the outer vessel the fluid flows by a pipe at the bottom into a receiver. In working this machine sugar combined with the fluid is fed into the machine by a funnel and pipe above.

[Printed, 10*d.* Drawings. Repertory of Arts, vol. 11 (*enlarged series*), p. 291; London Journal (*Newton's*), vol. 32, p. 358; Patent Journal, vol. 4, p. 551.]

A.D. 1847, November 9.—No 11,954.

CLAYPOLE, Henry Krebs.—(*A communication.*)—" Certain " improvements in the process, apparatus, and machinery for " making sugar." These are, first, the application of two separate generators of steam, provided each with thirty-two tubes filled with water, except about eighteen inches at the top, which is left empty for the generation of the steam. The fire passes under the teache and three boilers before it reaches the steam generators.

Second, "the use of a current of high-pressure steam in a " chimney to facilitate and promote the egress of the atmospheric " air by the top, and producing a perfect vacuum in a sugar " furnace." On the top of the generators are two sheet iron chimneys, into the bottom part of each of which is a jet of high- pressure steam.

Third, " the process of clarifying sugar by the application of an " iron tube to a double-bottom clarifier, and of increasing rapidly

s. G

" the temperature by high-pressure steam." This high-pressure
steam into the iron tube to a double-bottom clarifier is " obtained
" from the generators by means of a connecting pipe."

Fourth, " the application of a canal to the teache for the running
" out of the sugar." At the bottom of the teache is a canal,
terminated by a mouth, shut by a cast-iron valve, moved by a
lever, provided with a proper weight, which keeps the mouth of
the canal shut.

Fifth, " the use of galvanized sheet-iron coolers, oblong in
shape ; they are portable by means of a handle at each of two
ends, and " do not contain more than fifty pounds of sugar
" each."

[Printed, 1s. 4d. Drawings. Repertory of Arts, vol. 12 (enlarged series),
  p. 65; London Journal (Newton's), vol. 32 (conjoined series), p. 320 ;
  Patent Journal, vol. 5, p. 25.]

A.D. 1847, December 8.—Nº 11,991.

SCOFFERN, JOHN. — " Improvements in the manufacture and
" refining of sugar.". These are said to be " the combined use of
" sulphurous acid with lead " as follows :—A " lead material"
may be made by adding to " twelve gallons of vinegar of five per
" cent. strength, heated in a copper vessel to one hundred and
" sixty degrees Fahrenheit," gradually with continual stirring,
" forty pounds of finely powdered litharge, and boiling with
stirring until the mass becomes thick, when the heat is " gradually
" decreased, and the last portions of moisture should be dissipated
" by a gentle heat." The sugar to be treated is dissolved, and
the liquid at the temperature of about 180° F., the " lead
" material " is incorporated with it ; the solution is then filtered,
and any excess of lead is precipitated by " passing through it
" streams of sulphurous acid gas, until a portion of the liquor
" properly tested " (by means of hydrosulphide of ammonium
and sulphuretted hydrogen, the mode of doing which is
particularly described) " shall yield no trace of lead."

[Printed, 4d. No Drawings. Repertory of Arts, vol. 13 (enlarged series),
  p. 37; London Journal (Newton's), vol. 33 (conjoined series), p. 196 ;
  Patent Journal, vol. 5, p. 130.]

A.D. 1848, January 18.—Nº 12,033.

NEWTON, WILLIAM.—(A communication.)—" Certain improve-
" ments in the manufacture of sugar from the cane." These are,

" extracting the sugar from the sugar cane or other substance
" containing saccharine matter, by first drying and pulverizing
" the cane or other substance, and afterwards extracting the
" sugar therefrom by passing water through it " in a series of
vessels. The cane, cut into slices or small pieces by a cane
cutter, &c., is placed in a rectangular metal frame (perforated)
with a series of shelves placed one above another, and placed on
a car on a rail, is run into a stove to dry and pulverized is
brought above the tubs when the tubs are filled with pulverized
cane. The door in the top of each tub is to be closed. Water
from a reservoir above is let into the bottom of the first tub by a
pipe, and it rises until the tub is filled, and by an arrangement it
escapes and is carried by a pipe into the bottom of the second
tub, and so on through several of the tubs until of the degree of
saturation 20° to 25° of Beaume, when it is either run into one of
two filters or into a reservoir. The liquor is then evaporated, as
is well understood. After each tub is exhausted it is cut off
without suspending the operation of saturating the water passing
through the apparatus, the exhausted cane is removed into a car
on a rail below, and the tub is filled as before " with fresh
" pulverized cane or other material."

[Printed, 10d. Drawing. London Journal (*Newton's*), vol. 33 (*conjoined series*), p. 88 ; Artizan, vol. 6, p. 249 ; Patent Journal, vol. 5, p. 232.]

A.D. 1848, February 8.—N° 12,058.

HEBERT, LUKE.—" Improved mechanism for reducing, grind-
" ing, and sifting bark, sugar, coffee, seeds, and other substances."
These are, first, a revolving grinder, consisting of " a metallic
" 'core,' containing only the coarser grooves, from about a six-
" teenth to a quarter of an inch in depth," and a " case " with
similar grooves, " each having at one end a flat grinding ring
" projecting from its periphery at right angles to and concentric
" with its axis."

Second, " making the core of such dimensions that when new
". and adjusted in the case in a working position it shall project
" beyond the grinding surface," so that the mill may be repeatedly
recut before the core shall cease " to grind against the grinding
" surface of the eduction end of the case." Also making the
" bearing or support of the smaller end of the core not at the
" extremity, but within the core," so as to " preserve an ample
" space for the feed."

Third, "the application of wrought iron rings for the fine "grooves to cast-iron conical mills."

Fourth, "fixing wrought-iron conical grinders into and over "cast-iron forms."

Fifth, "the combination of any conical mill, or of the grinders "referred to in the first claim" above-mentioned, "and a sifting "machine on the same horizontal axis."

Sixth, "the mode of mounting grinders of any kind in a "vertical mill," "with or without a sifting machine on the same "horizontal axis." A box or frame has a bar across it, in which is an adjustable step supporting the axis, on which is fixed a running grinder surrounded by a stationary grinder, over which is a hopper, across which is a bearing to the axis, on which is a winch; fly nuts fasten the hopper to the box. "By prolonging "the axis several inches below the running grinder, and fixing "the bar" named above "suitably low, a horizontal wire gauze "disc may be fixed above the bar, and brushes be attached to "the axis or revolving grinder."

Seventh, "a machine for sifting and reducing tea and other "substances," consisting of a hopper, attached to which is a feeding worm, and below is a conical sieve of wire gauze revolving on an axis supported by arms; knives project inside and outside and parallel with the sieve; as the sifter revolves, the fine tea is sifted away, and at length the coarse leaves are broken by the knives, and enter the drawer. Coarse wire gauze may be used in place of the knives, and a stiff brush may be employed to keep the sifter free from adhering leaves.

Eighth, "the combination of a shaking sieve with my tea mill "patented in 1843," No. 9596.

Ninth, another machine for sifting and reducing tea in which a sieve is agitated vertically and laterally.

Tenth, "the construction of sieves with flat sides, and of "elliptical sieves revolving or vibrating on horizontal or inclined "axes."

Eleventh, "the application of 'direction slips or blades' to "the internal surface of any sieve revolving on a horizontal or "inclined axis."

Twelfth, "the construction of any mortar or recipient con-"taining grooves, and any pestle or instrument containing edges "entering the said grooves, and operating by percussion and "rubbing, for reducing substances."

Thirteenth, making "compound metallic mortars of any kind," by fixing with solder or other means one within the other."

Fourteenth, making " a prominence or peak in the centre of " any mortar, in which a pestle is used for pounding and " levigating."

Fifteenth, " increasing and varying the pressure of edge " stones, &c." A vertical shaft attached to their axis passes through the bed of the mill, to the lower part of which is appended a series of weights.

Sixteenth, spice hand graters.

Seventeenth, "pressing of succulent roots and other substances " against a revolving grater or cutter, or the grater or cutter " against the substances, by means of a screw or screws."

Eighteenth, "the helical arrangement of breaking arms work- " ing through racks, and the use of iron and steel wire brushes " as follows, " in a raw sugar mill " :—An upright case or frame having a hopper, at the lower end of which are a series of aper- tures, through which a series of breaking arms revolving on an axis pass ; someway below these arms is a semi-cylindrical coarse wire gauze sieve on a frame, and above this sieve is an arrange- ment of bars of wood carrying steel wires. The sugar broken up by the arms falls on to the sieve, whence it is combed or brushed through into a drawer below.

Nineteenth, "the combination of breakers and of a sifting " machine on a vertical axis for reducing raw sugar." A hopper feeds the sugar into a cylindrical case, to which rings are fixed having projecting arms, "not arranged, however, under each " other in vertical lines, but so as all to occupy intermediate " equidistant positions in plan." A frame carries a wire gauze sieve, and above it are whalebone or wire brushes which with a series of arms radiate.

[Printed, 1s. 2d. Drawings. Patent Journal, vol. 5, p. 296.]

A.D. 1848, July 18.—N° 12,218.

STEINKAMP, JOHANN ARNOLD. — " Improvements in the " manufacture of sugar from the cane, and in refining raw " sugar." These are, using " any cotton or vegetable fibre cut " up or divided into small lengths " in " the cleaning, purifica- " tion, and refining of sugar cane juice and refining raw sugar." " In order to clear one hundred pounds of raw sugar two to two " and a half pounds of raw cotton is required," cut into " lengths

" varying from about an eighth to a quarter of an inch long "
are saturated by " pouring hot water upon it, turning it about,
" and thoroughly wetting it, allowing it to remain in the water
" about twelve hours; it is then ready for use." The apparatus
required is a vessel rather wider at the top than the bottom, at
which there is a tap; above this is a wood frame, over which a
coarse cloth is stretched. The prepared cotton is placed in the
filter; on this cotton is a coarse cloth, and on this cloth is placed
a frame crossed by thin strips of wood. In the refining of raw
sugar, to 1,000 lbs. of sugar 5 or 600 lbs. of water are added, one
to two pounds of chalk in powder, and although not absolutely
necessary half pound of starch, boiling for 8 to 10 minutes,
stirring, skimming off the froth, passing through a sieve, and
passing through the filter prepared as above.

[Printed, 4d. No Drawings. Repertory of Arts, vol. 13 (*enlarged series*),
p. 178; London Journal (*Newton's*), vol. 34 (*conjoined series*), p. 35;
Mechanics' Magazine, vol. 50, p. 91; Practical Mechanics' Journal, vol. 1,
p. 278; Artizan, vol. 7, p. 133; Patent Journal, vol. 6, p. 166.]

A.D. 1848, November 21.—N° 12,335.

CLEMENT, WILLIAM HOOD. — (*Partly a communication.*) —
" Certain improvements in the manufacture of sugar, part of
" which improvements is applicable to evaporation generally;
" also improved apparatus for preparing the cane trash to be
" used as fuel." These are, first, " partly of improvements on an
" apparatus " described in No. 11,312. A mill is described
consisting of four rollers travelling at an equal speed by the usual
arrangement of cog wheels on their shafts. One roller is above
the three others, and the centre of the last roller is higher than
that of the first. The cane passing through these rollers gets
" three distinct bites," and passes out between the last two rollers
nearly in a vertical position, causing the juice to run more freely
from the crushed cane. The cane juice is directly conducted by a
trough on to a filter composed of an endless surface of fine wire
cloth travelling over a pair of rollers mounted at the sides of a
small receiver and under a drum which turns in bearings in the
sides of a tank placed immediately under the receiver and
containing water. As the juice flows on to the moving filter it
percolates into the receiver, while the fecculent or refuse matters
are carried by the cloth into the water tank, in which are revolving
brushes against which the cloth rubs and by which the fecculent
matters are brushed off. The filtered juice runs from the receiver

into an arrangement of heaters heated by steam and by a naked fire; lime is added, and the scum rises to the surface, and rolling over the edge of one of the troughs is carried by arrangements of pipes, &c. to the still house. To prevent the liquor rising in foam a flat perforated plate is fixed over the surface of the evaporators. There are arrangements for agitating the liquor, and a paddle wheel skims off the foam and pushes it over the inclined flanch of the evaporator. There are further arrangements for evaporating and concentrating the saccharine solution until it reaches what " is known as the ' striking point,' " when it is run " into crystal- " lizing and curing cases," consisting of boxes with compartments having divisions capable of being removed so as to drain off the molasses or liquid parts from the crystallized sugar. " To facili- " tate this separation of the molasses from the crystallized par- " ticles of sugar," introducing into the lower compartments a jet of steam, and this is repeated from time to time. The " heat to the " steam boilers for generating the steam to grind the cane, and " for other purposes," is applied as follows :—The flames pass from the furnace under the sugar evaporators and are conducted into a flue under an upper cylindrical steam boiler placed directly over another such boiler and connected to it by three large tubes. The flames are then carried downwards all around this lower boiler and also through an excentric cylindrical flue in this boiler until they reach the chimney shaft.

Second, " apparatus for the preparaton of cane-trash or ' bagass ' " to be used as fuel." " The bagass from the mill " is spread " on an open space round the works, and dried by tossing with a " machine consisting " of an arrangement of endless chains, bars, and hooks. This machine also conveys it to the furnace to be consumed.

Third, " processes of clarification ;" and first, what is termed " the high pressure clarification " the solution is put into a vessel with a safety valve and thermometer and heated to a high tem- perature; the vegetable albumen is thus completely coagulated and may be separated by filtration or by " scumming." " The " lime used to aid the clarification may be added to the liquor " either before or after it is submitted to this process." Second, precipitating the impurities from such solutions by causing the base of a salt or an oxide of a chemical compound to combine with the impurities in the solution through the instrumentality of electricity, and applying electricity to separate from combination

any salt or oxide which results from the process of clarification.
Third, " circulating a current of electricity through a bed of char-
" coal to promote the different combinations which take place
" when a partially clarified saccharine solution is in contact with
" the charcoal."

[Printed, 1s. 6d. Drawings. London Journal (*Newton's*), vol. 34 (*conjoined
    series*), p. 305; Mechanics' Magazine, vol. 50, p. 490; Patent Journal,
    vol. 7, p. 79.]

A.D. 1849, February 28.—N° 12,491.

CROSLEY, Henry.—" Certain improved modes or methods of
" and apparatus for heating and lighting, for drying substances,
" and for employing air in a warm or cold state for manufacturing
" purposes." These are, in reference to this subject, first, "the
" method of heating sugar pans by means of a moveable fur-
" nace mounted on wheels," which run upon rails fixed upon the
ground. The truck, &c. carrying the furnace can be drawn
backwards or forwards "from underneath the teache when it is
" charged with syrup, or when the boiled syrup is discharged
" therefrom." "The fire is placed and acts under the pan next
" to the teache, and the heat is conveyed by the flue under the
" other pans of the battery."

Second, "an improved whirling apparatus for drying sub-
" stances," in "so far as regards the means therein employed
" to give a reciprocating rotary motion to the hollow cylinder,
" that is to say, a rotary motion first in one direction and then in
" another." The cylinder is mounted on a hollow axis, round
which is wound a cord, which is pulled, so that it may work the
cylinder in either direction.

Third, apparatus "for warming or heating air, and applying it
" in a warm or hot state," "more particularly for the evaporation
" and concentration of liquid solutions." If hot air is used, air
is passed through a coil of pipes heated in a furnace when a low
temperature is required, there is an arrangement for admission of
atmospheric air by means of an air-diffusing wheel, into pipes in
the liquid. A blowing cylinder, or a fan-blowing machine may
be used. Another mode of applying warm or hot air is, the syrup
passes through a series of holes in a vessel while heated air passes
through holes in pipes in the bottom of the recipient.

[Printed, 1s. 2d. Drawing. Mechanics' Magazine, vol. 51, pp. 210 and 235;
    Patent Journal, vol. 7, p. 236.]

A.D. 1849, April 17.—N° 12,578.

BESSEMER, Henry. — "Improvements in the methods of
" extracting saccharine juices from the sugar cane, and in the
" manufacture of sugar, as also in the machinery or apparatus
" employed therein." These are, first, combining "in one frame-
" work of the machinery or instruments for expressing the
" saccharine juices from the cane with the machinery or instru-
" ments for expressing the saccharine juices from the cane with
" the machinery for working the said expressing machinery or
" instruments, the employment of perforated tubes or cylinders,
" or other suitable vessels left open at the discharge end, to hold
" the canes when in the act of being pressed;" the cutting of the
canes into lengths adapted to the size of the said containing
tubes, &c., by means of the same pistons or rams, by which they
are immediately afterwards pressed; and "the peculiar mode in
" which the reciprocating movements of the steam-driving
piston are combined with those of the expressing pistons,
" whereby the latter are made to exert throughout each stroke
" a gradually encreasing pressure with a gradually diminishing
" velocity."

Second, " the expressing of the saccharine juices from canes,
" by passing them through perforated tubes or other containing
" vessels open at the discharge end, and affording, by the resist-
" ance of their interior sides to the passage of the canes, the
" necessary resistance to the expressing piston or ram, whether
" the cane forms part of the general combination of machinery,"
or " of any other construction of machinery for the purpose of
" extracting the saccharine juices from the cane."

Third, "attaching to such perforated tubes or other containing
" vessels open at the discharge end loaded valves, doors, or flaps,
" for the purpose of retaining the canes in the same under pressure
" while in the course of being pressed."

Fourth, imparting to the reciprocating pistons when used as
above a gradually encreasing pressure combined with a gradually
diminishing velocity, whether the same is employed as part of the
general combination of machinery, or " of any other combination
" of machinery for expressing the cane juice from canes, and
" whether also steam or any other first source of power is made
" use of."

Fifth, " the apparatus for heating and regulating the tempera-
" ture of cane juice while in its progress from the cane press to

" the defecating vessel." There is a pipe which conveys the juice in at the top of one end of an oblong vessel, and a tube at the bottom of the other end for conveying it to the defecating vessel; this pipe is provided with a cock which regulates the flow, so that the oblong vessel is always kept full. This vessel is heated by a steam jacket, a valve admits the steam into this jacket, a tube of mercury lies on the bottom of the oblong vessel passes out of it and curves upward to the steam valve; on the top of this tube is a piece of vulcanized india-rubber which yields by the expansion or contraction of the mercury, and regulates the influx of steam by the valve.

[Printed, 1s. Drawing. Mechanics' Magazine, vol. 5, pp. 380, 385; Patent Journal, vol. 8, pp. 41, 133, and 144.]

A.D. 1849, April 26.—No. 12,590.

OXLAND, ROBERT, and OXLAND, JOHN.—" Improvements " in the manufacture of sugar." These are, " the employment of " acetate of alumina for defecating saccharine liquors, and for the " removal of color in making and refining sugar." In preference, the raw sugar is dissolved in water with steam heat, acidity neutralized by carbonate of lime in fine powder, and the solution should be made to boil at a temperature of 220° F., and filtered through ordinary filter bags into a blow up pan, acetate of alumina is now mixed with the syrup, and the whole boiled as above until the steam passing off shows no acid with blue litmus paper. The acetic acid evolved may be condensed and used for the preparation of acetate of alumina. When nearly the whole of the acetic acid has been evolved " a solution of tannin in water is added until no " further precipitate is produced thereby," carbonate of lime is added in fine powder in sufficient quantity to neutralize all acidity. The syrup is then passed through bag filters, and when bright it is turned into the reservoir, from which the vacuum pan is supplied.

[Printed, 4d. No Drawings. Repertory of Arts, vol. 15 (*enlarged series*), p. 29; London Journal (*Newton's*), vol. 37 (*conjoined series*), p. 106; Mechanics' Magazine, vol. 51, p. 429; Patent Journal, vol. 8, p. 79.]

A.D. 1849, May 24.—No 12,617.

REECE, REES, and PRICE, ASTLEY PASTON.—" Improve- " ments in the manufacture and refining of sugar or saccharine " matters." These are, first, the use of the hyposulphites of lime, magnesia, strontia, baryta, either singly or in conjunction with either solutions of the acid sulphate or acetate of alumina or acetic acid, as defecators of sugar and of saccharine matters.

Second, the use of the hyposulphite of alumina for the above purpose.

Third, "the use of the hyposulphuret of the sulphides" of magnesium, calcium, barium, and strontium, and of the bisulphurets, and the sulphurets of magnesium, barium, and strontium, " as " precipitants of lead or any salts of the same, from solutions of " sugar, and from saccharine solutions."

Fourth, subjecting saccharine liquors containing sulphuretted hydrogen " to the combined action of heat from steam or other- " wise and a vacuum, and to boiling in vacuo." Also " the use " of sulphurous acid and of the acid sulphites and of the hypo- " sulphite of alumina, and of hyposulphites, which by being " decomposed by an acid or otherwise, produce or liberate sul- " phurous acid," to remove the excess of sulphuretted hydrogen or sulphurets which may have been applied as precipitants of lead or its salts, "from saccharine solutions or otherwise."

Fourth, the use of saccharates of lime, baryta, and strontia, as sources of carbonates of those bases, by passing carbonic acid gas into the solutions of these saccharates.

Fifth, the use of the saccharates of lime, baryta, and strontia as precipitants of sulphate and acetate of alumina, when used as defecators of sugar or saccharine matters.

Sixth, the use of the saccharates of lime, baryta, and of strontia as sources of the hydrated sulphides of these basis, by saturating solutions of these saccharates with sulphuretted hydrogen gas, and the use of these sulphides as precipitants of lead, or its salts from solutions of sugar and from saccharine solutions.

Seventh, the use of bicarbonates of lime and magnesia as defecators of sugar and saccharine matters.

Eighth, " the use of the soluble acid sulphites as defecators of " sugar ·and saccharine matters when used in refining of the " same." Also " the use of the soluble acid sulphites when " applied to the treatment of the sugar cane and beet root, either " as in the before-mentioned processes for extracting saccharine " matter from the sugar cane and beet root, or when otherwise " employed in the treatment of the same." Also " the use of the " soluble hyposulphites either alone or in conjunction with the " before-mentioned adjuncts, when applied to the treatment of the " sugar cane and beet root, either as in the before-mentioned pro-

" cesses for extracting saccharine matter from the sugar cane and
" beet root, or when otherwise employed in the treatment of the
" same."

[Printed, 6d. No Drawings. Repertory of Arts, vol. 15 (*enlarged series*),
p. 38 ; Mechanics' Magazine, vol. 51, p. 524; Patent Journal, vol. 9,
p. 68.]

A.D. 1849, August 1.—N° 12,718.

DE CAVAILLON, FLORENTIN JOSEPH.—" Certain improve-
" ments in obtaining carbonated hydrogen gas, and in applying
" the products resulting therefrom to various useful purposes."
These are, in reference to this subject, as follows :—" The follow-
" ing materials may be used, either as a substitute for, or in
" combination with pit coal (namely), bones, kitchen stuff, graves,
" the residuum of suet, tallow, or other animal waste, the resi-
" duum of seeds or other oleaginous matters, spent bark, saw-
" dust, and also sawdust or pulverized or reduced wood that has
" been used for the purification of oils, also peat or turf, either
" pulverized or in fine powder." For gas making, " any of these
" materials may be taken separately, or all of them may be
" mixed together in greater or less proportions (that is to say)
" without being mixed in any definite proportions (except that
" fifty per cent. of the compound should consist of coal, and the
" remainder of the other substances either by equal or unequal
" proportions, or of some of them only if the others cannot be
" conveniently procured). These materials are consolidated or
" caused to adhere together, either by means of some gummy or
" resinous matter, or by some empyreumatic oil, or the molasses
" of sugar," and when thoroughly mixed they are operated upon
as coal in ordinary gas works. The products are gas, animal
charcoal, animal and vegetable charcoal in powder, &c. The
remainder of the Specification treats principally of the purification
of gas.

[Printed, 4d. No Drawings. London Journal (*Newton's*), vol. 36 (*conjoined series*), p. 257; Mechanics' Magazine, vol. 52, p. 115; Patent
Journal, vol. 8, p. 262.]

A.D. 1849, August 1.—N° 12,730.

MURDOCH, JAMES.—(*A communication.*)—" Certain improve-
" ments in converting sea water into fresh, and in ventilating
" ships and other vessels, applicable also to the evaporation of
" liquids, and to the concentration and crystallization of syrups
" and saline solutions." There are, in reference to this subject, in

apparatus "as applied to the concentration and crystallization " of syrups," consisting of an " evaporating vessel composed of a " series of compartments forming a continuous zig-zag channel, " and inserted in a casing;" steam is admitted at one part into the casing, and the water of condensation is carried off by an inverted syphon connected with another part; on the evaporating pan is a close fitting cover, to the under side of which is attached a perforated pipe, communicating with a cock, and over an aperture in the cover is fixed an exhausting fan. The liquid to be evaporated is introduced at an aperture at one corner of the vessel. " If the crystallization is to be effected in the vessel the lid is " lifted when the crystallization is sufficiently advanced, and the " crystals are withdrawn." The vessel can be wholly emptied by a cock at the bottom. In place of the evaporating vessel being formed in deep compartments, "it may be formed simply as a " flat-bottomed shallow vessel of any suitable figure. The steam " also may be generated in the casing, in lieu of employing a " separate boiler for the purpose, or the evaporator may be heated " by a naked fire."

[Printed, 10*d.* Drawings. Mechanics' Magazine, vol. 52, p. 115; Patent Journal, vol. 8, p. 259.]

A.D. 1849, August 16.—N° 12,742.

BROOMAN, RICHARD ARCHIBALD. — (*A communication.*) — " Improvements in machinery, apparatus, and processes for ex- " tracting, depurating, forming, drying, and evaporating sub- " stances." These are, in reference to this subject, as follows : first, in centrifugal machines introducing steam into the space between the outer and inner drums, and also the space between the inner drum and the strainer, and " causing the steam to act " from the outside towards the inside, and retaining the steam in " the apparatus during the process of extraction."

Second, "transferring the power which has been accumulated " in the working of one machine to work or assist in working the " other machine." The two machines are on one bed plate, and connected together by bevilled gearing to the main shaft, which is put in motion by an oscillating steam cylinder between the two machines, and by an arrangement of an endless cord over four rollers and around a pulley. Motion is given to the endless cord by a handle " which raises the guide fork of one machine while it " lowers that of the other."

Third, "discharging or unloading centrifugal machines." A bag fits the inner periphery and bottom of the cylinder and also the dome, and when the machine is at a low speed a nut causes it to rise up the shaft and the materials are thrown out by the centrifugal action into a suitable receptacle placed outside.

Fourth, strengthening and supporting centrifugal machines by means of "a combination of vertical and diagonal supports" (straps of metal). Also fixing the wire gauze or other strainer so that it may be readily taken out or fixed.

Fifth, "forming or moulding and depurating substances," as follows :—A cylindrical receiver, open at top, receives the sugar ; a series of sugar moulds, with their broad ends outermost, their small ends opening into the receiver, are disposed round the receiver. The broad ends have each two lids, the inner being a frame covered with wire gauze, the outer being impervious ; crystallized sugar is conducted into the machine by a pipe, on motion the sugar fills the moulds, the machine is then kept at a slow motion until the mass is set, when the impervious lid of the mould is slackened, and the liquid is driven out and liquor is supplied for purifying the sugar in the moulds.

Sixth, "apparatus constructed on the same principle as the " preceding for giving shape and form to plastic substances."

Seventh, depurating substances in a liquid state by the combined action of steam, or of gas and centrifugal agency. The liquid is revolved in a perforated cylinder, and falls in showers into the steam or into the gas in an outer cylinder ; the gases named are carbonic acid or sulphurous acid.

Eighth, the employment of filter bags, as follows :—Each filter is a bag of some woven stuff, enclosing a bed or layer of charcoal, or some filtering medium, covering entirely "the internal periphery " of the cylinder."

Ninth, for drying and evaporating purposes using an arrangement of steam tight coiled pipes, through which steam and air or hot air are made to circulate, as also the centrifugal arrangements described under the first head.

[Printed, 1s. 4d. Drawings. Mechanics' Magazine, vol. 52, p. 141; Patent Journal, vol. 8, p. 281.]

A.D. 1849, August 23.—N° 12,749.

NEWTON, ALFRED VINCENT.—(A communication.)—"Improve- " ments in manufacturing and refining sugar." These are, first,

" the use of bisulphites or acid sulphites (particularly the acid
" sulphite of lime) as preservative agents against fermentation, and
" as depurators of vegetable juices containing crystallizable sugar."
The solution of bisulphite of lime as a weak solution may be
applied over the cylinders of the crushing mill, the rasping
machine, or to the portions of cane as they fall from the slicing
machine, or otherwise.

Second, besides the above substances the use of the following
antiseptic substances as preventatives to fermentation in saccharine
fluids, namely, all the acid sulphites and neutral salts, " provided
" they be soluble in sulphurous acid, and provided also that the
" acid set at liberty can exercise no destructive action upon the
" sugar in solution," thus the phosphate of lime of bones dissolved in sulphurous acid may be used, and " mustard, creosote,
horseradish, nitrous acid gas either alone or in combination with
the salts of iron or other salts, tannin, lamp black saturated with
creosote, the products of the distillation of certain bituminous
matters, tars, or wood essence of turpentine, aldehyde, and analogous bodies, chlorine either in the gaseous state or in solution or
in combination with oxygen, may all be employed as antiseptics.

Third, " the employment of baryta, strontia, and of the sul-
" phurets of barium and strontium, of lime and other metallic
" oxides, for the precipitation and separation of the crystallizable
" sugar in the state of insoluble saccharates."

Fourth, " the method of producing the above-mentioned sac-
" charates, and of separating the sugar in a state of purity." In
using baryta the saccharine fluid " depurated " is well mixed with
hydrate of baryta " in the proportion of 50 to 60 parts of caustic
" baryta to every 100 parts of crystallizable sugar," and brought
up to boiling and kept so for a few minutes, the precipitate
saccharate of baryta is allowed to subside, is collected, dried by
pressure or by centrifugal apparatus, and decomposed by the
addition of small quantities of sulphuric acid, by carbonic acid, or
by sulphurous acid. When lime is employed as above the saccharate of lime being insoluble at the boiling point, is then
filtered by a filter known as Taylor's filter, or the liquid is
removed by a heated press. The precipitate is treated with carbonic acid or sulphurous acid, and in both cases the saccharine
matter is thus separated from the lime salt, " and is ready at once
" to furnish, by the usual treatment, refined sugar fit for com-
" mercial purposes." What was said of baryta is applicable

to strontia although the use of strontia is inferior, more being required, and the saccharate being soluble in pure water. When oxide of lead is employed " it requires to be kept for a considerable " time in contact with the juice, in order to produce the precipi- " tation of the sugar." "What has been already said of the " decomposition of saccharate of baryta by carbonic acid applies " equally to the oxide of lead." Any " of the known processes " may be employed in the production or regeneration of the " hydrate of baryta as the basis of the operations."

Fifth, " the employment of the bisulphites and acid sulphites " (particularly the acid sulphite of lime) as defecators of fluids " containing crystallizable saccharine matter." The liquid after the removal of mechanical impurities is treated with a solution of the salt, allowed to rest, and filtered and evaporated.

[Printed, 6d. No Drawings. London Journal (*Newton's*), vol. 36 (*conjoined series*), p. 229; Mechanics' Magazine, vol. 52, p. 178 ; Patent Journal vol. 8, p. 285.]

A.D. 1849, October 12.—N° 12,808.

FINZEL, CONRAD WILLIAM.—" Certain improvements in the " processes and machinery employed in and applicable to the " manufacture of sugar." These are, first, " the mode of applying " steam or liquids to machines used for separating syrups or fluids " from sugar by means of centrifugal force, for the purpose of " clearing and keeping clear the meshes or apertures in the peri- " phery or revolving cylinders of such machines." In the outer case of an ordinary centrifugal machine is a narrow recess about the same height as the revolving cylinder, a perforated box is placed in this recess, the box is connected with a supply of steam, so that when it is let on, it projects jets of steam when required against the periphery of the cylinder as it revolves, and so cleans it.

Second, " preparing such sugars as require mixing with liquid " before being operated upon in the centrifugal acting machines " described above. A vessel has a step in the centre of its bottom in which a shaft revolves ; there is a wire sieve fixed on the upper part of this shaft, and below this sieve are perforated steam pipes. On the revolving shaft below the sieve is a small centrifugal drum. The revolving shaft is hollow and has perforations opening into this small centrifugal drum, " in order that the liquid may " pass from the former into the latter." The sugar broken is put into the sieve, the centrifugal action of the machine throws the sugar through the sieve into the steam issuing through the perforated pipes, and the sugar is thus prepared to receive the syrup which is thrown from the centrifugal drum, and completes the mixing.

Third, apparatus acting as follows :—The vapour arising from a vacuum pan passes through the condenser (a series of tubes) flows into a receiver in a state of weak solution of sugar, in which are steam pipes, and to which an exhaust is applied until concentrated.

[Printed, 1s. 6d. Drawings. London Journal (*Newton's*), vol. 37 (*conjoined series*), p. 322 ; Mechanics' Magazine, vol. 52, pp. 319, 321 ; Patent Journal vol. 9, p. 34.]

A.D. 1849, November 14.—N° 12,844.

COWPER, Charles.—(*A communication.*)—" Certain improve-" ments in the manufacture of sugar." These are as follows :—" The juice having been obtained in the usual way " the temperature is raised from 120° to 167° F. Slacked and sifted lime from 15 to 50 lbs. for each 100 gallons of juice are well stirred in and the mixture is heated to 185° or 194° F. The clear liquor is now drawn off and the residues pressed to obtain the liquor from them. Any excess of lime is thrown down as carbonate from the saccharine solution by carbonic acid, produced by the combustion of charcoal or other species of carbon, and washed by passing it through water, excess of carbonic acid is expelled by boiling a few minutes, and filtered and concentrated gives a syrup sufficiently pure to be convertable into loaves, called loaves No. 1. If a still finer description of sugar is required the syrup after concentration to about 1.23 is filtered a second time. The drainings or green syrups which run out of the loaves of sugar made by the above improved process are sufficiently pure to be concentrated again and converted into loaves No. 2, but the following is the mode of operating :—The syrups diluted to 1.07 to 1.16 are heated to 150° F., from 20 to 30 lbs. of slaked lime for 100 gallons of syrup added with stirring, and then from two to four pounds of clay or siliceous loam made with water into a kind of cream, are mixed in and agitated and the temperature raised from 140° to 176° F., " after which the lime is saturated by a " current of carbonic acid gas, as already described for the first " juice. The liquor is now filtered through animal charcoal," and evaporated, &c. An apparatus for making carbonic acid gas and forcing it into the saccharine solutions is described.

[Printed, 8d. Drawing. Mechanics' Magazine, vol. 52, p. 397; Patent Journal, vol. 9, p. 82.]

A.D. 1849, November 20.—No 12,856.

COWPER, CHARLES.—(*A communication.*)—" Certain improve-
" ments in the manufacture of sugar." These are said to be,
first, " the recovery of baryta from carbonate of baryta formed in
" the manufacture of sugar by decomposing the same with sul-
" phuric acid," as follows :—" Carbonate of baryta is decomposed
" in close vessels by means of sulphuric acid and the carbonic
" acid, which is liberated is employed in the decomposition of the
" saccharate of baryta formed by boiling cane juice, beet root
" juice or other saccharine liquid " when mixed with a solution of
caustic baryta. " The sulphate of baryta which is formed is
" collected and converted into sulphuret of barium by charcoal,"
&c. The carbonic acid thus evolved may be collected or used as
above. A solution of the sulphuret of barium is made, and is
boiled with oxide of copper, which produces " sulphuret of copper
" and a solution of hydrate of baryta." By roasting, the sulphuret
of copper is converted into oxide and may be used over again.

Second, " the recovery of baryta from carbonate of baryta formed
" in the manufacture of sugar," as follows :—The carbonate of
baryta obtained by decomposing the saccharate of baryta as above,
by means of carbonic acid, is decomposed by double decomposition " by all or nearly all the sulphates, in the wet way with
" some and in the dry way with others, and with many of them
" in either way." " In some places there is an advantage in
" employing the sulphates of soda or of potash, for these sul-
" phates may thus be converted into carbonate or into caustic
" soda or potash." The process is the same with both these
salts, only " when sulphate of potash is employed 218 parts should
" be used in place of every 178 parts of sulphate of soda " to 123
parts of carbonate of baryta. These are, together with water in
preference boiled together, and an oxide added (lime) to take the
carbonic acid as it is separated from the carbonate of baryta, the
result is sulphate of baryta, carbonate of lime, and caustic alkali
in solution. The sulphate of baryta and lime are decomposed
as above.

[Printed, 4d. No Drawings. Mechanics' Magazine, vol. 52, p. 419; Patent Journal, vol. 9, p. 118.]

A.D. 1849, December 12.—N° 12,888.

BIRKMYRE, WILLIAM.—"Improvements in the manufacture
" and refining of sugar." These are, first, "the decolorizing of
" cane juice and syrups of raw sugar by adding to them a mixture
" of pounded chalk or limestone, and the substance (sulphate of
" alumina and silex) derived from the action of sulphuric acid
" and heat upon china or pipe-clays." To every 100 gallons of
cane juice of density 1,200 or 40° of Twaddell's hydrometer are
added a mixture of 14 lbs. of ground sulphate of alumina and
silex and 11 lbs. of ground chalk, and the mixture boiled, filtered,
run into cones, &c.

Second, "the decolorizing of cane juice and syrups of raw
" sugar by precipitating alumina" in the same, "from sulphate
" of alumina by lime or ground chalk, or limestone, or by a
" mixture of lime with chalk or limestone." " When the pure
" sulphate of alumina is used 11 lbs. of finely ground chalk
" should be thrown alternately into the 100 imperial gallons of
" concentrated cane juice with 9 lbs. of sulphate of alumina either
" in powder or solution." The manner in which the sulphate of
alumina is recommended to be used for raw sugar is given.

Third, the use of the substance precipitated as above collected
in the filters and washed "as a fertilizer, or converted by heat into
" a species of charcoal fit for decolorizing cane juice or syrups."

[Printed, 4d. No Drawings. Repertory of Arts, vol. 16 (*enlarged series*),
p. 38 ; Mechanics' Magazine, vol. 52 p. 496; Patent Journal, vol. 9,
p. 154.]

A.D. 1850, January 17.—N° 12,930.

COWING, HENRY.—" Improvements in obtaining motive power,
" and in steam and other ploughs, in land carriages, in fire-engines,
" in raising water for draining and other agricultural purposes,
" and in apparatus for evaporating saccharine and other liquors."

[No Specification enrolled.]

A.D. 1850, February 21.—N° 12,977.

SCOFFERN, JOHN.—"Improvements in the manufacture and
" refining of sugar, and in the treatment and use of matters
" obtained in such manufacture, and in the construction of valves
" used in such and other manufactures." These are, first, for
the above purpose in using the acetates of lead, the juice is

gradually raised to about 210° F. skimming all the time until the juice has acquired one degree (Beaume) of density or thereabouts over the original density of the cane juice, the heat is then removed and when the juice is just below boiling the subacetate of lead made into a thin paste with water is incorporated well by stirring, usually one-sixth per cent. of lead salt will suffice.

Second, "the manufacture of a pigment by employing sulphate " of lead," which is thrown down by sulphurous acid from sac- charine solutions where acetate of lead have been used as above.

Third, in "the construction of a valve," "suitable for the " pumps used when forcing sulphurous acid gas into syrups, and " such valves are also applicable to pumps when the fluid is to " be forced in one direction and prevented returning." This consists of a tube closed at its lower end, it is perforated for some distance with several holes, a band or tube of vulcanized india- rubber fits tightly around the tube and covers these holes; in this way, "a valve is made which will allow gases and other fluids " being forced through from the interior outwards, but the elastic " band will immediately close and prevent any return."

[Printed, 6d. Drawing. Repertory of Arts, vol. 16 (*enlarged series*), p. 233 ; London Journal (*Newton's*), vol. 37 (*conjoined series*), p. 96 ; Mechanics' Magazine, vol. 53, p. 159 ; Patent Journal, vol. 9, p. 251.]

A.D. 1850, February 27.—N° 12,981.

GWYNNE, GEORGE.—"Improvements in the manufacture of " sugar." These are, first, "the use and application of saccharide " of lead," whether prepared as follows, "or in any other manner, " of lead paste, and of oxide of lead in the manufacture of sugar," and the use of suitable chemical agents for rendering insoluble the lead left after the application of such substances. Also "the " use of bag or other filters for separating insoluble lead contained " in the bright liquor or juice when such lead has been introduced " through the use of hydrated oxide of lead." "Saccharide of " lead" is made by moistening litharge with water, grinding to a smooth paste, adding twice its weight of refined sugar, and grind- ing until no oxide of lead falls to the bottom, when a small portion is stirred up in a wine glass with some water. This agent is "worked up with water to the consistence of cream and " passed through a fine sieve into the "blow-up' or other clari- " fying vessel. The raw sugar, with the necessary quantity of " water, is then to be added," and the whole mixed up and heated and "filtered through a bag or other filter or other suitable "filtering apparatus." After filtration the excess of lead in solution may be removed by sulphuric or oxalic acid by hydro-sulphuric acid or sulphurous acid gases, acidulated ferrocyanide of potassium, phosphate of lime, binoxalate of potassa, &c. A preparation is made with bone earth, sulphuric acid, and carbonate of soda, called "phosphate of soda and lime," an acid preparation which is considered the best agent for rendering insoluble the lead in saccharine solutions.

Second, "preparing and using basic acetic of lead. To the lead paste prepared under the first head acetate of lead or an equivalent proportion of acetic acid is added, so as to form the highest basic acetate (hex acetate), but less acetate of lead may be used. The grinding up of these ingredients is not necessary, as they may be put into the blow-up, &c. vessel separately.

Third, "rendering insoluble the lead contained in 'the bright' "liquor or juice when the basic acetates of lead" have been em-ployed to purify the raw sugar, cane juice, &c. The agent which is preferred "for this purpose is prepared in the same manner as "the phosphate of soda and lime" mentioned under the first head, "with this difference, that it is made slightly alkaline instead "of slightly acid."

Fourth, "improvements in the vacuum pan." To the vacuum pan is attached a receiver, by means of a wide pipe or neck; a long pipe (thirty-four feet) leads from the bottom and dips into water in a small box, beyond the receiver and communicating with it by an opening near the top is a condenser, in the upper part of which is an injection cock for the admission of cold water; in the bottom of the condenser is a pipe similar to that in the receiver, and also dipping into water in a box; from the upper part of the condenser is a pipe worked by an air pump. In place of the long pipes attached to the bottom of the receiver and condenser small pumps may be applied.

[Printed, 8d. Drawing. Repertory of Arts, vol. 20 (enlarged series), p. 158; Mechanics' Magazine, vol. 53, p. 178; Patent Journal, vol. 9, p. 272.]

A.D. 1850, March 23.—No 13,014.

CURTIS, WILLIAM JOSEPH.—"Improved machinery and appa-"ratus adapted for the manufacture of sugar." This consists, first, in the application of a hammer or hammers, stamper or stampers, for breaking or crushing the canes before they pass to the squeezing rollers.

Second, the apparatus whereby the proportions of bisulphite of lime and cane juice are regulated. There is an overshot water wheel to which motion is given by the juice and bisulphite falling into the buckets, as in an ordinary water wheel, a governor, as in a steam engine, works a lever which rises or falls as the governor revolves quicker or slower, and by an arrangement opens and closes a cock in connection with a reservoir of bisulphite of lime.

Third, depriving sugar of its molasses by placing it in a revolving perforated cylinder surrounded by a jacket, which receives the molasses; from this jacket is a pipe by which the molasses flows off into a suitable receiver. There are two arms or orifices through which the cured sugar escapes into a hopper, by means of which the sugar is delivered into the hogshead below. There are rammers for ramming down the sugar and apparatus for regulating the discharge of the sugar from the cylinder.

Fourth, machinery for making bisulphite of lime, arranged so that three sets of bellows being set to work forces air in a receiver by means of a pipe, and from there it passes into the sulphur chamber, the sulphur being kept in a state of combustion, by means of the air blown upon it; the sulphurous fumes are then let off into an agitator and kept in contact with the cream of lime until it acquires a density of $10°$ Beaumé, when it is let off into a vessel and a fresh supply of cream of lime is renewed from a reservoir.

Fifth, the machinery for dipping or transferring the liquor or syrup from pan to pan. A shaft is worked along the centre of a series of pans, a crank works a pump. A main suction pipe is placed over the pans and communicates with each pan by means of branch or offset pipes, which are connected to the main pipe by means of moveable joints; these pipes, when they are lifted out of the liquor, are turned upwards and serve as cocks to shut off the communication.

Sixth, apparatus for skimming the coppers. One method is by scoop wheels turned by a vertical shaft, and another method is by an oscillating paddle. In each case the skimmings are swept into a receptacle at the side of the pans.

Seventh, "shielding or protecting the teach from the action of " fire whilst the teach is being struck or discharged." A shield is formed of cast iron, and its outer surface is faced or protected with fire clay, which burnt is as hard as a fire brick. The shield is arranged in a frame supported upon three wheels running upon rails and in such a manner that it can be placed between the teach and the fire and removed when necessary.

Eighth, " burning the liquid pitch or the inflammable oil from " the pitch-lake or other places in the furnace of a steam engine " boiler, likewise of a force or injecting pump for the same " purpose." There is an engine furnace, above it is the steam boiler. A strong vessel higher than the steam boiler contains the liquid asphaltum or pitch ; from the bottom of this vessel is a pipe, which may be carried along each side of the fire and across the bridge, in order to heat the pitch before it passes into the furnace by a nozzle like a gas burner. A steam pipe from the boiler enters the upper part of the strong vessel containing the liquid asphaltum or pitch, the steam from which drives the oil from the nozzle.

Ninth, burning the liquid pitch, &c., in the furnace of a sugar pan ; likewise, the use of a force or injecting pump for the same purpose. This arrangement is somewhat similar to the last, only the vessel containing the pitch is in an offset of the chimney and partly surrounded by flame and heated vapours.

Tenth, employing " electrotyped and enamelled pans for the " boiling of syrup."

Eleventh, " a web or band and the use of hot air," for " con-" centrating the sugar at a low temperature." An endless web or band dips into the teach and then passes successively over a series of drums. The band is set in motion by the pinion, moved by any first mover, taking into the spur wheels fixed on the drum shafts.

[Printed, 2s. 8d. Drawings. Mechanics' Magazine, vol. 53, p. 257 ; Patent Journal, vol. 10, p. 21.]

A.D. 1850, March 26.—N° 13,023.

ROTCH, THOMAS DICKASON.—(A communication.)—" Improve-" ments in separating various matters usually found combined " in certain saccharine, saline, and ligneous substances." These are, in reference to this subject, first, to diminish the oscillation of the drum of centrifugal machines, attaching a weight by means of a hook to the lower end of the socket, at the bottom of which is the brass step on which the lower end of the vertical axis of the drum turns."

Second, " to prevent all oscillation of the vertical shaft," three small friction rollers are placed under the pivot of the vertical shaft of the drum, and two bearings are placed above the friction rollers.

Third, the introduction of syrups from large coolers, say from 2,000 to 3,000 quarts, and of 30° to 35° Beaumé, and from 120° to 125° F., into the centrifugal machines, " whereby a much more " perfect and valuable effect is produced by the said machines " than heretofore."

Fourth, "the application of woven fabrics and sponge " in place of clay, as follows :—On the large end of the sugar loaf, when in the mould when prepared as usual, for clay placing a thick piece of cloth, over which a piece of sponge the size of the end of the loaf, and from 1½ to 2 inches thick, the loaf having previously been twice liquored, then pour from a watering pot over the sponge as much pure water as it will absorb, and continue to " moisten the sponge for two days, gradually decreasing " the quantity of water poured over the sponge."

[Printed, 10d. Drawing. London Journal (*Newton's*), vol. 37 (*conjoined series*), p. 229; Mechanics' Magazine, vol. 53, p. 275; Patent Journal, vol. 10, p. 5.]

A.D. 1850, June 1.—N° 13,093.

HILLS, Frank Clarke, and HILLS, George.—" Certain im- " provements in manufacturing and refining sugar." These are " the employment of sulphite of lead, or carbonate of lead, or, " precipitated protoxide of iron to deprive saccharine solutions of " sulphuretted hydrogen or of hydrosulphurets of the earths which " may have been employed for their purification of such solu- " tions." " When sulphuretted hydrogen is the agent which has " been employed to precipitate the lead or other metals," and all the metal is precipitated, sulphite of lead (made into a cream with water) is added, the liquid being about 150° F. until a filtered solution will not darken " by adding a solution of acetate of lead." " The saccharine solution is then filtered and boiled down in the " usual manner for crystallization." When hydrosulphurets of the earths have been employed in place of sulphuretted hydrogen, their excess may be removed by the addition of either the sulphite or carbonate of lead or the precipitated protoxide of iron ; "and " also, if sulphuretted hydrogen be used as a precipitant, provided " the saccharine solution be rendered perfectly neutral or charged " with an excess of base, the carbonate of lead or the precipitated

" protoxide of iron may be employed in lieu of the sulphite of
" lead to abstract the sulphuretted hydrogen."

[Printed, 4d. No Drawings. London Journal (*Newton's*), vol. 38 (*conjoined
series*), p. 182 ; Mechanics' Magazine, vol. 53, p. 439 ; Patent Journal, vol. 10,
p. 117.]

A.D. 1850, June 24.—N° 13,147.

MACFIE, ROBERT ANDREW.—" Improvements in manufactur-
" ing, refining, and preparing sugar; also improvements in
" manufacturing and treating animal charcoal."

[No Specification enrolled.]

A.D. 1850, July 17.—N° 13,178.

VARILLAT, JEAN JULES.—" Improvements in the extraction
" and preparation of colouring, tanning, and saccharine matters
" from various vegetable substances, and in the apparatus to be
" employed therein."

[No Specification enrolled.]

A.D. 1850, July 31.—N° 13,202.

BESSEMER, HENRY.—" Certain improvements in apparatus
" acting by centrifugal force in the manufacture of sugar, and
" other improvements in the treatment of saccharine matters by
" such apparatus." These are, first, treating " of the saccharine
" juice of the cane immediately after it has been expressed, and
" in which state it is found to be mixed to a greater or less extent
" with small fragments of the cane, and the arrangement of centri-
" fugal filtering apparatus for separating these particles from the
" juice."

Second, the cane juice having been filtered in the centrifugal
apparatus " it may be defecated by any of the usual methods now
" practised, when a certain quantity of coagulated matters will
" be found suspended in the juice." A centrifugal filtering apparatus is now applied at " this stage of the manufacture, in order to
" separate such coagulated or other solid matters from the saccha-
" rine juice."

Third, the juice having been " thus clarified and filtered by the
" two preceeding operations, it is then heated in an open pan, in
" order to evaporate the aqueous parts of the fluid." To facilitate
this evaporation the juice is lifted up " (by centrifugal force) in a
" tube placed in the centre of the boiling vessel, the upper part of

" such tube being pierced with numerous small holes in order
" that the juice raised within it may by the rotation of the tube
" be dispersed in a shower over the entire area of the vessel."
Evaporation is thus conducted at a lower temperature.

Fourth, " apparatus for applying centrifugal force to the filtering
" of syrups in the refining of sugar," and consists in two modifications of the centrifugal filter above used " for operating upon
" the saccharine fluid, and separating therefrom the coagulated or
" other solid matters after the process of ' blowing up,' and
" preparatory to passing such syrups through the charcoal filter."

Fifth, the application of certain improved modifications of
centrifugal apparatus to the manufacture of sugar when separating
the crystals from the fluids and other matters with which they are
mixed. A portion of these improved modifications are intended to
reduce the vibratory motion communicated by such descriptions
of machines to the buildings in which they are worked," while
others " are in the mode of feeding and discharging the matters
" operated upon in such machines, and also in the methods of
" applying power for driving such descriptions of machines, and
" of the construction of the perforated drums used therein," by
grooving them internally and externally at right angles, and the
holes conical, are kept open by a wire or other suitable brush.

[Printed, 2s. 10d. Drawings. Repertory of Arts, vol. 17 (*enlarged series*),
pp. 295, 321; Mechanics' Magazine, vol. 54, p. 136; Engineers and Architects' Journal, vol. 14, p. 15; Patent Journal, vol. 10, pp. 227, 240.]

A.D. 1850, October 17.—N° 13,286.

SHEARS, DANIEL TOWERS.—(*A communication.*)—" Improve-
" ments in the manufacture and refining of sugar." These are,
first, employing in the interior of the vacuum pan " a rouser or
" displacer" of any convenient form.

Second, introducing into the pan when the sugar has been
boiled to dryness or nearly so a quantity of thick liquor to mix
with the sugar, so as to admit of its discharge from the pan.
Another improvement consists in introducing cold air into the
evaporated liquid after the vacuum is destroyed to aid crystallization.

Third, converting beet root or other saccharine juices into
" refined sugar at one operation, instead of first making it into
" raw sugar and then refining it, as at present practised." The
first method is to boil, crystallize in vessels, in preference, as large

as convenient, and extract the syrup "by the centrifugal or pneu-
" matic processes." The dry sugar is dissolved by heat with
" liquor or other liquid to the consistency required" for moulding,
and when hard the moulds are placed in a " centrifugal apparatus,"
and the sugar washed. The sugar is dried in the usual way or
heated air dries it in the machine. The second mode is "the
" juice having been purified and crystallized in any manner not
" calculated to produce loaf sugar," the sugar is reduced "to the
" consistency and temperature needful for loaf sugar" and
moulded and treated as usual.

Fourth, " purifying saccharine solutions by means of the com-
" bination of an alkali and of an acid gas." Employing by
preference, " an alkaline material composed of prepared lime and
" a West Indian plant, 'Wassama,' or the ingredient No. 1,
" specified" in No. 11,370. A solution is made of either of these
materials, in preference, in cold water, of density "about 15°
" Baumes saccharometer, it is then applied to the saccharine
" solution" till it becomes strongly alkaline, when carbonic or
sulphurous acids are passed into it until the solution is very
slightly alcaline. The solution is then heated and boiled for a
few minutes, the impurities allowed to settle before filtering, or
" it may be clarified in any of the usual ways, then filtered."

Fifth, " reducing West Indian molasses, rhoar (or other
" saccharine substances) to a thin solution, and purifying and
" treating them," by means of " the ingredients No. 1 and No. 2
" specified in" No. 11,370, "though the ingredients prepared
" with 'Wassama' have been declared in the West Indies to be far
" preferable." The proportions and methods of using these ingredients are given. This purified solution of molasses is " employed
" to dissolve the sugar intended to be refined." " Crystallizing
" to dryness of solutions of West Indian molasses, rhoar, or such
" like materials, either individually or in combination with each
" other or with raw sugar, by combining the evaporation after the
" formation of the crystals."

Sixth, the arrangement of "apparatus for regulating the steam
" for vacuum pans and apparatus employed in the manufacture
" and refining of sugar." The steam from a boiler enters a pipe,
passes through parts when they are uncovered by a slide valve
through a pipe to the vacuum pan ; on this pipe is a cylinder into
which a piston is fitted. The piston is loaded to any pressure
required. The action of steam or vapour on the under side of the

piston will always keep the slide valve parts so adjusted as to admit such a quantity as to keep an unvarying pressure on the vessels or apparatus attached to the pipe, modifications of parts of this apparatus are given.

Seventh, the combination of apparatus. There are three boilers, the juice is put into the first and boiled by steam from a boiler, or waste steam from high pressure engines. The steam in the interior of the first boiler passes into a series of tubes " of " the next compartment or boiler," and " thence by a similar set " of passages into the boiler of the third boiler," " to serve as heating medium for the second and third pans, or if this is not sufficient, a further quantity is admitted by a pipe direct from the boiler. The juice in the first boiler " is concentrated to about " ten degrees Beaumé, and is constantly drawn over to the second " boiler by the vacuum in the second boiler to the constant level " of the pipe connecting them." In the second pan the juice is evaporated to the density of about 25° or 27° Beaumé, from thence it is drawn by a pump to be filtered if necessary, and is then returned into the third pan or boiler " to be concentrated " to the crystallizing point, as is usual by the ordinary methods at " present in use."

Eighth, " crystallizing cane juice or other saccharine solutions " to dryness, when boiling in open vessels subject to the pressure " of the atmosphere," by causing " the process of evaporation to " be continued after the liquid has been been brought to ' proof,' " and then to dryness, " taking care, however, to keep down the tem- " perature," particularly after " proof," so as to prevent burning.

[Printed, 1s. 4d. Drawings. Repertory of Arts, vol. 18 (enlarged series), p. 294; Mechanics' Magazine, vol. 54, p. 338; Patent Journal, vol. 11, p. 49.]

A.D. 1850, December 19.—N° 13,416.

NIND, PHILIP.—(A communication.)—" Improvements in the " manufacture of sugar, and in cutting and rasping vegetable " substances." These are, first, in the manufacture of beet root sugar, the beet washed is cut into slices, about 1,000 lbs. is then allowed to fall into the first of eight vessels containing water saturated with sulphurous acid, after an hour a communi- cation is to be opened with the second vessel, into which has been previously introduced 1,000 lbs. of cut beet root. Fresh water is then introduced into the upper part of the first vessel all over the surface, which forces the liquor previously therein into the upper part of the second vessel. The water in the first vessel is to be impregnated with sulphurous acid as before. The liquor remains in the second vessel about an hour, when it is forced into the third vessel, in which there is a charge of beet root, by a charge of fresh water introduced at the top of the first vessel as above, and the liquor impregnated as before. This process is repeated until the liquor first displaced from the first vessel has passed through the eight vessels, when the first vessel will have to be opened, and a fresh charge of beet root introduced. The liquour is removed into an evaporating apparatus in which is a ventilator or fan in motion, while gas is forced through a perforated pipe within the apparatus. The evaporation and clarification now takes place by the aid of the gas generated by throwing sulphate of ammonia into the fire, the alkaline gas or vapour from which coagulates the albumen, the liquor should be concentrated to about 30° Beaumé. Cold water is now added, the whole is mixed and filtered and evaporated and crystallized as is well understood. The same improvements will be applicable when manufacturing sugar from the cane. Another apparatus consists of a series of vessels in which screws work, and the slices of cane are made to pass through each of these vessels until the water becomes saturated with the saccharine matter.

In centrifugal apparatus " arranging the outer reticulate cover-" ing so that it may be readily removed with the sugar," and replaced by another. In another improvement in centrifugal machines, the frame is elastic in all its points, by means of india-rubber and other springs. The lower bearing is elastic, and the upright shafts are coupled by elastic tubes or springs. The cutting machine is a roller, which has affixed thereon a series of blades moved by a fly wheel, as the ends of the cane are fed up to the rollers by hand they are cut off in short lengths. These lengths are further reduced by rasping, but the rasping machine is not claimed.

[Printed, 3s. 6d. Drawings. Repertory of Arts, vol. 19 (*enlarged series*) p. 280; Mechanics' Magazine, vol. 54, p. 516; Patent Journal, vol. 11 p. 161.]

A.D. 1850, December 27.—N° 13,428.

FRASER, JOHN MATHISON.—(*A communication.*)—" Improve-" provements in the manufacture of sugar." These are, the treatment of " cane or beet root juice " as follows :—" Operating

" upon cane juice, which is expressed between cylinders in the
" usual manner, and conducted by an open channel or gutter into
" a proper tank." At the extremity of the gutter is a sieve or
strainer. A mixture is made of cane juice and quicklime in
certain proportions, and the mixture is added in certain proportions, " with continuous stirring whilst the juice is running into
" the tank. But as soon as the tank is about half full" there
is added a saturated solution of sulphurous acid in certain
proportions, and the whole mixed together, then left for about an
hour for the sediment to subside, the clear liquor is drawn off
" into the usual evaporating pans, and as the evaporation proceeds carefully skimming. The materials used in preference for
making the sulphurous acid are common sulphuric acid or oil of
vitriol of commerce and charcoal, these are put into a flask-shaped
vessel, heat is applied, and the gas is washed in a wash bottle
before passing into the water to be saturated.

[Printed, 6d. Drawing. Mechanics' Magazine, vol. 55, p. 17; Patent
Journal, vol. 12, p. 38.]

A.D. 1851, February 3.—N° 13,490.

ALLIOTT, ALEXANDER.—" Certain improvements in cleaning,
" dyeing, and drying machines, and in machinery to be used in
" sugar, soap, metal, and color manufacturing." These are, in
reference to this subject, first, in centrifugal cleaning and drying
machines, constructing the cylinders thereof so as to diminish the
labour of emptying and working the same, and facilitating the
escape of the liquid. A cylinder is fastened on a spindle (driven
from below) ; the periphery of this cylinder is of perforated metal,
or of framework to carry off the liquid ; an inner perforated cylinder fits loosely and made very light fits inside. the first cylinder,
and can be easily "lifted out and emptied when used for operating
" upon substances of a cohesive nature," such as sugar. Also
the revolving cylinder has bottom openings made by a flange
fitting loosely on the dome of the cylinder, and a couple of spring
bolts for preventing this flange rising, or lids fitting in recesses
on the dome may be used. There is an opening in the pan surrounding the revolving cylinder. For drying substances, making
perforated channels in the cylinder, " which project inwards or
" towards the interior of the cylinder."

Second, this is said to be improvements in the " forming
" machine" described in No. 12,742. The forms, sugar loaf

shaped, are fitted into bearings in the revolving cylinder, so that each revolves upon its own axis, the inner bearings are formed in a ring which is attached to the cone of the cylinder by joints. The lids, at the outer or large end of the forms, have suitable perforations, and are fastened by catches. When the sugar is set in the forms, and a syrup or other liquid is passed through the loaves, while the centrifugal action drives the liquid endwise, the revolution of the forms upon their own axes causes a dispersion of the fluid outward towards the side of the loaves. The moulds may have small perforations in their sides.

[Printed, 10*d*. Drawing. Mechanics' Magazine, vol. 55, pp. 136 and 161; Patent Journal, vol. 11, p. 239.]

A.D. 1851, March 20.—N° 13,560.

BESSEMER, HENRY.—" Improvements in the manufacture and
" refining of sugar, and in machinery or apparatus used in pro-
" ducing a vacuum in such manufacture, and which last improve-
" ments are also otherwise applicable for exhausting and forcing
" fluids." These are, first, "clarifying, evaporating, and concen-
" trating saccharine fluids in open pans or boilers" different "in
" form and arrangement, so as to suit the kind of fuel to be
" employed, and the nature of the previous or succeeding opera-
" tions which the fluid has been or is afterwards to be subjected to,
" but in all of which forms of apparatus the heat for producing
" evaporation of the saccharine fluid is transmitted thereto
" through the agency of steam generated in another compart-
" ment of the same pan or boiler in which the saccharine fluid is
" contained."
Second, "forming a partial vacuum in pans or vessels used for
" concentrating saccharine fluids by means of the vapour of
" steam arising therefrom, which is made to act on a piston in
" connection with an air pump and condenser, and thus supplies
" the motive power for its own exhaustion, and also in another
" form of apparatus, in which rotary motion is produced by the
" emission of steam and water in the condenser, such steam being
" produced by the evaporation of the saccharine fluid in the
" vacuum pan, and which apparatus is otherwise applicable for
" exhausting and forcing liquids." Also, "in heating such
" vacuum pans by means of steam generated in a lower and sepa-
" rate compartment of the same pan or boiler."
Third, "in apparatus for producing a partial vacuum in pans

" or vessels used in the manufacture and refining of sugar, where
" the motive power required in the production of such partial
" vacuum is to be supplied from a fall of water or other con-
" venient source of motive power, and not from the vapour of
" saccharine fluid " as under the second head, and "which appa-
" ratus is more particularly applicable to those cases where a
" small amount of exhaustion is required, as in the filtration of
" saccharine fluids, and in the ' curing ' of sugar," and "which
" apparatus is otherwise applicable to the exhausting and forcing
" of liquids."

Fourth, "apparatus for ' curing sugar,' by which apparatus the
" crystals of sugar are separated from fluids." There is a cast
iron bed plate forming the main framework of the apparatus, and
having a semicylindrical cavity or trough, extending all the way
along over this trough is a frame filled with rings, with interstices
between them, forming as it were a perforated tube ; at one end of
this tube is joined a three-way cock, on the top of which is fixed a
hopper ; on the other end is a force pump. The perforated tube is
tightly filled with dry sugar, and a cover is put over it. The sugar
from the coolers having been well stirred up is put into the hopper.
The force pump is worked by motion applied to a cylinder ; the
semi-fluid mass flows in, "the fluid portion cannot be kept under
" pressure in a tube having so many openings extending around
" it, but will flow through these openings while the solid matters
" are retained and undergo a further squeezing as they progress
" along the tube, and are finally discharged at its open end " by
means of a spiked roller into a hopper. The crystals, if desired,
may be still further cleansed by "liquoring " or "by mixing the
" sugar as it is discharged from the machine with the liquor, and
" subjecting it to a second operation." When the crystallization
of the sugar is imperfect, and the crystals minute, in preference,
using an arrangement of a cistern with a mixture of crystallized
sugar and fluid matter well mixed ; in this is slowly revolved a
perforated and otherwise constructed cylinder, which is exhausted ;
a crust of sugar forms on the drum, and becomes "dryer as it
" rises in consequence of the fluid parts being drawn into the
" drum. On arriving at the top it will be subjected to the action
" of a great number of minute jets of liquor forcibly projected
" upon it for cleansing the crystals of the more viscid matters that
" may be adhering to them," and the drum travelling on to a
scraper the crystals are removed by it.

Fifth, apparatus for separating in Muscavado sugar the large crystals from the smaller ones, and from impurities; for this purpose an arrangement is described in which the sugar passes from a hopper between two revolving fluted rollers; beneath the rollers is a spout which conveys the sugar into the interior of an inclined wire drum, where it is acted upon by wire brushes mounted on a shaft "similar to the mode by which flour is separated from the "bran." The smaller crystals are brushed through the upper part of the drum, and pass into a vessel with warm water, in which they are dissolved, while the larger crystals are brushed through the lower part of the drum into a chamber, and while the small lumps and impurities are discharged at its lower end. The best crystals are mixed "with a solution of sugar or 'liquor'" and submitted to either of the processes described under the fourth head. The inferior sugar is "heated in the 'blowing-up "'pan'" and filtered.

Sixth, "apparatus for making into lumps or loaves crystals of "sugar by subjecting such loose crystals of sugar to the action "of heat and pressure in the moulds." The apparatus is arranged so that the action is as follows:—The crystals of sugar are damped with water and put into a hopper, above one of the moulds of which there are a series fixed all round a table, "steam "is turned on so as to heat the mould to the desired extent, the "operator will move round the table so that the mould which "has been filled will pass under the piston," he will then turn the handle of a small steam cock to admit steam to the upper side of the piston at half stroke, so as to exert a powerful compressive force on the crystals of sugar; "the pressure may be retained for "about a minute, when the cock will be reversed, and the piston "raised from the mould; the machine will continue in the position "until the table is turned round by the workman so as to bring "the mould that was being filled under the hopper during the time "the pressing operation last described is going on," and so on.

[Printed, 6s. 2d. Drawings. Repertory of Arts, vol. 20 (*enlarged series*), pp. 1 and 73; Mechanics' Magazine, vol. 55, p. 269; Patent Journal, vol. 12, pp. 21, 35, and 45.]

A.D. 1851, March 24.—N° 13,562.

HERRING, MATTHEW.—"Improvements in the manufacture "of sugar and rum, parts of which improvements are applicable "to evaporation generally." These are, in reference to this subject, a "defecator," consisting of a series of hollow metal plates heated "by steam, or hot water or hot air," &c., beneath these is a receiver for the syrup after passing over these plates. A syphon, at the apex of which is a stop-cock, at the end of each leg is a stop-cock, the one at the end of the short leg being connected with the one at the apex "by a stout wire, so that the opening of "the one shuts the other, and vice versâ." At the end of the short leg a flexible tube may be attached, and to this is attached a float (a square box), with means of straining. This is used for conveying the liquid into the reservoir over the "defecator." The juice from the receiver may be made to pass through a blanket or any reticulated substance. A stratum of animal charcoal, or if the liquor is acid, a stratum of pounded limestone or coral or shells well washed." In order to make greater draft a fan is placed over the hollow metal plates. "The cleanser," this is "a "vessel which may have either a flat or circular bottom, having a "reticulated false bottom" on which the crystals to be cleansed are spread evenly. The liquor is poured on the top, and a pump set in motion, "the liquor is drawn through the sugar at the "bottom and returned over the top, the attrition of which cleanses "the crystals of their superficial impurities." When the sugar is sufficiently cleansed the liquor is drawn off, and the pump is used for drying the crystals, in which hot air may be used. The liquor may be passed through animal charcoal before it is again returned to the crystals. Should the centrifugal machine be used for cleansing, it is arranged so that "should the shaft swerve "from its centre it will press on the internal edge of the flanch, "and be restored by the springs under the circumference." A small quantity of low pressure steam keeps the sugar moist at the lower part of the periphery. The remainder of the Specification refers to a continuous still.

[Printed, 10d. Drawings. Repertory of Arts, vol. 19 (*enlarged series*), p. 83; Mechanics' Magazine, vol. 55, p. 279; Patent Journal, vol. 12, p. 5.]

A.D. 1851, March 31.—No 13,577.

GWYNNE, JOHN.—"Improvements in machinery for pumping, "forcing, and exhausting of steam, fluids, and gases, and in the "adaptation thereof to producing motion to the saturation, "separation, and decomposition of substances." These are, in reference to this subject, first, "a vacuum sugar boiling apparatus "as fitted" with the Patentees "centrifugal pump for maintain-

" ing the vacuum therein." The pump is bolted down to the bottom of the water cistern in which it is immersed, the suction pipe from the pump passes into the dome on the top of the vacuum pan, the driving shaft carrying the pump disc is driven by a pulley carried in an external bearing, being passed through the upper side of the cistern by a stuffing box. As the pump revolves the hot air and vapour are drawn through the suction pipe into which a jet of cold water is injected, " and condenses it, when the " combined fluid and vapour is carried forward into the pump " and finally discharged into the cistern."

Second, what the patentee terms his " variable pressure centri- " fugal seperator as applied for the cleansing, washing, and " separation of raw or unrefined sugar," as well as for the treat- ment of " various other matters to which centrifugal force is " applicable." A conical copper vessel contains the sugar during the process of separation; its sides are perforated with conical appertures, the narrower sides of the apertures being on the inner side of the vessel. This vessel is keyed to a vertical shaft, there is an inner smaller conical vessel, to the lower or inferior surface of which is attached a curved scraper working in contact with the surface of the first conical vessel. The driving shaft has a helical screw attached to it, above which is a hopper. There is likewise a screw feeder into this hopper. There is an iron vessel below the first conical vessel to receive any liquid matters which may pass through, and from thence they pass away by a pipe. There are arrangements for supplying fluid, if washing or cleansing is required, and also for supplying steam, air, or gases, when neces- sary. A rotatory motion is given to both vessels by two pairs of bevil buff or surface wheels working in connection with a horizontal driving shaft over head.

Third, " centrifugal apparatus for moulding refined sugar. A " vertical shaft is arranged, driven by the " fast and loose pulleys " arrangement," beneath which is keyed on it a friction disc. " Immediately above the collar bearing, and on the extremity of " the shaft is keyed a malleable iron chamber," round " the cir- " cumference of which are a set of circular holes to receive the " inner ends of the conical sugar moulds," the outer ends being held up by straps. The moulds are separately filled with the sugar from the coolers, the hole in the apex stopped till the mass is solidified, when the plug is withdrawn and it is placed on a pot to drain; after draining, the moulds are placed in the centrifugal

machine in one or more rows or stories, and the machine is made to
rotate at a high speed so as to discharge the fluid from the moulds
into a surrounding chamber, from whence it is drawn by a pipe.
The complete washing is effected by a supply of decolorized syrup to
the central chamber from above, and rotating is pursued as before.
This machine is actuated from below, but it is preferred to do so
from above.

Fourth, " centrifugal apparatus," " for drying, filtering, wash-
" ing, separating, and pressing various substances, such as peat
" or charcoal." A perforated vertical pipe is attached to a driving
shaft by a collar at its upper end, its lower end working on an
upright branch of a horizontal pipe, which delivers into the perforated cylinder the liquid to be filtered or decolorized. The
perforated cylinder is surrounded with cages filled with sponge,
felt, animal charcoal, or other filtering medium, and the fluid
" emerges by the perforations, and is made to impinge by the
" centrifugal force against the inside of the filtering cages, and
" being driven through the layers of material therein, is filtered
" or decolorised as may be intended."

[Printed, 3s. 10d. Drawings. Mechanics' Magazine, vol. 55, p. 299 ; Practical
Mechanics' Journal, vol. 4, pp. 107, 121, 126, and 147 ; Engineers and
Architects' Journal, vol. 14, p. 474 ; Patent Journal, vol. 12, p. 37.]

A.D. 1851, April 15.—N° 13,591.

SCHRODER, HERMAN.—" Improvements in the manufacture
" and refining sugar." These are, " the combination of moving
" or revolving discs or surfaces with pans or vessels heated by
" pipes or tubes or passages containing steam or other heated
" fluid." The pan is preferred to be of the shape of a segment of a
cylinder, the pipes for heating are coiled round the vessel to the
bottom ; steam enters at the top of these pipes, condensed water
runs from the pipe through the bottom. One large disc is shown
revolving in the pan.

[Printed, 6d. Drawing. Repertory of Arts, vol. 19 (enlarged series), p. 102;
London Journal (Newton's), vol. 40 (conjoined series), p. 355 ; Mechanics'
Magazine, vol. 55, p. 319 ; Patent Journal, vol. 12, p. 52.]

A.D. 1851, May 15.—N° 13,634.

OXLAND, ROBERT, and OXLAND, JOHN.—" Improvements
" in the manufacture and refining of sugar." These are, employing " phosphoric acid in a combined state," as afterwards
described for " defecating saccharine liquids or solutions of sugar,

" and for removing their color." In a former Patent, No. 12,590, Old Law, acetate of alumina was used for refining sugar, and lime for separating the alumina, but lime does not effectually remove the alumina, and to remove it it is proposed after the addition of the lime to add a small quantity of either of the superphosphates of alumina or of lime, boiling for two or three minutes, carefully neutralizing any excess of acid by adding " aluminate of lime, " saccharate of lime, lime water, or milk of lime," the aluminate of lime preferred, and completing the process as in No. 12,590. In place of the foregoing process phosphates may be employed directly, as phosphate of soda ($1\frac{1}{2}$ lbs. to the ton of sugar), bringing the solution to the boiling point, carefully neutralizing any acidity with aluminate or saccharate of lime, milk of lime, or lime water, and " then passing the syrup thus obtained " (of sp. gr. 25° to 30° Beaumé)," through the ordinary bag filters; this liquor may be evaporated, &c., but a further amount of color may be removed by using hydrate of alumina dried at 212° F., diffused through the water, thus dispensing with animal charcoal. When superphosphate of alumina is used (6 lbs. of alumina dissolved in phosphoric acid for the ton of sugar), the syrup at 25° or 30° Beaumé is brought to the boiling point, and the process proceeded with as above. Modes of preparing superphosphate of alumina and aluminate of lime are given.

[Printed, 4d. No Drawings. Repertory of Arts, vol. 19 (*enlarged series*), p. 168; London Journal (*Newton's*), vol. 40 (*conjoined series*), p. 27; Mechanics' Magazine, vol. 55, pp. 419, 435; Patent Journal, vol. 12, p. 86.]

A.D. 1851, July 7.—N° 13,689.

MIRRLEES, JAMES BUCHANAN.—" Improvements in machi-
" nery, apparatus, or means for the manufacture or production
" of sugar." These are, " constructing or arranging sugar mills
" by combining the engine with the mill in such a manner that
" the same framing shall answer both for the engine and mill,
" or in other terms, that the standards for the rollers shall also
" carry the engine shaft."

Second, " constructing or arranging sugar mills, whereby the
" standards themselves shall serve as the elevated support for
" raising the rollers or other heavy details of the mill."

Third, " constructing or arranging sugar mills so that the weight
" of the mill may add stability to the engine." " The framing
" of the mill being also the framing of the entire engine and

" gearing, the combined arrangement is less liable from strains
" or the subsidence of the building or foundation than existing
" plans."

Fourth, "the system or mode of heating the cane juice by
" passing the waste steam of the engine or other steam or heated
" air through a hollow base plate formed to receive the expressed
" juice." The upper surface of the base plate is concave, the
waste steam is made to pass into the interior of the base plate, and
thus "the juice is kept hot in its way to the boiling apparatus."

Fifth, "the application and use of malleable iron to the con-
" struction of the framing of sugar mills."

Sixth, "the general arrangement of the various parts of sugar
" mills and engines combined as delineated in the drawings."

[Printed, 10d. Drawing. Mechanics' Magazine, vol. 56, p. 60; Practical
Mechanics' Journal, vol. 4, p. 275, and vol. 5, p. 52.]

A.D. 1851, August 5.—N° 13,709.

DE MORNAY, EDWARD. — "Improvements in machinery for
" crushing sugar canes, and in apparatus for evaporating saccha-
" rine fluids." These are, first, "single and double-acting roller
" mills." The single mill gives but one effective squeeze as in the
common three-roller mill, but in this mill one or more small feed
rollers are substituted for the large lower roller on the feed side
of the mill, by these means dispensing with the trash returner,
and by increasing the number of feed rollers, the two horizontal
expressing rollers can be brought as near as desired to the same
level, which would obviate in a great measure the defect observ-
able in the common mill of the second roller carrying on some of
the juice past the pressing parts, where it becomes absorbed by the
relieved megass. The double mill gives two effective squeezes by
" two pair of rollers being so placed with one or more feed rollers
" that the megass ascends after each separate pressure between
" the two pairs of rollers. In this mill it may be advisable to
" place a table or trash returner between the first and second pair
" of rollers to prevent the small trash from falling into the juice
" below; but as the canes would be very much crushed at the
" first squeeze, this would not be objectionable, as in the common
" three-roller mill."

Second, an improvement in No. 12,578. In expressing juice
from cane in No. 12,578, the canes are forced "by means of a
" plunger through a parallel tube throughout, or a parallel tube

" for some distance, and contracted towards the end where the
" trash is ejected." It is proposed " having a tube whose sides
" shall diverge some distance towards the extremity where the
" megass is expelled, and that afterwards it shall be either parallel
" or slightly contracted." In No. 12,578, Old Law, all the cane is
" introduced into the tube in the same direction, so that all the
" pieces lie parallel to each other," but it is proposed to introduce
" the pieces of cane into the tube in such a manner that they lie
" in alternate layers having the fibre of the cane of any other layer
" nearly, if not quite, at right angles to the intermediate ones."
It is also proposed to construct " the tube in two pieces, which
" will be bolted together, the larger orifices being in contact, and
" the smaller orifice of the piece of tube nearest the plunger
" placed so as to receive the end of the plunger as it cuts off and
" crushes the canes. If the other half of the tube is also tapered,
" the small end will be furtherest from the plunger, and where
" the megass will be rejected, it may be found by experience that
" this half the tube should be made parallel."

Third, in evaporating pans, having them long and shallow, and
the bottom, in preference, formed like an inverted arch. The juice
is admitted at one end, there is a moveable partition or partitions
fitting closely which runs on wheels on a tramway so as to prevent
the fresh juice mixing with the other ; a tap is at the other end
for drawing off the concentrated liquor. Another pan is constructed with a series of wires running across it near the bottom ;
the pan set in an inclined position, the juice enters at one end,
and runs out at the other deprived of a portion of its water.

[Printed, 1s. Drawings. Repertory of Arts, vol. 19 (*enlarged series*), p. 144;
Mechanics' Magazine, vol. 56, p. 155.]

A.D. 1851, September 11.—Nº 13,745.

VARILLAT, WILLIAM JEAN JULES.—" Improvements in the
" extraction and preparation of colouring, tanning, and saccharine
" matters from various vegetable substances, and in the apparatus
" to be employed therein." These are, first, in a machine for
cutting dye woods, in which a log of wood (maintained in a vertical
position) is placed between two cutting cylinders, having four flat
blades and four fluted blades, the fluted blades which project more
than the flat blades, at each turn make fluted shavings, and the
flat blades, coming round immediately after, cut off all the edges
the fluted blades have not carried away; the logs are then raised

higher up by a block and pulley as the shavings are cut away until there remains but a very short length, when another log is placed between the cylinders ready to be cut in a similar manner.

Second. In placing the shavings which have been cut into tin cylinders filled with about 14 times there own weight of distilled water for extracting the dye; these vessels are placed in a large reservoir of water at 90° heat, and when the liquid has been stirred for about 15 minutes, it is made to pass from one vessel to another by means of a pump placed at one side of the reservoir, and thus the cylinders are emptied and filled; when the spent shavings in the first vessel are sufficiently stirred to extract the dye they are lifted out by a strainer placed at the bottom of each vessel on which the shavings rest. Thus the wood is taken out every half hour from the vessel at the end of the series, and when the vessels are in operation receives 18 weaker and weaker baths, the hot water passing through all the vessels, by which the matters contained therein become richer and richer and more dense during 18 changes. This mode of extraction or dissolution may also be applied to cane or beet root, sugar, tannic acid, extracts of liquorice, and from all kinds of vegetables.

Third. The extract thus obtained is next run into a cylinder to be evaporated, in which a vacuum is formed by means of air pumps and a condenser, the cylinder is mounted upon hollow trunnions in connection with a coil fitted inside the cylinder. Steam is let into the trunnions and into the coil, and when the coil on the hollow trunnion revolves it will set in motion the extract contained in the cylinder, which in turning is covered with a thin coating of liquid renewed at each revolution. This extract is run in at one end of the cylinder, and when it becomes sufficiently dense is run out at the other into a vessel for the purpose, causing an intermittent evaporation, and allowing the steam being emp.oyed at a temperature of less than 100°.

Fourth. The extract is to be finally concentrated in an apparatus which consists of two vessels, one placed on the other, the upper one being in the form of a dome, the lower one a basin, an air-tight partition dividing the two. The basin is provided with a steam coil, water being let into the basin until it rises above the coil to " about 80°," which heats the partition forming the bottom of the upper vessel, and on which a small quantity of the liquid to be concentrated is let in. At the upper part of the dome is a tube communicating with a condenser and air pumps, a valve in the tube being opened a vacuum is produced when the liquid is turned into a pulp, and if required to be solidified is stirred by a rotating scraper, which in its traverse sends out bubbles from the mass; an air pipe is then opened to destroy the vacuum, and a valve in the partition opened to allow the pulpy mass to fall into a trough, beneath which are placed a series of rectangular boxes lined with paper, which are to be filled one after the other, and when concentrated and cooled may easily be taken out of the boxes and packed in cases ready for sale.

[Printed, 2s. 8d. Drawings. Mechanics' Magazine, vol. 56, p. 257.]

A.D. 1851, December 8.—N° 13,846.

BROOMAN, RICHARD ARCHIBALD. — (*A communication.*) — " Improvements in the manufacture of sugar, in the preparation " of certain substances for such manufacture, and in the machi- " nery and apparatus employed therein." These are, first, for evaporating saccharine fluids, employing " a series of centrifugal " apparatuses, by means of which the liquid is passed through a " succession of evaporating operations until sufficiently concen- " trated." One one shaft are fixed a series of these drums revolving in a large pan which is divided horizontally into as many compartments as there are drums by means of circular plates, in the upper part of which is a space left for the escape of vapor. The liquid is introduced through a pipe at the top, and flows over the circular plate into a heating apparatus, "from " which it passes into the centrifugal drums and is dispersed over " the next plate, flowing back into the next heating apparatus and " drum, and so on through all of them in succession."

Second, a method for producing a vacuum for evaporating saccharine fluids. The pan is in communication with a box, which has two-valves fitted to it, two cylindrical vessels communi- cate with a boiler by two steam pipes, and with the box by the two valves. The two cylindrical vessels have two taps at the ends opposite to the steam pipes for the escape of air, and are further provided in the bottom with valves and pipes for the outlet of water. The degree of vacuum is ascertained by indicators on the pan.

Third, unloading and working the drums of centrifugal machines. This is effected by lowering a moveable casing which envelopes the machine so as to bring it below that of the fixed portion of the periphery made fast to the bottom. The speed of the machine being reduced, a hand wheel is then taken hold of, which causes the loose top and loose bottom to be drawn up, by which means the lining will be raised and turned inside out, thus discharging the drum of its contents.

Fourth, cleaning the charcoal used in the manufacture of sugar previous to re-burning, by passing water through it when submitted to rapid rotation in a suitably constructed centrifugal machine. "The charcoal is afterwards dried as far as convenient " before removing it from the machine."

Fifth, "constructing the forms of moulds used in making " sugar loaves " with an extension at the small end, the ordinary plug hole being at the extremity of the extension." "The " hydraulic pressure of the liquid will cause the sugar to be " cleared lower down than is effected by the methods now in " use."

[Printed, 1s. 4d. Drawings.]

A.D. 1852, January 24.—N° 13,916.

FONTAINEMOREAU, Peter Armand le Comte de.— (*A communication.*)—" Certain improvements in lithographic, " typographic, and other printing presses, which improvements " are also applicable, with certain modifications, to extracting " saccharine, oleaginous, and other matters, and to compressing " in general." These are, in reference to this subject, "the " apparatus consists of a number of plates of iron or steel, " forming the links of two endless chains, which pass between a " series of five or more pairs of rollers, which receive the matters " to be pressed. When beet root, turnips, or other similar " matters have to be compressed they are previously reduced to " pulp." The cast-iron rollers turn upon wrought-iron axes, and communicate motion to one another by means of cast-iron cog wheels. "The plates of the lower chain differ from those of " the upper ones, the former having a flange on each side running " in the direction of the length of the chain," and also having a number of conical holes, through which the liquid escapes and falls into a receiver beneath the machine. The material to be pressed falls from a rasp or mill fixed above, "(this fall being " regulated by any suitable means) forms a thick and even " coating upon the plates of the lower chain which begin to be " pressed when it reaches the first pair of rollers," after passing

between the last pair of rollers the chaines separate and the lower
ones let fall the residue into a receiver.

[Printed, 1*s.* 6*d.* Drawings. Mechanics' Magazine, vol. 57, p. 115; Engineers and Architects' Journal, vol. 15, p. 312.]

A.D. 1852, January 29.—N° 13,936.

SMITH, WILLIAM.—"Improvements in apparatus for cutting
" or breaking lump sugar and other vegetable substances."
These are, first, the "general arrangement of machinery or appa-
" ratus." A wooden framing carries at its upper part a table,
upon which is fixed a rectangular frame of cast iron; to this frame
two stationary knives or cutting edges are bolted.

Second, "mounting the moving knife or knives of such ma-
" chinery on a stem working in a fixed guide or socket, or on a
" pair of stems similarly fitted."

Third, "communicating the reciprocating motion of a treadle
" to the moveable knife or knives of such machinery by means
" of a jointed connection of which a crank pin is the centre of
" motion."

Fourth, to facilitate the placing of these pieces of sugar so that
they shall be cut into portions approaching to uniformity of size,
a ledge or rest carried by the above cast-iron frame is provided
for supporting the sugar while under the action of the knives, and
a stop plate " stands out from the back of the framing for deter-
" mining the distance that the sugar may be thrust beyond the
" line of the knives." "The capacity of the machine may be
" doubled by affixing to the traversing stem a knife in the form
" of a cross in plan view and providing fixed knives corresponding
" thereto on the table of the machine."

[Printed, 8*d.* Drawing. London Journal (*Newton's*), vol. 41 (*conjoined
series*), p. 268; Mechanics' Magazine, vol. 57, p. 134.]

A.D. 1852, February 3.—N° 13,954.

TORR, GEORGE.—"Improvements in re-burning animal char-
" coal." These are, in constructing and setting the revolving
retorts or apparatus used for the above purpose. The retort
" has an open end in front, and a closed end with a hollow axis
" at the back;" the back end of the retort is within the masonry
or setting, and can be heated. A fixed tube passes through the
axis in front, bending at right angles as it enters the retort, and

opening into the upper part of the retort, through which any vapours evolved in re-burning the animal charcoal escape, and are conveyed away by a pipe connected to this pipe outside of the retort. To turn over the matters under process, a long plate is fixed by means of arms within the retort. The retort is revolved by pulleys at each end. The fire is underneath, and the heat passes into a chamber within which the retort revolves, and thence into the chimney.

[Printed, 1s. 2d. Drawings. Repertory of Arts, vol. 20 (*enlarged series*), p. 147; Mechanics' Magazine, vol. 57, p. 157.]

A.D. 1852, February 24.—N° 13,988.

BESSEMER, HENRY.—"Improvements in expressing saccharine " fluids, and in the manufacture, refining, and treating sugar." These are, first, "constructing the cranks or eccentric shafts of " cane presses so that the plungers actuated by them are each " brought into successive operation."

Second, the construction of cane pressing tubes. These tubes have plungers fitted into them, they are rectangular in cross section, and of much less length than those hitherto in use. In preference they are cast, and of "Muntz's metal."

Third, "the direct action of the plungers of cane presses by " connecting one end of them to the crank or eccentric shaft."

Fourth, "guiding and giving a parallel motion to the plungers " of cane presses." A portion of the bed plate of the press is planed true, and between it and the under side of the cap of the cane tube, the enlarged end of the plungers are fitted so that the space between the surfaces of the bed plate and of the cap "forms " a guide for the plungers, and insures their parallel motion " backward and forward when actuated by the crank."

Fifth, the construction of double acting cane presses; the general details and arrangement of these apparatus are the same as before described for the single presses.

Sixth, "the use of wrought iron tension rods in the construc- " tion of cane presses" for staying up the machinery.

Seventh, constructing clarifiers. A cast iron cylindrical vessel having a nearly flat bottom consisting of hollow chambers with vertical radial partitions. A cylindrical jacket fits on to the vessel with a rust joint at top and bottom.

Eighth, the use of glass in the construction of subsiding vessels,

Ninth, the forcing of heated air upon revolving or moving surfaces partly immersed therein.

Tenth, " the evaporation of the aqueous portions of saccharine " fluids by the joint action of heating media below 212° F., used " in combination with currents of hot or cold air brought in con- " tact with such fluid on the surface of apparatus revolving or " moving, partly immersed therein."

Eleventh, " the use of a 'jacket' or double bottom to pans as a " means of heating saccharine fluids by steam, when used in . " combination with currents of heated air brought in contact with " such fluids on the surface of apparatus revolving or moving " partly immersed therein."

Twelfth, " the use of a spiral blade or screw for scraping off or " preventing any accumulation of solid matter at the bottom of " pans or vessels used in the concentration of saccharine fluids, " and also for the purpose of exposing a large surface of fluid in " order to facilitate its evaporation."

Thirteenth, forcing " air through a central air pipe or drum in " apparatus revolving or moving, partly immersed in the fluid to " be evaporated."

Fourteenth, " the use of large hollow axes for the purpose of " increasing the firmness or rigidity of screws or series of discs " used in the concentration of saccharine fluids."

Fifteenth, " the evaporation of the aqueous portions of saccharine " fluids, by forcing heated air in contact with such fluids or " surfaces which move into and out of the fluid to be evaporated."

Sixteenth, evaporating such fluids by " forcing heated air in " contact with thin films or coatings of such fluids on fixed " surfaces not heated by any other means."

Seventeenth, rapidly heating concentrated syrups.

Eighteenth, a cooler made of wood secured by iron hoops ; the bottom is conical, with a plug. It has a handle and wheels.

Nineteenth, separating fluid matter from crystals of sugar by spreading them on a table with a pervious material, and a partial vacuum beneath.

[Printed, 2s. Drawings. Repertory of Arts, vol. 21 (*enlarged series*), p. 1 ; Mechanics' Magazine, vol. 57, p. 218, and vol. 58, p. 43.]

A.D. 1852, March 22.—N° 14,030.

SYMINGTON, William, FINLAYSON, Charles, and REID, John.—" Improvements in flues, and in heating air, and

" in evaporating certain liquids by heated air." These are, in
reference to this subject, " for heating air and evaporating saccha-
" rine liquids, a series of tubes are placed in a furnace and
" heated; an inlet pipe admits air which passes amongst and
" between these pipes becomes heated," from thence it passes into
a pan containing the saccharine fluid to be evaporated, in which
revolves a spindle on which are about forty discs, " kept three-
" quarters of an inch apart by collars." The flue from the
apparatus in which the pipes are heated passes under two
evaporating pans, and likewise under a clarifier. The peculiarity
of this invention it is said is to have " the same means or
" furnace to heat the air-heating apparatus, and also the pans."

[Printed, 1s. 6d. Drawings. Repertory of Arts, vol. 20 (enlarged series),
p. 309; Mechanics' Magazine, vol. 57, p. 277.]

A.D. 1852, May 25.—No 14,141.

WALKER, JOSEPH, junior.—(A communication.)—" Certain im-
" provements in vacuum pans for the evaporation and crystalliza-
" tion of saccharine or other solutions." These consist in " such
" a combination of rectilinear tubes that their collective surfaces
" may be made to bear any desired proportion to the contents of
" the vacuum pan for which they may be intended, or in other
" words, such that the amount of heated surface in a pan of any
" given size may be infinitely increased beyond that which can
" be obtained by any of the methods as now used." A vacuum
pan is shown consisting of a bottom, top, and body; the bottom
part is traversed by a series of vertical pipes; there is a pipe for
the admission of steam, and a pipe for discharging the water
formed by condensation of the steam.

[Printed, 6d. Drawing. Mechanics' Magazine, vol. 57, p. 417.]

A.D. 1852, May 29.—No 14,143.

VON HERZ, ADOLPHUS CHARLES. — " Improvements in
" treating, preparing, and preserving roots and plants, in extract-
" ing saccharine and other juices from roots and plants, in the
" treatment of such juices, and in the processes, machinery, and
" apparatus employed therein."

[No Specification enrolled.]

A.D. 1852, June 12.—N⁰ 14,168.

BRANDEIS, JOSEPH.—"Improvements in the manufacture of " raw and refined sugar." These are, precipitating lead from saccharine solutions by means of hydro-sulphuret or sulphuret of potash, soda, or ammonia, ascertaining whether all the lead is removed by adding to a small portion of liquor "hydrosulphuric " acid, when if all the lead is removed it will make no further " precipitate or change its color." The excess of the hydro-sulphuret is removed by adding a small quantity of some of the insoluble salts of lead as "phosphate, tartrate, sulphate, &c., or " some of the insoluble salts of manganese as phosphate, &c., " which will abstract the excess of hydrosulphuret" which has been added.

[Printed, 4d. No Drawings. Mechanics' Magazine, vol. 57, p. 496.]

A.D. 1852, July 20.—N⁰ 14,233.

EGAN, JOHN FRANCIS.—(A communication.)—" Improvements " in the manufacture of sugar." These are, first, in apparatus ; a spout with a lid is placed in front of the fourth and fifth coppers into which the thick gelatinous and other matters that separate from the cane juice are brought by means of a kind of rake into the spout and flow entirely away. During its further ebullition it is cleaned by what are generally called brushes which convey the impurities into canals also with lids between the second and third and third and fourth coppers. The coolers have divisions of boards or planks and plugs at the bottom for drawing off the uncrystallized saccharine matter.

Second, the application of a defecating agent consisting of ten gallons of juice of the plantain tree, seven pounds quicklime thoroughly mixed and the clear liquor drawn off and one ounce of flowers of sulphur added to every six gallons. "After the first " thick scum coagulated by the action of the heat has been " removed in the fourth and fifth copper from the cane juice two " or three quarts of the above liquor are added and immediately " large amount of impurities rise to the surface and are carried " rapidly and efficiently away by the rake or instrument used for " the purpose."

[Printed, 1s. 2d. Drawings. Repertory of Arts, vol. 21 (enlarged series), p. 150; Mechanics' Magazine, vol. 58, p. 116.]

A.D. 1852, July 24.—N° 14,239.

BESSEMER, HENRY. — "Improvements in the manufacture,
" refining, and treating sugar, part of which improvements are
" applicable for evaporating other fluids." These are, first,
" the use of close or covered vessels for 'clarifying' and elevating
" the liquor" as follows:—"A cast iron cylindrical vessel the
" interior of which terminates downwards in an obtuse cone" has
a steam jacket covered by felt and wood; it has a dome or cover
in the centre of which is a hole capable of being closed for the
admission of the fluid to be clarified. The liquor is run into the
vessel; lime or other matters are then added, and steam admitted
into "the jacket will soon produce ebulition which may be
" continued for a few minutes, the lid is then to be closed and
" screwed down" and a cock is opened into a pipe which leads
into an elevated receiver into which it is desired to raise the
fluid by admitting steam by a pipe into the top of the iron
cylindrical vessel. Two such vessels are preferred.

Second, "heating by hot water, hot air, or other heated media
" below 212° Fahrenheit the vessels used for the defecation or
clarification of saccharine fluids." It is preferred to use for this
purpose an open jacketted vessel.

Third, "discharging the saccharine fluids from filtering drums."
This is said to be an improvement upon No. 13,560. A drum is
composed of a number of circular plates mounted on an axis,
these plates are covered with wire gauze or pierced metal, in these
plates are holes through which bent pieces of metal pass forming
inclined gutters which at their lower ends are connected by
openings with one of four passages formed in the brass axis of
the drum. This bent piece like the gutter extends from end to
end of the drum and forms a sort of bucket which by the rotation
of the drum conveys the filtered liquor into the slanting gutters
which carry it away.

Fourth, working two screws in one pan or vessel by "the use
" of gearing so arranged as to cause the said screws to give a
" circulation to the fluid therein contained." The screws are
provided with hollow perforated axes for the purpose of passing
air over the screws as described in No. 13,988.

Fifth, in screws for evaporating the use of two or more threads.

Sixth, "scraping the threads of the screw, and rendering such
" threads or discs more rigid by making them thicker at the

" central part, and becoming gradually thinner towards their
" outer edge." A scraper is placed between the threads of the
screws. Screws are made with a greater number of threads than
one. Discs are made thicker "at that part where they join the
" axis and gradually becoming thinner at their peripheries."

Seventh, "the use of steam to displace overheating media used
" to evaporate saccharine fluids and using such steam to heat
" the partially granulated sugar before discharging it from the
" pan."

Eighth, in cleansing or airing sugar. By No. 13,988, Old Law,
the whole of the sugar on the table "was exposed to the pressure
" of the air above it and to a partial vacuum below." By this
apparatus it is proposed to expose only a portion of the annular
channel which is covered with sugar to exhaustion.

Ninth, "the filtration or separation of dust, soot," &c. from air
" brought in contact with saccharine or other fluids " by passing
them "through any woven or felted fabric."

Tenth, "treating and combining albuminous matters with
" charcoal." These are evaporating albuminous matters " at so
" low a temperature as not to coagulate them and then to mix
" them with powdered or granulated charcoal."

[Printed, 2s. Drawings. Repertory of Arts, vol. 24 (*enlarged series*),
p. 105; Mechanics' Magazine, vol. 58, p. 135; Engineers and Architects'
Journal, vol. 16, p. 185.]

A.D. 1852, September 18.—N° 14,293.

MACINTOSH, John.—"Improvements in manufacturing and
" refining sugar." These are, "evaporating and concentrating
" saccharine fluids in the manufacturing and refining of sugar"
as follows :— The liquor to be evaporated is in a pan, a travelling
surface passes through the liquor and over and under a series of
rollers. This apparatus is enclosed in a chamber heated at or
near the bottom, or hot air is passed in at the bottom and out
at or near the top ; or the saccharine fluid is passed over the fixed
surfaces into "any suitable vessel for further operations." "A fan
" or other apparatus may be applied for withdrawing the air so
" that the atmosphere of the room will be constantly changing."

[Printed, 6d. Drawing. Mechanics' Magazine vol. 58, p. 278.]

A.D. 1852, October 7.—N° 14,318.

BROOMAN, Richard Archibald. — (*A communication.*) —
" Improvements in the manufacture of sugar and in machinery

" and apparatus employed therein. These are, first, employing
" in presses for expressing saccharine juice from substances
" containing the same, of friction rollers arranged round a main
" roller or cylinder in combination with a fixed bed or in com-
" bination with a revolving bed."

Second, "the employment for such purpose of a plain cylinder
" or roller in combination with an endless chain of revolving
" friction rollers or with a bed of rollers revolving only on their
" axes."

Third, a "carbonic acid generating and impregnating appa-
" ratus" for freeing saccharine solutions " of any excess of alkali
" they may contain; and at the same time of decoloring them."
The gas generator is connected with the cistern in which is the
fluid to be treated by means of a branch pipe "on which is
" affixed a conical bearing upon which the generator is free to
" revolve." There is a ball valve opened by means of the
pressure of the gas. In the generator is placed a large glass or
earthenware tube filled with acid and chalk is carefully packed
in; the lid "ground air-tight or rendered so by vulcanized india-
" rubber packing" is fastened on and by turning the vessel downward the acid is spilt amongst the chalk and the carbonic acid
evolved opens the valve and enters the solution under treatment.

Fourth, apparatus for cleansing and purifying sugar, consisting
of a revolving hollow cone of wire gauze or of finely perforated
metal terminating in a pipe; there is an outer cone of similar
material; diverging from its apex is a hopper in which is a screw
or screw blades. Below the hopper are two chambers for holding
" the cleansing or decoloring agent or agents to be employed.
" A partial vacuum is established on the lower part of the
" apparatus. Any coloring or liquid matter passes by the pipe
" into the cistern and the sugar is delivered upon a trough.
" Another apparatus consists of a trough divided into three com-
" partments; at each end of the trough are pulleys upon which
" an endless web or band of any textile fabric or material travels.
" A series of friction rollers are fixed in the trough for supporting
" the band. At one end is a hopper terminating in a box con-
" taining a screw for delivering the raw material on the endless
" web or band; in the trough an exhaust pipe in connexion
" with an air pump and pipes below for drawing off the molasses.
" The cleansing and purifying agents whether steam alone or
" steam and heated air combined" are admitted by an apparatus

placed over the first compartment of the trough, and the employment of "the endless bands for carrying the sugar to be
" cleansed and purified " and " of steam combined with heated
" air for bleaching and purifying sugar " are claimed.

[Printed, 1s. 6d. Drawings. Mechanics' Magazine, vol. 58, p. 317, and
vol. 59, p. 21.]

A.D. 1852, December 21.—N° 14,354.

GALLOWAY, ROBERT. — "Improvements in manufacturing
" and refining sugar." These are, first, "employing lime com-
" bined with lead or saccharate of lead in such forms as to act
" in a similar manner to plumbite of lime in refining sugar, or
" saccharate of lead when an acetate of lead is used, to act on
" saccharine matter, and employing lime or magnesia before or
" after the use of acetate of lead."
Second, "using an acetate of lead twice, employing another
" or other processes immediately, and using bicarbonate of lime
" or bicarbonate or carbonate of magnesia, or mixtures acting
" similarly to separate the lead."
Third, "recovering back the lead employed in such processes "
by after washing the scum of saccharine matter, and digesting it
in acetic acid, "thus obtaining acetate of lead for further use,"
drying and igniting the insoluble residue, oxide of lead remains
behind which is again used. Or the washed scum dried and
ignited yields oxide of lead.
Plumbite of lime is made by dissolving 112 lbs. litharge and
28 lbs. quicklime in boiling water or by adding litharge to boiling
lime water. Saccharate of lead is made by adding powdered
litharge to a boiling solution of sugar so long as it is dissolved.
Saccharate of lime is made by adding slacked lime in excess to a
cold solution of sugar. Saccharate of magnesia is made by
substituting caustic magnesia for the above lime. In each case
the solution of sugar is preferred to be of density 26° Beaumé.

[Printed, 4d. No Drawings. Repertory of Arts, vol. 22 (enlarged series)
p. 288; Mechanics' Magazine, vol. 58, p. 517.]

PATENT LAW AMENDMENT ACT, 1852.

1852.

A.D. 1852, October 1.—N° 26.

MACINTOSH, JOHN.—" Improvements in evaporation." These are " combining and arranging apparatus for evaporating fluids " as follows :—Within a chamber are fixed a series of parallel plates which may be inclined or vertical, the water or other fluid is introduced into the top of the chamber by a coiled pipe, perforated with numerous small holes, upon three or more surfaces of fine wire cloth ; there are openings at the bottom of the chamber for air and the products of combustion to pass through, over which may be placed slide valves. Two pipes force air in below the fires. There is a safety valve at the top, and a pipe near the top to carry off the vapours. " If steam without the " products of combustion is desired," the vessel has " a close " bottom or a valve large enough to cover all the openings, and " in order that no grit, dust, or other substantive matters may " pass," the products from the fire are filtered through a filter composed of asbestos. When the apparatus is to be used for concentrating or evaporating other fluids the vessel is made with a close bottom, and air is introduced above it and flows against the stream of fluid, or the air is dispensed with and a partial vacuum obtained in the vessel and the concentrated fluid flows out " freely at the bottom by suitable outlets."

" The saccharine or other fluid that may be evaporated is but " a short time exposed to heat."

[Printed, 6d. Drawing.]

A.D. 1852, October 1.—N° 29.

EBINGRÉ, JOHN DANIEL.—" Improvements in the manufacture " of animal charcoal." These are, " rendering the dust of animal " charcoal (whether produced in the manufacture of the ordinary " granular charcoal or by reducing animal charcoal into powder) " into paste, and then into a granular form," as follows:—" Take " 100 lbs. of animal charcoal in powder, 1·30 lbs. of clay, 4 lbs.

" of mineral pitch, 100 lbs. water," and soak them together for four or five days in a tub, stirring from time to time, when they are poured into a vessel having in it a stirrer or agitator put in motion by a strap acting on a drum ; from this vessel the mixture passes through an aperture closed by a suitable sluice or valve, in which it is similarly agitated ; it then passes into another vessel in which are two grinding stones, and when well ground it is passed into another vessel below, from which it is " ladled out " and " placed in brown paper spread over a tray formed by parallel " bars," and is dried in a " stove heated to about 104° F., and " when dry the matters are to be divided by a cutting instrument." The matters thus prepared are carbonized with about 124 parts by weight of bones deprived of their fat to 76 parts by weight of the composition above described ; in like manner to bones and the calcined product is granulated " as heretofore when treating " calcined bones only, and the same will be fit for like uses as " granulated animal charcoal."

[Printed, 6d. Drawing.]

A.D. 1852, October 1.—Nº 85.

BRANDEIS, Joseph.—" Improvements in the manufacture of " sugar and saccharine solutions." These are, " the removing " the salts of lead from sugar or saccharine solutions by filtering " such solutions through a bed or beds of granulated charcoal," in preference, " using animal charcoal previously immersed in " dilute acid, and then burned in the usual way in a charcoal " retort, but other kinds of granulated charcoal, such as animal, " peat, or wood charcoal will answer for the removal of lead from " saccharine solutions."

[Printed, 4d. No Drawings.]

A.D. 1852, October 1.—Nº 90.

ASPINALL, John.—" Improvements in evaporating cane juice " and other liquids, and in apparatus for that purpose." One apparatus consists of " a cylindrical vessel, placed vertically and " having a perforated false bottom," under which is a chamber ; there are a number of vertical tubes, connected at bottom to this chamber, and at top to another chamber on which is a short pipe with a valve. The whole of this apparatus is placed in a pan containing water or other fluid, heated by a fire or steam or heated air. The pan may be covered over so as to form a jacket or casing round the apparatus. The cylindrical vessel is closed at the top, and is provided with an air pump or other exhausting apparatus. The air is drawn through the saccharine solution placed in the cylindrical vessel. Another apparatus consists of a pan in which two rollers are partially immersed and an endless band of wire gauze or other suitable material is passed round the rollers, which are enclosed or otherwise. The liquid raised on the endless band is exposed to a current of warm or cold air produced by a fan or air pump, &c. There is a scraper attached, " whereby the solid and liquid matters, or a portion of them, are " removed from the band, and prevented from returning into the " vessel from which they have been raised by the band." Such machines may have several bands and scrapers. The rollers need not be immersed in the liquid as the band may " bag down " between them."

[Printed, 8d. Drawing.]

A.D. 1852, October 1.—N° 113. (* *)

HARCZYK, BERNHARD.—"An improved preparation or com-
" position of coloring matter to be used in washing or bleaching
" linen and other washable fabrics, and in the manufacture of
" paper and other substances."

The object of this invention is to supersede the use of the " blue
" balls or stones," usually employed by laundresses in washing
linen, by substituting a soluble preparation of indigo, when previously coated " on sheets of paper, linen rags, or other substances
" capable of receiving the preparation." In practise, a piece of
the blued paper or rag is put into the water employed. The preparation of indigo is prepared by dissolving it in strong sulphuric
acid, which is afterwards neutralized by an alkali, and the deposit
washed and filtered. The Patentee states, that colouring matter
so prepared may be used by " sugar refiners to whiten their
" sugars," or by " paper manufacturers, when a fine blue tinge is
" required."

[Printed, 4d. No Drawings.]

A.D. 1852, October 5.—N° 220.

BROWN, DAVID STEPHENS. — "An improved apparatus or
" instrument for evaporating or distilling liquids." This con-

sists "of a tube open at both ends, made in the form of a
" horseshoe, the three sides of a square, or the two sides of a
" triangle," "it is placed with its two ends downwards, and
" having a vacuum inside;" the two ends dip into two open
buckets, in one of which is the liquid to be evaporated, while
some liquid is in the other or condensing one. The air is exhausted
from the condensing tube by means of a tube placed inside it.
The liquid inside the other tube is heated by means of a coil of
pipe, &c. in it, while it is agitated by a revolving stirrer, which
may be hollow, with hot air or steam passing through it. The
vapour is condensed in the condensing tube by various means
according as the condensed liquid is valuable or otherwise. A
modification of this apparatus is described, in which a tube is
attached to the lower end of the distilling limb for the bucket
and a stack or chimney of a fire passes through the distilling
limb instead of a heated worm tube, &c., and the lower part of
the condensing tube is enlarged to receive dry sand or other
substances when they are used as condensers. "The apparatus
" is to be employed for concentrating and crystallizing saccharine
" and other solutions, purifying water, distilling spirits, and
" condensing the steam of steam engines.".

[Printed, 6d. Drawing.]

A.D. 1852, October 13.—N° 366.

NASH, JOSEPH.—"The treatment and refining of sugar." This
consists, first, in the application of "muriate, the proto-muriate,
" the nitro-muriate" of tin, &c. in precipitating extractive and
other colouring matters.

Second, the application to syrup of chloride of lime, magnesia,
alumina, the alkaline chlorides, or chlorine water or gas, or the
separate or combined use of the same.

Third, "the separation of lead from sugar or the displacement
" of oxide of lead from sugar by the replacement of tin or its
" oxide, or any other body applicable as an equivalent agent for
" the same purpose, and specially the sulphite of tin," the
" manifest excess of lead" having been previously removed "by
" any of the ordinary agents which have been or might be used
" for that purpose," also "the concurrent use of albumine (by
" boiling with the alkaline solution of same) for fixing and com-
" bining with the displaced lead, and for the total or partial

" separation of the agent of replacement;" also the use of ammonia as " a solvent for animal albumine," but no claim is made to "the use of any of the salts of lead in the treatment " of sugar otherwise than in connexion with the principle of " substitution and the agency described," nor for the "use " of albumine generally in the treatment of sugar" otherwise than as above.

Fourth, " the special use of phosphoric acid, the neutral and " super-phosphate of lime, the acids and alkalies, and salts," (oxalic and sulphurous acids and their soluble salts, bone ash " dissolved in muriatic or other acid may be added to be after- " wards precipitated by the subsequent neutralizing re-agents " employed,") and the syrup thus treated is neutralized and " boiled with the alkaline solution of albumine;" but no claim is made " to the general or exclusive use " of these substances otherwise than as described.

Fifth, " the use of ammonia generally as a solvent of sugar, " and for its extraction from the sugar cane or other natural " sources of sugar, and also as a solvent of albumine, and also " for cleansing and restoring animal charcoal;" also the use of " the alkalies potash and soda, in part, for some of these purposes " in so far as they are applicable thereto, as the natural equivalents " for ammonia."

[Printed, 6d. No Drawings.]

A.D. 1852, October 16.—N° 418.

JOHNSON, John Henry.—(*A communication.*)—(*Provisional protection only.*)—" Improvements in the manufacture of sugar." These are "in centrifugal mechanism used in the manufacture " and refining of sugar," driving each machine " direct by a small " engine attached to the framework thereof, a disc being fitted on " to the extremity of the axis of the revolving drum," and " fitted with a stud pin, which serves as the crank pin of the " connecting rod and actuates the valve spindle also. The disc " is made heavy and serves for a fly-wheel and friction disc for " stopping the machine when the steam is cut off, a friction " strap being attached for that purpose." The speed is regu- lated by a governor attached to the engine. The centrifugal drums larger than ordinary ones, are fed by perforated boxes fitted round them. " In place of a vertical machine with one drum, " the axis may be horizontal and fitted with a rotary drum at

" each extremity." " By another arrangement, a hollow shaft
" may be used, the machine being thereby suspended from
" the upper extremity of a fixed vertical shaft, in place of
" working on a footstep." " A bent rim may also be fitted over
" the edge of the revolving drum," to prevent the molasses, &c.
being thrown out and allow the machine to be open during
working.

[Printed, 4d. No Drawings.]

A.D. 1852, October 27.—N° 544.

YOUNG, JAMES HADDEN.—(*Provisional protection only.*)—
" Improvements in expressing juice or fluid from the sugar cane
" and from other matters." These are, pressing such matters
" between two unyielding surfaces, one being perforated or open
" to some extent" combined with the use of a partial vacuum,
with or without the flow of air or fluid. The best arrangement,
" it is believed, is a cylinder and piston with a perforated bottom
" arranged suitably for introducing and removing the canes or
" other matters."

[Printed, 4d. No Drawings.]

A.D. 1852, November 2.—N° 614.

ARCHIBALD, CHARLES DICKSON.—(*A communication.*)—(*Provisional protection only.*) — " Improvements in machinery and
" apparatus for crushing grinding and triturating refractory and
" other materials, and for washing and separating ores and metals
" from earthy and other substances." These are, in reference to
this subject, the use of " balls or spheres rotating in a circular
" groove, and which are loaded by superincumbent pressure to
" increase the crushing or grinding power." This grinding apparatus, it is said, " may be also used for crushing sugar canes and
" other like substances, and for triturating pigments, drugs, and
" other materials."

[Printed, 4d. No Drawings.]

A.D. 1852, November 19.—N° 795.

BESSEMER, HENRY.—" Improvements in apparatus for con-
" centrating cane juices and other saccharine solutions, and in
" the treatment of such fluids."

These are, first, "the formation of a gutter or channel at the
" bottom of the pan when working with discs or screws," for
" the purpose of facilitating the outflow of the concentrated
" syrups."

Second, the arrangement or combination of apparatus for
" applying currents of air partly round the hollow axis carrying
" discs or screws, and the bringing the axis of the disc or screw
" in contact with the fluid, for the purpose of preventing a direct
" communication between the inlet and outlet air passages."

Third, "preventing the formation of dry concrete sugar on or
" near the axis of the screw or discs, by causing the axis to be
" partly immersed in the fluid."

Fourth, "heating the bottom of the pan with the same air
" which is used to evaporate the syrups carried up by the discs or
" screws, whether such air be applied to the bottom of the pan
" before or after its use in the pan;" and "conveying the air to
" and from the pan by means of channels below the pan."

Fifth, "preventing the bending or warping of discs or screw
" blades by staying or connecting them together."

Sixth, "obtaining access to a covered pan" for evaporating
saccharine fluids, "having therein revolving discs or screws, by
" the use of moveable curved plates covering an opening made on
" one side of the pan a few inches above the level of the fluid,
" and also the absorption of a portion of the heat from the air
" which escapes from the pan and has acted on the fluid raised by
" the discs or screw by other portions of air which are afterwards
" to be used therein to promote evaporation."

Seventh, "the use of a circulating channel and strainers on the
" side of the pan."

Eighth, "the use of a toothed scraper or comb in evaporating
pans.

Ninth, "bleaching or decoloring of saccharine solutions by
" bringing them in contact with the vapours arising from the
" combustion of sulphur."

Tenth, "the use of revolving discs or surfaces for the purpose
" of exposing saccharine solutions to the action of any gases or
" vapours (other than atmospheric air)," as sulphurous acid or
carbonic acid, for "the purpose of effecting chemical changes
" in such solutions."

[Printed, 8d. Drawing.]

A.D. 1852, November 19.—N° 796.

BESSEMER, HENRY.—"Improvements in the crystallization
" and manufacture of sugar. "These are, substituting steam or
" other motive power for much of the manual labor" consequent
on the use of a number of small moulds "in the production of
" 'crushed lump'" sugar, also slicing the sugar more perfectly
without its previous removal from the crystallizing vessel; to
ensure bold coarse crystals by using large deep vessels in place of
the cones now used. The chamber or trough is rectangular and
holds from one to two tons of sugar, the sides are vertical, the
bottom sloped to an angle of about 45°, each side is hinged to the
bottom, at the centre of the bottom is a row of holes. The sides
of the trough are moved for facilitating the movement of the
crystalline mass along the trough. Any number of these troughs
may be placed parallel to each other, and along each end of them
are two beds similar to the beds of a large lathe. One of the
beds has a sliding headstock carrying a piston. The other bed
has a cutting wheel which revolves nearly in contact with the end
of the trough. Using " a screw or other equivalent in apparatus
" for sliding sugar, whereby a concrete mass of sugar may be
" advanced with a regular motion up to the cutter, such motion
" having a fixed or definite proportion in speed with the cutter."
" Placing the crystallizing troughs or vessels herein described
" about a central point that lines drawn horizontally through
" the centres of them shall be radii of a circle." The "use of a
" floating piece of wood, or other slow conductor of heat, when
" placed on the surface of saccharine fluids during their crystalli-
" zation." The sugar after it is sliced is passed between a pair
of horizontal metal rollers separated by springs and finally between
a pair of rollers covered with vulcanized india-rubber, to separate
" small concrete lumps."

[Printed, 8d. Drawing.]

A.D. 1852, November 19.—N° 797.

BESSEMER, HENRY.—"Improvements in the treatment of
" washed or cleansed sugars." It is stated that the cleansing
of sugars is generally effected by a solution of white sugar called
liquor or by water alone, but the sugar becomes dry and these
improvements are to render it "permanently moist," and are first

combining saline matters (chloride of sodium, &c.) with such
solutions.

Second, gelatine dissolved in water and sugar.

Third, decolorized grape sugar or other glucose matter.

Fourth, "the combination of decolored molasses or syrups
" rendered uncrystallizable by heat."

Fifth, "the combination of saline and gelatinous, matters with
" glucose or grape sugar."

Sixth, "the mixture of sugar with uncrystallizable syrup before
" such sugars are removed from the machines in which they are
" washed or cleansed."

[Printed, 4d. No Drawings.]

A.D. 1852, November 19.—N° 799.

BESSEMER, HENRY.—"Improvements in apparatus for con-
" centrating saccharine fluids." These are, first, "the use of
" hot water or other heated fluids, at a temperature not exceeding
" 212 degrees Fahrenheit, for the purpose of transmitting heat to
" saccharine fluids contained in vacuum pans."

Second, "the use of steam at a pressure not exceeding that
" of the atmosphere, or below a temperature of 212 degrees
" Fahrenheit, for the purpose of transmitting heat to saccharine
" fluids contained in vacuum pans."

Third, "the use of air or other aëriform fluids, at a tempera-
" ture below 212 degrees of Fahrenheit," for the above purpose.
Also "regulating the pressure of steam when used at a pressure
" below that of the atmosphere" by establishing a communication "between the discharge end of the steam coil and the
" condenser of the vacuum pan;" and upon the pipe which
forms this communication placing "a loaded valve, similar to an
" ordinary safety valve, one side of which will be pressed upon
" by the steam in the coil, while upon the opposite side of it
" there will exist a vacuum, or nearly so." The pressure desired
is regulated by the weight applied to the loaded valve.

[Printed, 4d. No Drawings.]

A.D. 1852, December 3.—N° 941.

BANFIELD, THOMAS COLLINS. — (A communication.)—"Im-
" provement in the process of, and apparatus for, extracting
" saccharine and other juices from beet-root or other roots and

"plants." These are, first, "by subjecting several portions
"thereof successively to the same water or other solvent in a
"vessel or vessels from which the air is excluded, for the purpose
"of obtaining concentrated solutions of such juices."

Second, "the combination of two or more vessels, furnished
"with tubes and other apparatus," for "the purpose of causing
"portions of water or other solvent to be applied successively to
"the maceration of several portions of material, in several vessels,
"so as to extract saccharine and other juices therefrom, and make
"concentrated solutions of such juices, and also for the purpose
"of causing such portions of water or solvent to be transferred
"from one vessel to another without being exposed to the action
"of the atmosphere."

In preference, a system of twenty vessels are recommended, and
the substance, sliced, chopped, dried or otherwise, is introduced
into them by a mouth at top, which is then closed. There is a
similar opening at the bottom for removing the material after it
is exhausted. There are pipes with valves to regulate he flow of
the fluid from the lower part of one receiver to the upper part
of the adjoining one. The hot water or other solvent is fed from
a reservoir into the first vessel, macerates the material, and, aided
by the pressure of an air pump flows into the second, and so on
through the other vessels.

[Printed, 8d. Drawing.]

A.D. 1852, December 3.— 953.

BROOMAN, RICHARD ARCHIBALD. — (*A communication.*) —
"Improvements in the manufacture of sugar." These are, first,
the mode of feeding the materials into the presses by means of
wire gauze bands on rollers, so that they form a passage between
them, gradually diminishing in width, and so that the materials are
subjected to preparatory pressure. The application of a vacuum
behind the bands, by which the juice is drawn off at the same
time that the pressing operation is performed. "The mode of
"arranging the pressing-boxes in pairs, so that the entrance of
"the piston into one box shall discharge the pressed matters
"from the opposite box, and vice versâ;" and the application of
a liquid (water) forced through the matters undergoing pressure,
for more thoroughly obtaining the juices therefrom. The water
is forced through pipes between the plates of the boxes, and is

conducted to receivers through orifices in the bottom of the pistons
and boxes ; and whether employed in combination or separately,
or any two or more of them combined with other arrangements
of pressing machinery.

Second, "the application of centrifugal force for spreading
"saccharine liquids over revolving surfaces, in apparatus for
"evaporating such liquids." There are two drums, and between
the envelopes of each are two spiral channels, one of which is to
receive the liquid to be concentrated, and the other the steam by
which the liquid is heated, &c. "Also the combination with such
"apparatus of arrangements for heating the air employed for the
"purpose of aiding or effecting the evaporation of the saccharine
"fluids, by means of the saturated hot air or vapour escaping
"from the apparatus." This apparatus may be employed with
or without a vacuum.

Third, "the construction of the drums of centrifugal sugar
"machines with spirally corrugated peripheries, and without
"perforations," except "an opening of sufficient size at the
"bottom to throw off the extracted liquid."

[Printed, 1s. Drawings.]

A.D. 1852, December 4.—N° 956.

MANIFOLD, John Thornborrow, and LOWNDES, Charles
Spencer.—(*Provisional protection only.*)—"Improvements in the
"method of extracting juice from the sugar-cane." This, it is
said, "relates to a novel arrangement or adaptation of the hydraulic
"press," and consists in the use of "two presses, the tables of
"which are each connected by connecting rods and links to the
"ends of a vibrating beam or lever overhead," for "the purpose
"of raising one plate while the other is being depressed, the
"cylinders and rams being above the plates, which consequently
"work downwards. The cane carriers pass right through the
"presses, travelling beneath the pressing plates, and are stopped
"and driven on again as they are alternately pressed and released
"by their respective plates above them. As the pressing plates
"ascend after each downward stroke their respective cane carriers
"traverse forward the extent of the pressing plate, and are then
"thrown out of gear, and consequently stopped by self-acting
"mechanism for that purpose."

[Printed, 4d. No Drawings.]

A.D. 1852, December 6.—Nº 968.

DE DOUHET, GUILLAUME FERDINAND. — (*Provisional protection only.*)—"Improvements in the manufacture of alcoholic,
" saccharine, and starch products." These are said to be " com-
" bining the constitutive elements of these different bodies, viz.,
" carbon, hydrogen, and oxygen, in different quantities." For
this purpose sulphuret or sulphide of carbon, obtained from distilling sulphur over red hot charcoal, is employed, and mixed with
milk of lime or a solution of any caustic, alkali having an affinity
for its sulphur; and this mixture shut up and heated in a still
with precaution, the lime or alcali unites with the sulphur, the
carbon liberated unites with the hydrogen and oxygen of the
water, and forms, " according to the proportions, several ethers
and alcohols of the different degrees." " To have amylaceous or
" saccharine products it is requisite to mix sulphuret of carbon
" with an alkaline solution more concentrated than for alcohol,"
and the mixture, secured in a close vessel, is " gradually heated,
" left to get cool, after which the mass is lixiviated, poured off,"
and the lime in solution separated " by carbonic acid, and the
" saccharine or amylaceous products are produced, which are
" afterwards treated by means known." " It is not necessary to
" indicate the various proportions of the substances, as the
" reactions herein-before described will always take place, what-
" ever may be the proportions."

[Printed, 4d. No Drawings.]

A.D. 1852, December 11.—Nº 1031.

DIXON, GEORGE.—(*A communication.*)—" Improvements in the
" manufacture and refining of sugar." These are, " evaporating
" saccharine fluids by causing them to descend in a divided state
" through vertical tubes, around which steam circulates in a close
" vessel, and up which tubes streams of air (caused by a fan or
" blower) pass at the same time." The apparatus consists of an
upright box or chamber, with a plate near the top and near the
bottom, in which numerous tubes are fixed; between these plates
steam is admitted; the saccharine liquid flows down these tubes;
at the bottom is a chamber to receive the saccharine liquid, through
which streams of air are forced by a fan or other blower, and passing
up through the tubes, is drawn off with the steam from the upper
part of the chamber.

[Printed 6d. Drawing.]

MOINIER, JEAN BAPTISTE, and BOUTIGNY, CHARLES
CONSTANT.—"Improvements in concentrating syrups and other
" solutions, and in distillation." These are, the arrangement or
combination of apparatus for the above purpose, as follows :—
" A vessel, heated by means of steam externally, in which are a
" number of hollow metal spheres, or other forms. The liquid is
" allowed to descend in a divided state from the upper part
" of the apparatus amongst the spheres, at the same time air is
" admitted at the lower part, and drawn by an air pump."
" The syrup or more concentrated fluid is drawn off from below,
" and the vapours, passing off when desired, are condensed in a
" suitable condensor."

[Printed, 6d. Drawing.]

NESMOND, PIERRE CHARLES.—"Improvements in machinery
" applicable to the manufacture of ice and to refrigerating pur-
" poses generally." This consists " in constructing apparatus for
" affecting the alternate compression and dilatation of atmospheric
" air and other permanent gases for the manufacture of ice, and
" for refrigerating purposes generally." The apparatus for the
purpose of " purifying sugar from cane and beet roots " is constructed as follows. The foundation plate of wood or sheet or
cast iron is mounted on small rollers, and has an oblong upright
casing fixed upon it made of non-conducting substances ; in the
interior of this casing is a condenser in which the pressure of the
air is effected ; of metal or other material capable of resisting ;
above this condenser is a cooling reservoir (at the bottom of which
is a tap), in which is water retained by a sheet iron or galvanized
iron partition ; on the top of this reservoir is a wooden lid,
externally lined with sheet iron. There is a tube which forms a
communication between this condenser and what is called the
dilatation chamber. This dilatation chamber in the lower part of
an oblong upright casing made of non-conducting materials, called
the refrigerator, is divided into an upper and lower compartment
by means of an iron plate, but these compartments communicate
with each other by means of a tube. The tube of communication
between the condenser and dilatator has two taps united together
by levers, by which they are prevented from remaining simul-

taneously open. "The vegetable from which the sugar is to be " extracted (beet root for instance) is cut into thin slices," and put in the upper division of the refrigerator; an air pump is attached to the condenser, and the recepient is filled and emptied several times with cold air. The materials are then withdrawn and placed in the vessel above the condenser, and about four times their water at a temperature not above 104° F. added. When the liquid appears to have separated a sufficient quantity of saccharine matter, it is drawn off, and replaced by an equal quantity of pure water at 32° F. By congealing briskly saccharine solutions from beet root the useless salts are in solution.

[Printed, 8d. Drawing.]

1853.

A.D. 1853, January 15.—N° 106.

VION, HIPPOLYTE CHARLES. — " Certain improvements in " apparatus for refrigerating." These are, in absorbing latent heat by a vaporized gas, in which tubs or vats lined with gutta percha are employed to contain the liquid to be refrigerated; above these vats are sheet iron cylinders for holding any liquified gas, fitted with lids, also with thermometers, manometers, and man-holes. These cylinders are alternately supplied by pumps through means of suction and forcing cocks fitted into pipes, which unite at their extremities in the liquid gas cylinders. The first cylinder is filled with about six and a half gallons of liquified gas, and the other three with thirty-two gallons. Suppose the gas in the first cylinder at the temperature of 20°, and indicating by the mano-meter 1 atmosphere of pressure; the second cylinder at the tem-perature + 12°, indicating by the manometer three atmospheres; the cocks should be arranged so as to evaporate in the second cylinder, and compress in the first cylinder; the tubs or vats into which the cylinders are alternately precipitated (by means of crabs or winches) being filled with liquid. When a body changes its state it gives out caloric, and thus the first portion of gas absorbed by the pump ought to convey to the uppermost parts of the liquified gas a certain quantity of caloric necessary for effect-ing the change in condition of the liquified gas. As the equili-

S.   L

brium of the caloric becomes established rapidly by contact the
lower portions of the liquified gas become progressively cooled.
If the absorption be continued until the manometer of the second
cylinder indicates one atmosphere, a certain quantity of caloric
will be taken from the liquid, and so on. This apparatus may be
used for taking heat from any body whatever, and among other
purposes which are named for obtaining crystals of sugar from
solution in water.

[Printed, 1s. 6d. Drawings.]

A.D. 1853, January 24.—N° 174.

KNAB, DAVID CLOVIS.—" Improvements in the process of, and
" apparatus for, distilling certain vegetable and mineral matters,
" and also animal bones and flesh." These are, in reference to
this subject, as follows :—In distilling animal, &c. substances,
employing " a temperature suitable to the nature of the operation,
" and at a lower degree than hitherto practicable," also the apparatus employed therein. " A cast iron retort, similar to those
" used for manufacturing gas," is provided with a tube at the
top for the passage of the volatile products to the condenser.
The retort is set in a cast iron trough, between which and the
retort is a space forming a bath of molten metal, the metal varying according to the degree of heat required. " Boiled down
" bones, flesh, bone black which has already been used," &c.
" should be treated at a temperature of 680° F., when they begin
" to give up a notable quantity of carburetted hydrogen. This
temperature may be varied according to the nature of the volatile
" products to be obtained. The bath may range up to 795° F.
" Bones produce bone black as residue, and essential oils and
" ammonia as volatile products, and acetic acid."

[Printed, 8d. Drawings.]

A.D. 1853, February 17.—N° 414.

PIDDING, WILLIAM.—" Improvements in the treatment and
" preparation of saccharine substances, and in the machinery or
" apparatus connected therewith." These are first, " the use of
" bituminous substances (that is to say) animal or vegetable tar
" or pitch, margaric, oleic, or fat acids, wax, stearine, spermacetti,
" and greases," prepared by placing them " in an iron vessel and
" kept boiling by means of steam jets ; clear hot water should be

" constantly kept flowing into the vessel, and the waste water
" carried off," as long as any taste or odour is imparted to the
water, " when combined with the hydrate of alumina, precipitated
" silica or oxide of iron." The black rosin, or resin of colophony
prepared as above is mixed in the proportion of about 18 per
cent. to the dry alumina contained in the hydrate, they are mixed
to the consistence of cream by water. When it is required to
reduce the melting point of the rosin, which is about 140°, it is
mixed with the other fatty acids, &c., mentioned above.

Second, the combination of margaric acid, wax, and grease
prepared as above with rosin previous to its admixture with
hydrate of alumina, silica, or oxide of iron, for the foregoing
purposes.

Third, " the combination of the vacuum pan and filter." There
is a pipe between the vacuum pan and filter, and in this pipe is a
tap. " The temper " it is said is prevented mixing with the
charcoal by means of a coarse cloth.

Fourth, treating " sugar after crystalization by drawing off the
" residual syrup by means of vacuum, both before and after it is
" placed in moulds." The sugar arrived at the proper state of
crystallization, a slide is removed and a jet of steam introduced
and a vacuum formed. The crystalline mass put in moulds, the
moulds are acted upon by a vacuum caused by a jet of steam or
otherwise. "

[Printed, 1s. Drawings.]

A.D. 1853, February 17.—N° 420.

HAWES, WILLIAM.—(*Provisional protection only.*)—" Improve-
" provements in the manufacture and refining of sugar." These
are, defecating " saccharine juices and solutions by employing salts
" or oxides of metals," for which purpose using "bismuth, tin,
" zinc, or manganese," and subsequently filtering " through
" charcoal or other filter."

[Printed, 4d. No Drawings.]

A.D. 1853, February 19.—N° 431.

HILLS, FRANK CLARKE, and HILLS, GEORGE.—" Certain
" improvements in refining sugar, and in preparing materials
" applicable to that purpose." These are first, " the employment
" of phosphate of lime, steamed or calcined bones, sawdust,

" asbestos, pumice stone, or other analogous absorbent material,
" except animal or other charcoal, to abstract lead or salts or other
" combinations of lead from saccharine solutions." The saccharine solution previously defecated by the lead compound is passed
through a filter of the above materials, about fifteen feet deep,
and " preferably of about four or five feet diameter, having a false
" bottom to support the materials so as to constitute a filter."
The filter should be maintained at about 150° F., preferably by
means of steam in a coiled pipe placed among the materials. The
bones, &c., should be reduced to a coarse powder, and the sawdust
washed with caustic soda or potash and afterwards with water.

Second, " the employment of acetic, muriatic, or nitric acid to
" remove lead or salts or other combinations of lead " from filters
as above, " which may have been employed in abstracting such
" lead or salts or other combinations of lead from saccharine
" solutions."

[Printed, 4d. No Drawings.]

A.D. 1853, February 26.—N° 487.

BRANDEIS, JOSEPH.—" Improvements in the manufacture and
" refining of sugar." These are said to be " the application of
" shale or schist charcoal, for separating the excess of lead, tin,
" zinc, or bismuth from solutions of sugar which have been
" defecated by salts or compounds of such metals," as follows :
to a solution of the raw sugar dissolved in water, from one to
two per cent. of " the sugar or subacetate of lead " dissolved in
water is added, the precipitate removed by a bag filter, and the
filtered liquid passed " through a bed or beds of charcoal from
" five to six feet deep (more or less), obtained by carbonizing
" shale or bituminous schist," until the solution tested in the
ordinary way shows no traces of lead, when it is crystallized in
the ordinary manner. " The process is the same when other salts
" of lead are employed, or when salts of tin, zinc, or bismuth are
employed to defecate the solution of sugar."

[Printed, 4d. No Drawings.]

A.D. 1853, March 14.—N° 633.

HOWARD DE WALDEN AND SEAFORD, CHARLES AUGUS-
TUS, Lord. — (Provisional protection only.)—" Whitening and
" cleansing sugar by the application of steam and hot air in a

" centrifugal sugar machine." There is " a furnace in which a
" coil pipe, coming from any steam pipe of the factory, is heated
" to a high degree of temperature and expansion. A pipe
" extends from this coil pipe to the centrifugal machine," passes
through an opening in the cover and projects the steam into the
centre of the revolving drum, while the proper quantity of air is
admitted through a hole in the cover, " to form, with the hot
" steam, the proper temperature and moisture for cleansing the
" sugar." To force the hot air vertically through the sugar
" there is a concave annular ring fixed on the exterior plate of the
" border of the revolving drum, in which is suspended nearly to
" the bottom of this concave ring, a circular leather strap fixed
" to the immoveable upper part of the receiver, which forms a
" flexible joint, and prevents the steam and air from passing over
" the revolving drum without acting on the sugar."

[Printed, 4d. No Drawings.]

A.D. 1853, March 14.—N° 639.

SCOTT, JOHN, junior.—"Improvements in the treatment or
" manufacture of animal charcoal." These are, " arranging the
" kilns, ovens, or furnaces used for re-burning and cleansing
" the animal charcoal used in sugar refineries, that the burned
" material may be cooled by a continuous process without the use
" of water." The kiln is arranged with a series of vertical tubes
surrounded by fire or hot air flues, and into these tubes the charcoal is placed to be burned. Immediately beneath this set of
tubes or retorts is a second set of smaller content, also set vertically, the axis of each tube of this lower set being coincident with
that of a corresponding retort. When a charge has been burned,
the attendant opens a slide in the bottom of the retort, and discharges about one-third of the charcoal in it, letting this quantity
fall down into the tube below. When this discharge is effected,
the lower tube being filled, or nearly so, the operator shuts up the
tube, and allows the contained charcoal to cool by an external
circulation of air. The upper part of the retort then receives a
fresh supply of the unburned charcoal to fill up the emptied
space, and thus the operation goes on throughout the whole series
in the kiln. A good current of cold air is kept circulating round
the cooler tubes, and this air is then passed up in its partially
heated state to the retort furnaces above, where its heat is made

available for securing further economy in its action upon the
" furnaces from which the original re-burning or charring heat
" is derived."

[Printed, 10d. Drawings.]

A.D. 1853, April 8.—N° 851.

ROBINSON, HENRY OLIVER.—" Improvements in machinery
" for crushing sugar canes." These are, first, " constructing a
" steam engine with the gear to connect it to a sugar cane mill in
" such a manner that the engine and gear are placed together
" upon a base plate common to both of them, and also in such a
" manner that the end of the said base plates may be connected
" with the base plate of any sugar cane mill by screw bolts and
" nuts, whether the latter be specially adapted to it or not," so
that a steam engine and gear so constructed "may be attached to
" any sugar mill already in existence by screw bolts and nuts,
" whether to replace a worn-out steam engine, a water wheel, or
" other motor."
    Second, a "coupling or clutch, consisting of a disc of iron
" formed with a groove on each of its flat sides at right angles to
" each other; the one groove receives the end of the connecting
" gear shaft, which is formed in a tongue to fit into it, and the
" other groove receives a tongue formed upon the shaft of the top
" roller of the mill."

[Printed, 6d. Drawing.]

A.D. 1853, April 12.—N° 878.

GREENWOOD, THOMAS. — " Improvements in evaporating
" saccharine fluids." These are, " arranging or combining and
" working the heating apparatus of vacuum pans used for boil-
" ing saccharine fluids," "in such manner as to admit of being
" brought into action in succession as the pan becomes more and
" more filled." For this purpose it is preferred "to use a series
" of coils of pipes fixed one above the other in a vacuum pan,
" each such having its own stop cock or valve for the admission
" of steam, and its own stop cock for the exit of con-
" densed water, so that it may not become a heating apparatus
" till the charging of the vacuum pan brings the saccharine fluid
" to cover it." In place of coils of pipes, " arrangements of

" heating pipes or other surfaces may be employed." In carrying
out the arrangements as above, the vacuum pan is made with a
jacket.

[Printed, 4d. No Drawings.]

A.D. 1853, May 2.—N° 1062.

BELLFORD, AUGUSTE EDOUARD LORADOUX.—(*A communication.*)—" Improvements in the extraction and manufacture of
" sugar and of saccharine matters." These are, " the application
" of the apparati " afterwards described " or of any other similar
" apparati, to the clarifying of sugar, and the extraction of
" saccharine matters contained in beet root, either fresh or dried,
" as well as the application of the same apparati to the extraction
" of malt for the manufacture of beer." " These apparati con-
" sist of vessels of a conical form," in which is placed the
material to be exhausted. This vessel is supported by axles or
bearings, on which it turns to facilitate the operations, it is
furnished with a perforated metallic sheet, fit for retaining the
crystals of sugar, at its upper part a forcing pump is attached of
a tube several yards in height, is furnished with a cock for the purpose of bringing upon the mass a column of clarifying matter,
and afterwards to force through it air from above. Other
" apparati " are described which differ little from this.

[Printed, 8d. Drawings.]

A.D. 1853, May 7.—N° 1131.

FINZEL, WILLIAM CONRAD.—"An improvement in refining
" sugar." This consists "in the melting of sugar in vacuo." An
ordinary vacuum pan is employed and there is no limit to the
form or size of the pan, and it is said that " a much less quantity
" of treacle is produced, and the sugars are of a better color than
" those melted in the ordinary manner."

[Printed, 4d. No Drawings.]

A.D. 1853, May 19.—N° 1243.

MANIFOLD, JOHN THORNBORROW, LOWNDES, CHARLES
SPENCER, and JORDAN, JOHN.—" Improvements in the method
" of extracting the juice from the sugar cane." These are, " reduc-
" ing the raw canes to the condition of saw dust or minute particles,

" disintegrated by the action of circular or other saws or by other
" convenient reducing agents," and extracting " the juice from
" the solid portions of the canes " by " continuous or duplex
" action hydrostatic presses," or otherwise, and " to aid the dis-
" charging action, steam is passed through the reduced material."

[Printed, 8d. Drawing.]

A.D. 1853, May 25.—N° 1278.

HIGGINSON, George Irlam.—" Improvements in machinery
" or apparatus for evaporating or concentrating liquids." These
are, " the general arrangement of apparatus," the application and
use of rotatory or partially rotatory screw spiral steam chambers
or discs, " and the system or mode of evaporating and concen-
" trating liquids by means of hollow heating screw blades or
" discs dipping into the evaporating liquid," as follows :—The
arrangement preferred, " is that of a hollow screw or spiral blade
" of great breadth, carried upon a horizontal tubular spindle
" revolving upon bearings above or upon the containing vessel
" of the liquid." The " steam is passed into and through
" it by means of a thoroughfare in the tubular shaft," and is
" finally discharged at the opposite end of the shaft into an
" external pipe, which conveys it away to the external casing
" or steam heating chamber of the evaporating vessel." This
apparatus is " specially applicable to the concentration of cane
" juice or syrup." " The constant dipping into and emerging
" from the evaporating liquid of the heated surface of screw
" threads or discs presents a very extensive evaporative area."
" Hollow discs or surfaces of other forms may likewise be used
" instead of screw blades," and instead of revolving " con-
" tinuously, reciprocatory rotation may be employed," whilst the
steam may be first passed " through the outer casing of the
" containing vessel, and afterwards through the screw blades."

[Printed, 6d. Drawing.]

A.D. 1853, June 3.—N° 1363.

GOSSART, Ferdinand Louis. — " A system of permanent
" circulation of caloric intended to produce and overheat steam,
" gas, and liquids." The applications it is said of this apparatus
are numerous, among which is named, in reference to this subject,
" the manufacture of sugar," and it is said that " the apparatus

" is composed of three essential parts " namely :—" a heating pipe,
" receiving the waste heat by the condensing tube, through which
" comes with a certain pressure the liquid that is to be converted
" into steam, or the gas to be heated. Secondly, one or several
" pipes, cylinders, boilers, or generating tubes, or heaters,
" receiving directly or indirectly the action of the fire. Thirdly,
" a condensing tube, in which the steam or gas is cooled by
" communicating its heat or caloric to the heating pipe." " The
" principles upon which this invention are based," are, first, to
multiply in the interior of the tubes the heating or condensing
surfaces by the use of conducting surfaces (fragments of metal).
" Secondly, to establish a continuous circulation of the caloric by
" using, either by double pipes or by conducting liquids serving
" as intermediate agents, the heat that the steam or gas retains
" when it arrives in the condensing pipe, for the heating of the
" liquid, steam, or gas which passes through the heating pipe to
" arrive into that part of the apparatus where it receives directly
" or indirectly the action of the fire."

[Printed, 1s. Drawings.]

A.D. 1853, June 3.—N° 1364.

MAYELSTON, James.—" Certain improvements in the manu-
" facture and refining of sugar." These are, first, " the employing
" of atmospheric pressure, induced by exhaustion or suction, in
" vessels with false bottoms or in moulding apparatus combined
" or used therewith," " for the purpose of facilitating the draining
" of the mother syrup from crystals of sugar, which along with
" their mother syrup have been admitted into such vessels or such
" moulds when in or about the state in which sugar is usually
" moulded, that is to say, in a state of sufficient fluidity whilst
" hot to admit of being poured into vessels, and of sufficient
" consistency to become fixed on cooling, and also for the purpose,
" when desired, of subsequently washing or liquoring the said
" crystals in such vessels or moulding apparatus."
Second, " suspending of pneumatic draining and cleansing
" apparatus in such manner as to admit of being readily inverted
" to facilitate the discharging or unloading the same."
Third, " retaining saccharine fluids between given degrees of
" temperature," by attaching " to the vessels containing such fluids
" any apparatus which, acting on the well-known principle of the
" expansion and contraction of materials, or their unequal expan-
" sion and contraction by change of temperature " so as to impart

motion to and open or shut any valve or apparatus " whereby heat
" is admitted to or excluded from such vessel or such fluid, so as
" thereby to raise or maintain the temperature of such fluids
" above the limits at which fermentation is most liable to take
" place without exceeding another given limit," and for this
purpose preferring to maintain the fluid " between 120° and 140°
" of Fahrenheit."

Fourth, " moulding, draining, and cleansing of sugar by the aid
" of pneumatic pressure in hexagonal moulds, so united or placed
" together in one common vessel, or within one common boundary,
" as to admit of the sugar in them being drained or washed in
" common."

[Printed, 6d.   No Drawings.]

A.D. 1853, June 20.—N° 1510.

GALLOWAY, ROBERT.—" Improvements in manufacturing and
" refining sugar."   These are said to consist " in employing
" tannic acid (usually called tannin), or gallic acid, or a compound
" of one of these acids with a base, as potash or ammonia," " for
" the purpose of precipitating lead used in defecating saccharine
" fluids."   In the Provisional Specification " pectine or pectic
" acid " are also named.   The juice or solution of sugar after
treating with " saccharate of lime, saccharate of lead, or saccharate
" of magnesia, or plumbite of lime, and afterwards neutral acetate,
" or one of the basic acetates of lead," as described in No. 14,354,
Old Law, is treated, either before or after filtration, with " a so-
" lution of tannic or gallic acid, or some tannate or gallate," until
only a slight precipate takes place in some of the liquid filtered.
The whole is then filtered and " a sufficient quantity of sulphurous
" acid or some bisulphite " is added, " to remove the last traces
" of lead, and the solution is " again filtered to free it from the
insoluble sulphite of lead thus produced."   The solution of tannic
acid is made " by infusing two pounds of crushed valonia in from,
" two to three gallons of boiling water, and using the clear liquor."
" Gallic acid is prepared by exposing a solution of tannic acid to
" the air for some time."

[Printed, 4d.   No Drawings.]

A.D. 1853, July 15.—N° 1687.

BESSEMER, HENRY.—" Improvements in the manufacture and
" refining of sugar."   These are, separating " the whole or greater

" part of the glucose matter, or uncrystallizable coloured syrups,
" from sugar," prior to manufacturing it into " loaves, lumps, or
" concrete masses, instead of separating such glucose matters after
" the lumps of sugar has been formed." In carrying out this,
" the raw sugar may be dissolved in water, ' blown up,' and passed
" through animal charcoal," concentrated to 45° Beaume, in
preference, " by bringing the fluid in contact with currents of warm
" dry air, at a low temperature," as described in No. 795, A.D.
1852, or " the vacuum pan may be used in lieu thereof." " The
" syrup when sufficiently dense is heated in the usual manner to
" a temperature of about 170° to 180° Fahrenheit, and then put
" into large crystallizing vessels," which may be of wood, preferably of a circular form, capable of holding a ton each, or it may be
put into common sugar moulds; as soon as the sugar has completely crystallized the whole matters are " removed from the
" vessels in a ' green state,' and are then sliced or broken up so as to
" render the sugar suitable for the process of washing or cleansing in
" a machine," in preference in apparatus described in No. 795, A.D.
1852, " or other machines, such as the centrifugal curing machine,"
and when cleansed the sugar is suitable " for the formation of a
" loaf or concrete mass of sugar." The crystals thus obtained are
distinguished " by the term of product No. 1, and the fluid matter
" separated therefrom I call product No. 2," and so on, using
other numbers to distinguish the different products. In dealing
with No. 1 it is dissolved by means of heat in a minimum of hot
water or syrup (called No. 3), the solution having a density of
about 46° Beaumé, is heated quickly to 180° F., put into moulds
placed in pots, and when the sugar is perfectly crystallized the
plugs are withdrawn, and after draining the loaves are " ' cleaned
" off' and dried in a stove." If the color is not quite so good as
desired, " ' water clay or liquor ' may be used in such quantity as
" may be found necessary to produce the required color." The
drips may be mixed with No. 3 product and be worked up, or they
may be returned to the evaporating pan. No. 2 product may be
concentrated to about 45° Beaumé, moulded, crystallized, washed,
and cleansed. This product is No. 4, and the syrup is No. 5.
The crystals No. 4 may be either mixed with water or with syrup
No. 3, so as to yield a semifluid about 46° Beaumé; this mixture
on crystallizing yields loaf sugar of a good quality, or No. 4 and
No. 1 are mixed and treated as before directed in reference to
product No. 1. Whenever the syrups No. 3 or the drips acquire

too deep a color they are put into the evaporating pan and form part of a new charge. No. 5 is evaporated to 46° or 47° Beaumé, crystallized and pressed, the product No. 7 is small crystals of " bastard sugar in a nearly pure state, and may be sent to market," but, in preference, it is dissolved in a fresh solution of filtered raw sugar, or it may be returned to the evaporating pan and form " part of a future first product." The last or eighth product is the treacle, obtained by pressure " in a fit state for sale." In refining sugar, to more nearly equalize the value of the products, " the inferior sugar, such as ' pieces ' and ' bastards,' after cry- " stalization, are washed, cured, or pressed, dissolved in a filtered " solution of raw sugar ; or the sugar may be put into the vacuum " pan or other concentrating apparatus, whereby the crystals are " blended with the new portions of sugar, and form part of the " loaf." In some cases sugars that have been cleansed by machi- " nery may be added or mixed with the concentrated syrups as " above.

[Printed, 6d. No Drawings.]

A.D. 1853, July 15.—No 1689.

BESSEMER, HENRY.—" Improvements in the manufacture and " treatment of bastard sugar and other low saccharine products, " such as are obtained from molasses and scums." These are, " first, the mode described in the after claims " of treating bastard " sugar and low saccharine products, obtained in processes of " refining sugar herein mentioned."

Second, " the combined agency of heat and pressure for the " purpose of separating the solid from the fluid parts of bastard " sugar and other low products obtained in the manufacture and " refining of sugar."

Third, " the encreasing the density of sugar solutions " " three " or four degrees Beaumé beyond the usual density," after which they " may be let down into large crystallizing vessels " about one ton each; cooling the syrup " down to about 120° or 110° degrees " Fahrenheit," putting it into bags of a textile fabric, these bags are piled in a hydrostatic or other press and pressed between corrugated plates.

Fourth, when " the syrups are brought up to a greater density " than usual and are intended to be cooled as much as possible," shallow trays are used for crystallizing cakes of sugar that are afterwards to be pressed for the purpose of separating the glucose matters, both by cold and by hot pressure.

Fifth, " dissolving pressed cakes of bastard sugar in a filtered " solution of raw sugar, in order that such bastard sugar may " form a part of the refined product " by concentrating, crystallizing, cooling, and pressing as above.

[Printed, 6d. No Drawings.]

A.D. 1853, July 15.—N° 1691. (* *)

BESSEMER, Henry.—" Improvements in the manufacture and " refining of sugar ;" these are, constructing " cylindrical sugar " moulds having a moveable bottom, so that the syrups may " drain off from a surface whose area is equal to the body of the " mould." " The mould may in consequence be made of much " greater height," because this increased area of outlet will allow " the syrups to drain off quickly, which the hydrostatic pressure " of a tall column also materially assists." " This increased " capacity of the mould will render it " too heavy " to be handled " by the workmen in the usual way," " therefore they are made " fixtures or moveable only with revolving apparatus somewhat " like a turntable ; and instead of detaching the loaf from the " mould by a blow," " an hydraulic press or other suitable " mechanical force " is employed " to push out the loaf from the " mould." " In carrying out the above," " the moulds have a " vertical slit or opening made on one side of them, which extends " from end to end ; on each side of this slit angle flanges are " rivetted ; there are projecting pieces or lugs formed on the angle " iron through which a screw passes. A piece of vulcanized india- " rubber is put in between the angle irons, so that when the screws " are tightened a close joint will be formed, through which the " fluid matters in the mould cannot find their way," and " on " the bottom of each mould a cap is fitted with a bayonet joint, " and having a piece of vulcanized india-rubber between it, a " sound joint is formed."

[Printed, 8d. Drawing.]

A.D. 1853, August 13.—N° 1900.

GWYNNE, John.—" Improvements in the preparation of a " black powder from coal, and in the application thereof to the " manufacture of paints, blacking, and various other purposes."

These are,—coal is carbonized and ground "in mills or by
" manipulations of like and suitable nature" reduced to small
fragments, and is applied to a number of purposes which are
named. The ground matter is either sifted dry or separated by
the wet method, namely, washing. "For the purpose of refining
" sugar, after 'Gwynne's patent solidified peat or coal' has been
" carbonized and reduced to powders of different degrees of
" fineness, I add to those powders certain portions, as required,
" of phosphates, sulphate and carbonate of lime, sulphate of soda,
" also alkaline carbonates, sulphate of alumina, sulphate of baryta,
" or silicious sand."

[Printed, 4d. No Drawings.]

A.D. 1853, August 13.—N° 1902.

GWYNNE, JOHN, and GWYNNE, JAMES EGLESON ANDERSON.
—"Improvements in the preparation of beet root for the manu-
" facture of sugar, which improvements are also applicable to the
" preparation of other vegetables." These are, the roots after
washing or otherwise preparing are cut into small pieces, which
are then passed through a series of inclined heated and perforated
cylinders fitted with rotating ribs so as to turn over and agitate
the roots as they pass through the cylinders. Another mode of
drying the roots is by passing "the cut roots over an endless band
" contained in a case heated by any convenient means, such
" band being of any suitable material, and mounted on a series
" of rollers." By similar arrangements to the above the roots
may be cooled by passing cold air over them. The roots after the
processes are completely dried, and "may be placed under com-
" pression so as to solidify the mass and save space in their
" stowage until required for use."

[Printed, 1s. Drawings.]

A.D. 1853, August 30.—N° 2011.

PICCIOTTO, JAMES.—(*A communication.*)—"Improvements in
" burning or reburning animal charcoal." These are, "con-
" structing apparatus for burning and reburning animal charcoal
" of tubes of fire clay made in several parts," "moulded to a
" form suitable to admit of several like pieces or bricks being
" brought together and built one on another, to produce the
" desired form of retort." These retorts or tubes "are built in

" an upright position, and several of them are combined into one
" stack." The fire-place or furnace is near the lower ends of the
retorts ; the heat and products of combustion pass from the fire
amongst the retorts, "which are set in a chamber, and then pass
" away by a flue at the upper part of the building." The lower
end of the tubes or retorts rest in sockets, and have " valves to
" close them, which, when opened, allows the burned or charred
" matter to descend in the tubes ;" care is " taken to prevent the
" atmosphere coming in contact with the charcoal till it is cooled
" down." The tubes or retorts are filled with the material to be
burned at the top.

[Printed, 6d. Drawings.]

A.D. 1853, October 12.—N° 2343.

MAUMENÉ, EDME JULES.—" Improvements in the treatment
" of lignite or wood coal, and in obtaining various useful pro-
" ducts therefrom." These are, in reference to this subject, as
follows :—The above coal is distilled in a retort, " a current of
" ordinary, or of overheated steam, may be introduced into the
" retort during the process of distillation or carbonization."
The residue in the retort is washed with dilute muriatic or other
acid, capable of dissolving the impurities and washed with water
and dried and ground to a fine powder. This substance may be
used "in lieu of animal charcoal for decolorizing solutions of
" sugar or other solutions. In some cases I calcine it with other
" suitable materials, such as refuse wool, tartar, or sawdust in
" order to increase or facilitate its decolorizing effect."

[Printed, 4d. No Drawings.]

A.D. 1853, October 13.—N° 2358.

WAY, JOHN THOMAS.—"Improvements in making and refining
" sugar, and in treating saccharine fluids." These are, " the use
" of soluble or gelatinous silicia (or of natural earths, mineral
" beds or strata, and other substances containing soluble silica in
" considerable proportions) for the purpose of neutralizing an
" excess of lime or other alkaline earth employed in the defecation
" or purifying of the juices of the cane and beet root, or of other
" saccharine liquids, and also the use of silicate of lime and the
" silicates of other alkaline earths for the same purposes." By
soluble silica is meant silica soluble in boiling caustic potash or

soda; it may be artificially prepared by adding an acid to solutions of silicate of potash, or soda. Certain earths, such as occur in "certain beds or strata of earth at the base of the chalk hills " occurring in parts of Surrey and elsewhere, contain large pro- " portions of silica in this soluble condition," and this is the earth or mineral preferred, and "when the proportion of silica is from " fifty to seventy per cent., one ton of the earth or mineral is " sufficient for one hundred tons of sugar," the lime added to the sugar not being "unnecessarily great."

[Printed, 4d. No Drawings.]

A.D. 1853, October 17.—N° 2388.

CHANTRELL, GEORGE FREDERICK. — " Improved apparatus " applicable to the manufacturing and the revivification of animal " or vegetable charcoal and other useful purposes." This consists of a row of furnaces " parallel to each other, over which are thrown " reverberatory arches with openings to allow the heat to pass " through them, which is then carried by au S or backwards and " forwards draught through a series of floors constructed one " above another, openings being left at alternate ends thereof to " allow the heat to pass upwards." The floors are "tiles " supported by fire lumps or fire blocks." Between each furnace are a number of narrow chambers to cause the draught to double. The retorts have hopper mouths at the top, and openings at the bottom to allow the charcoal, &c., when burned, to fall into suitable coolers. " These lower openings are covered by means of " a plate of iron with holes cut through to correspond with those " in the bottom of the chambers." "The structure is built on " pillars to raise it sufficiently above the ground to allow space " for working and coolers. The whole of the tile and brickwork " forming the structure where the heat acts upon it is laid and " run in with fire clay lute about the consistency of treacle." " By this arrangement the char burnt within the chambers " described is clear of iron until all humid matter is extracted " from it, and thereby obviating the objection made by sugar " refiners to the use of char made or reburnt when in contact " with iron, as char so treated is found to give sugar filtered " through it a pink or 'foxy' tinge."

[Printed, 1s. 6d. Drawings.]

A.D. 1853, October 21.—N° 2435.

CHALLETON, Jean François Felix. — "Certain improve-
" ments in carbonizing and distilling peat, coal, wood, and other
" animal, vegetable, and mineral substances." Among the
substances which are named to be distilled are bones, and the
improvements are in order to " carbonize and distil in a
" progressive and continuous manner " these substances so as to
obtain products therefrom. The retort is divided into compartments by sliding doors ; each compartment contains a waggon
rolling on wheels, made of metal or fire clay, in which is the
substance to be carbonized ; on one of the external side walls of
the furnace are fixed the bearings of several hand or fly wheels ;
" the shafts of these wheels pass through the wall, and carry
" each inside the furnace a spur wheel gearing into a rack
" attached to a carriage," " which can thus be moved backwards
" and forwards inside the retort." " The sole plate carries a line
" of rails, which is broken at each sliding door, and on which
" rolls the carriage." One or more compartments are " left
" outside the furnace for the purpose of extinguishing in them
" the matter under treatment. This is effected by a jet of
" alkaline vapour which is introduced into the extinguishing
" chamber." " These waggons are shoved forward progressively,
" and as they travel in the opposite direction to that of the heat
" they pass successively from a low temperature to a higher one
" until the carbonization is complete," upon which the waggon is
driven into " the extinguishing chamber." " The waggons are
" assisted in their forward motion by a slight incline given to the
" sole of the furnace."

[Printed, 1s. 2d. Drawings.]

A.D. 1853, November 19.—N° 2693.

DIMSDALE, Thomas Isaac.—"The use and preparation of
" certain solid and liquid substances for the defication, purifica-
" cation, and decolorization of saccharine juices and syrups or
" solutions, and for neutralizing, decomposing, and absorbing
" noxious and fetid gases." These are, in reference to this
subject, as follows :—" For decolorizing or purifying syrups or
" saccharine solutions, I propose to mix blood with peat earth,
" spent tan, sawdust, or any fibrous or woody matter, divided or

" powdered, and to char or convert the mixture into charcoal by
" any of the ordidary methods by which charcoal is made. If
" peat earth or spent tan be used, they must be well washed
" previous to being mixed with the blood, to cleanse them from
" all dirt or impurities. In like manner clay or other aluminous
" earths may be used for the above mentioned purpose, mixed with
" blood, and then roasted or burned in retorts, kilns, or by any
" of the methods by which charcoal is prepared. Bituminous
" shale, charcoal, or any description of charcoal washed and
" cleansed may be mixed with blood, and if the blood be fresh
" the mixture may be used in that state, and when so mixed the
" mass may be recharred. Either of these preparations can be
" employed precisely as bone charcoal is."

[Printed, 4d. No Drawings.]

A.D. 1853, November 29.—N° 2777.

MICHEL, Louis Alexandre.—"A system of apparatus for
" sawing and breaking sugar." This consists, first, of a frame
carrying two uprights which have grooves or long mortices cut in
them in which the ends of a beam slide up and down. This beam
is connected with a lower beam by two uprights, thus forming
a frame carrying the saw; this frame may carry any number of
saws, set apart by about the thickness which the pieces of sugar
are to be. The sugar loaf is carried on a carriage capable of
holding all the pieces of sugar till they are perfectly separated
from each other. This might be done with a kind of grate having
as many bars as there are pieces to be cut.

Second, "these layers are then brought under a breaking
" machine, which is so arranged that by putting a shaft to motion
" a knife falls on one or several layers of sugar, and breaks or
" cuts off one or several pieces. Whilst the knife is lifted up, the
" layer of sugar receives a motion that causes it to advance, and
" the knife descending again breaks one or several pieces of the
" sugar. It will be easily understood that by changing the motion
" given to the layers of sugar, the thickness of the pieces may be
" varied." The whitish appearance produced by the sawing is
removed by immersing the long pieces in water, or by steaming
them, then drying them thoroughly, "after which they pass
" through the breaking machine, and are ready for use."

[Printed, 6d. Drawing.]

A.D. 1853, December 2.—N° 2809.

REYBURN, ROBERT. — (*Provisional protection only.*)—" Im-
" provements in sugar refining." These are, " so arranging the
" apparatus that the portion of charcoal which is most exhausted,
" that is to say, that which is nearest to the point where the
" syrup enters, may be from time to time removed, and that fresh
" charcoal may be added at the end where the syrup escapes, so
" as to supply the place of that which has been withdrawn, by
" which means the syrups are at all times obtained as free from
" color as at the commencement of the operation."

[Printed, 4d. No Drawings.]

A.D. 1853, December 2.—N° 2811.

BESSEMER, HENRY.—" Improvements in the manufacture and
" refining of sugar." These are, first, " for increasing the size
" and purity of the crystals of raw sugar, is the solution and re-
" crystallization of it without the intermediate process of evapo-
" ration." This is effected in a vessel in the bottom of which are
a series of steam or hot water pipes, and an agitator, boiling
water is sprinkled over the sugar (about a ton) so as to have a
syrup of 40 to 50 degrees Beaumé. The syrup flows into a large
crystallizing vessel with a lid so as to cool slowly. Or a portion
of the sugar may be dissolved in a pan with "a steam jacket, and
" then adding by degrees as much sugar as will form a syrup of
" the required density," or better, dissolving sugar in water to
form " a liquor of about 28° to 34° Beaumé, and blowing it up
" with albuminous matter," filtering and using it in lieu of water
for making dense solutions for crystallizing ; in like manner
" refines liquor," &c. may be used as solvents. When the crystallization of the sugar has taken place it is removed to the washing
table or machine.

Second, this table it is said "forms the subject of a former
" patent granted to me," and "consists of a revolving horizontal
" table or sieve on which the sugar is spread in a thin stratum,
" while a partial vacuum is formed beneath, " but "instead of
" applying the fluid from perforated pipes above the sugar," it is
" better to allow the fluid to flow from a perforated box, the per-
" forated bottom of which is in contact with the thin stratum of
" sugar under operation." " Over the moving stratum of the
" sugar a series of small inclined blades " are fixed to "pass
" downwards through the sugar."

[Printed, 8d. Drawing.]

A.D. 1853, December 8.—Nº 2855.

BORDONE, Philippe Joseph Toussaint. — (*Provisional Protection only.*)—" Improvements in extracting and treating the " juice of beet root and other vegetables." These are, in reference to this subject, as follows :—The pulp fresh, or previously dried, is introduced into an upright cylindrical vessel, having hemispherical ends, the lower being perforated. The vessel has a lining or filter of woollen stuff, removable after each operation, and steam is introduced into it at a temperature from 125° to 130° centigrade, by which the saccharine particles will be dissolved and flow out at the bottom. " The juice on running from these vessels is " treated with the quantity of milk of lime necessary for its " defecation." It is protected from fermentation by a small quantity of oil floating on its surface, which prevents its " absorbing oxygen from the air." " If it be desired to produce " alcohol instead of sugar, 20 per cent. of lukewarm bran water " must be introduced to the pulp before the introduction of the " steam, which will cause fermentation much better than any " other fermenting agents." The remainder of the processes are for obtaining oils and making bread, &c.

[Printed, 4d. No Drawings.]

A.D. 1853, December 14.—Nº 2898.

BEANES, Edward.—" Improvements in the manufacture and " refining of sugar." These are, first, " an improved vacuum " pan." This is a cylindrical vessel, terminated at top by a dome, connected with an air pump ; into the floor of the pan on which the fluid is placed are screwed a number of vertical tubes of equal size and length closed at the upper end, open at the lower. Below the floor of the pan are two steam chambers, communicating with each other, one above the other, from the lower of which are a series of tubes open at both ends which pass up into the vertical tubes named above, and through which steam is driven into them. Tubes from the upper steam chamber " communicate with the " lower steam chamber," " through which, as the case may be, " steam may be introduced into, or the steam and water produced " by the condensation of steam withdrawn from the said chamber."

Second, " a filter for filtering sugar cane juice or other liquids " containing saccharine matter in solution." This consists of a vessel with two false bottoms, on each of which is laid a stratum of sand or powdered flints, preferring, "the washed sweepings of " such roads as are mended with flint stones." The upper layer being coarser than the lower one.

[Printed, 6d. Drawing.]

1854.

A.D. 1854, January 2.—Nº 1.

COLLETTE, Charles Hustings.—(*A communication.*)—" Im-" provements in the manufacture of sugar." These are, treating " saccharine juices and syrups for the purpose of obtaining sugar " therefrom " as follows :—The liquors in the " defecation pan " are treated with lime or lime water, " 30 or 40 per cent. of lime " is sufficient for this purpose," when " the lime has produced " the requisite effect upon the liquid," superphosphate sometimes called biphosphate of lime about 3 parts to 100 parts of the juice are added so long as red "litmus paper dipped into the juice is " turned blue," if an excess of superphosphate happens to be added it is neutralized by lime or lime water. The whole is then filtered, concentrated to 18° Beaumé, superphosphate of lime again added as above, and the whole again filtered, and the liquid concentrated, " the vacuum pan and crystallizing tubs may be " used in the usual way," to obtain " as much sugar as can " be separated." Sugar refined in this way may be dissolved and " again submitted to the process for the purpose of further " purifying it." The residual juice may be treated in a some-what similar manner so as to obtain a further quantity of sugar, and the residual juices from this may be likewise so treated for further quantities of crystallized sugar. In the Provisional Speci-fication it is said that " a superphosphate of an alkali," may be employed.

[Printed, 4d. No Drawings.]

A.D. 1854, January 11.—Nº 72.

TUSSAUD, Felix.—" An universal pump-press with continuous " action, called continuous producer." This consists, in the application " of one or more chains, or endless racks, or a wheel " or wheels with moveable teeth, gearing into the thread or

" threads whatever may be the form or dimensions of these
" threads. This screw being set in motion by any motive power,
" causes this chain, or these chains, or else the wheel or wheels, to
" move in either one or the other direction, *i.e.*, backwards or for-
" wards." This mechanical combination may be used " either as
" a continuous press, or as a force pump, or as a suction pump, or
" else as a double-acting pump. The endless chain or chains, or
" the wheel, or else, if required, wheels, after having traversed the
" cylinder in the longitudinal direction, come out through the
" bottom by an aperture, or apertures, in which they act like
" pistons as it were." " These endless chains may be made up of
" links in the same manner as plate chains, or in any other suit-
" able way. The joint may be in the space between the teeth, or
" in the middle of the tooth, or else there may be one joint for a
" certain number of teeth, provided the joints be so arranged
" that the teeth in the cylinder act like pistons." " The chain or
" chains may be made of metal, or of any other suitable sub-
" stance, such as india-rubber, gutta percha, or any other soft or
" elastic material, according to the uses which this mechanism is
" put to." This apparatus it is said " is applicable to all indus-
" trial operations where pressing, forcing, or pumping, is re-
" quired," a number of which are named, and it is said that
" this pump having no valves is particularly applicable to raising
" and lowering the cane juice and treacle in sugar factories,"
&c.

[Printed, 8*d.* Drawing.]

A.D. 1854, February 7.—N° 302.

TAYLOR, James, BROWN, Isaac, and BROWN, John.—
" Improvements in the charring of vegetable and animal sub-
" stances." These are, the use of a drying chamber, and of
" retorts with or without internal flues, for effecting the con-
" tinuous and economical carbonization of animal or vegetable
" substances;" also " of an air tight receiver or carriage in con-
" junction with the retort," and " the general arrangement and
" construction of drying chambers and carbonizing furnaces," as
follows :—The substance to be dried is passed upon trays of iron
or wire cloth, &c. attached to endless chains through a drying
chamber; it is " then put into a vertical retort enclosed in brick-
" work or a casing of segmental flags made of well burned fire-

" clay and hooped with wrought iron hoops, the heat being
" applied round and through the retort in any convenient
" manner." When the charge is " snfficienly charred it is with-
" drawn at the bottom whilst it is hot and received into an air-
" tight portable vessel to be removed to a suitable place to cool."
" When one charge is withdrawn another is at once put into the
" retort, the heat being thus regularly kept up and the charring
" process going on."

[Printed, 6d. Drawing.]

A.D. 1854, February 20.—N° 398.

ASPINALL, John.—(*Provisional protection only.*)—" An im-
" provement in machinery employed in the manufacture of sugar."
This consists, in machinery for separating molasses from sugar,
in which there are two cylinders, one surrounding the other, and
the sugar is supplied to the space between the two, while a vacuum
or exhaustion is applied, the sugar is found to clog the wire gauze
of the cylinders, and it is proposed to provide " the inside of the
" outer cylinder with a screw thread or threads either solid or in
" the form of a brush, whereby the sugar is not only caused
" to advance, but the wire gauze or screen surrounding the inner
" cylinder is kept clean."

[Printed, 4d. No Drawings.]

A.D. 1854, February 28.—N° 489.

WAY, John Thomas, and PAINE, John Manwaring.—
" An improvement in the manufacture of gas, and also of a
" charred product." This consists in " distilling a compound
" matter consisting of a stone or earth (largely composed of
" soluble silica, found in Surrey, and probably in other places),
" and tar or fat, or oil," but preferring " coal tar as the cheapest
" substance that can be procured; but under some circumstances
" molasses or waste sugar, and other vegetable or animal liquids,
" may be employed," and " thereby obtaining gas, for the
" purposes of light and heat, and a charred product suitable for
" making filters, and for decolorizing and deodorizing purposes."
The above mineral is " easily recognised " by reducing it " to a
" fine powder, boiling the powder in caustic soda, and afterwards
" adding to the solution formed an excess of hydrochloric acid
" which precipitates the silica " if the mineral contained soluble
silica.

[Printed, 4d. No Drawings.]

BOUR, Joseph. — "Improvements in evaporating saccharine.
" liquids." These are, "a combination of hollow revolving ves-
" sels suitably constructed for being filled with and heated by
" steam;" these vessels, "by their revolution and also by means
" of cups at their peripheries, carry up the liquid contained in
" the trough below and in which the vessels are partly immersed.
" The cups discharge the liquid over the heated surfaces of the
" hollow revolving vessels." "Each hollow vessel is composed
" of two portions of a large sphere, which are connected together
" at their peripheries. The several hollow vessels are connected
" together at their centres, and in order to obtain stiffness at the
" centres to act as an axis, two bent or trough-like strips of metal
" are connected together, and they form open troughs in the cen-
" tral part of the combined hollow vessels into which the water
" resulting from the condensation of the steam is raised by means
" of hollow arms with spoon-like ends, and the water flows off
" from the ends of these troughs through a hollow axis at one
" end of the apparatus. The steam to heat the apparatus is
" introduced at one end by a hollow axis and flows off at the
" other end of the apparatus." The apparatus revolves slowly,
and when the liquid in the trough is at the required degree of
concentration, streams of the liquid to be concentrated "are
" allowed to flow from above on to each heated vessel till the
" trough is filled" as desired. "The steam employed is the
" waste steam from a high pressure engine used for crushing
" the sugar cane."

[Printed, 8d. Drawing.]

WRIGHT, James.—(A communication from John Reid.)—"Im-
" provements in machinery or apparatus for 'curing' and 'liquor-
" 'ing' sugar by centrifugal force." These are, first, "the
" combination of the outer jacket and interior drainer in such a
" manner that both may partake of the rotary movement, and
" the one revolve with the other." The outer case or jacket is
attached to the curb or upper flanged part of the drainer, then
carried down about eight inches lower than the bottom of the
drainer, and flush rivetted to a bottom of its own, "thus leaving a
" space between the two bottoms amply sufficient to contain all
" the molasses from a charge."

Second, "the mode of exhausting the air as well as the dis-
" charging the contents of the machine through the bottom."
The vacuum is obtained " by inserting 2 two-inch pipes into the
" bottom of the case through stuffing boxes with trumpet-shaped
" or flanged elbows " moving freely up or down in the stuffing
boxes; when pushed up, their ends will project above the molasses
in the receptacle so that, none can run out, but when pulled
down the molasses will flow into and through them, and so be
discharged from the machine.

[Printed, 8d. Drawing.]

A.D. 1854, March 31.—N° 738.

COSTE, JEAN MARC GUSTAVE.—(*Provisional protection only.*)
—"Revivifying animal charcoal that has already been used, and
" obtaining by a peculiar process prussiate of potasse or soda
" from it." This consists as follows, employing "potass or soda
" about one part; sulphate of iron, about one part; and animal
" charcoal as above 30 parts. These ingredients or substances are
" to be well pulverized and exposed to the action of caloric until
" they are calcined. They are then mixed with about 60 parts of
" water, subjected to ebullition for about half an hour, and then
" carefully filtered and washed; this is twice repeated. After
" the last filtration the animal charcoal is to be dried in a proper
" stove, and the liquid is to be evaporated so as to produce the
" prussiates in the crystallized form."

[Printed, 4d. No Drawings.]

A.D. 1854, April 6.—N° 792.

NASH, JOSEPH.—"The manufacture and refining of sugar."
This consists, first, " in the use of india-rubber rings or bands
" with connected seats and loose covers for the ordinary sugar
" moulds;" also the general arrangements afterwards named,
excepting " the use of the hydro-extractor, but only the improve-
" ments therein as the separation of the bottom and sides of the
" draining sieve and the telescopic slide shaft " of the same, and
" the self-feeding draining cone." The neck of the cone sieve is
immersed in the supply chamber, which is fed with syrup by a
hopper near the bottom, the liquoring pipe comes into the cone
from above. " Using a suitable closed vessel for draining and
" liquoring loose crystalline sugars in bulk." " Arranging the

" charcoal filters in a circle or connected with series of closed
" vessels, for the application of pneumatic and hydraulic pressure
" thereto." Filling and emptying the hydro-extractor, without
stopping its motion, by means of a hollow shaft or tube for filling,
and the sides and bottom of the drum or sieve are made to
separate for emptying. A " self-acting feed and delivery may be
" obtained by using a cylinder partly conical in a horizontal
" direction or by a cylinder revolving at a suitable angle." Employing " carbonic acid gas or other suitable gas as a carrier of
heat to be blown through syrup in a close vessel, its condensation
being effected either by absorption or by the action of an
air-pump forced through a refrigerator, to be returned through
the heating apparatus in a circle of re-activity and re-condensation,
for the purpose of evaporation at a low or moderate temperature
for crystallizing sugar.

Second, the use of silex or silica alone or combined with
ammonia, or mixed with " rough sulphate, phosphate, carbonate,
" or other suitable salt of ammonia having the property of precipi-
" tating lime, or the same salts containing albumen in solution,"
made by dissolving these salts with blood previous to applying
it in variable proportions; the quantity required of blood thus
prepared " being much less than usual in its raw state, while the
" lime usually retained in the clarified sugar will be precipitated
" with scum waste from which the sugar is more readily and
" perfectly separated by the operation of draining and pressing."

[Printed, 10d. Drawings.]

A.D. 1854, May 1.—Nº 973.

ARCHBALD WILLIAM AUGUSTUS. — "Improvements in the
" manufacture of concrete cane juice and sugar." These are
first, "in reducing boiled cane juice or other saccharine matter
" into concrete," avoiding "the excessive heat of the tayche" by
completing "the boiling of the juice in a separate pan in which
" the temperature may be kept more under control," in preference, " a pan called by the French bascule," is used and after
evaporation; it is discharged into shallow coolers six inches deep.
In some cases employing "a composition of loaf sugar, of alum,
" and of Howard's finings."

Second, placing " over the vessel or tray before the cooling
" material is placed upon it a sheet or cloth, or even leaves to
" facilitate its withdrawal from the vessel in which it is hardened.

Third, regulating the temperature in concentrating it.

Fourth, drying the concrete in the sun or in a warm atmosphere. .

Fifth, employing concrete prepared as above " in the manufac-
" ture of sugar, or other manufacture in which saccharine mate-
" rials are employed." The concrete or concentrated syrup, in
preference from West India molasses, is clarified, evaporated, in
preference, in a vacuum pan, described in No. 13,286, Old Law,
just liquid enough to run out.

Sixth, making sugar from molasses as above.

Seventh, making sugar by introducing sugar into the centrifugal machine, in a bag, to separate the molasses or syrup from
the crystals.

Eighth, when the sugar has run from the pan under the fifth
head, it is mixed with " a quantity of water sufficient to reduce "
it " to a density convenient to put into the moulds for making
" refined sugar."

[Printed, 4d. No Drawings.]

A.D. 1854, May 23.—N° 1150.

REYBURN, ROBERT. — (*Provisional protection only.*) — "Im-
" provements in refining sugar." These are, arranging the apparatus used in filtering syrups of sugar through animal charcoal, so
that " the portion of charcoal which is most exhausted, that is to
" say, that which is nearest to the point where the syrup enters,
" may be from time to time removed and that fresh charcoal may
" be added at the end where the syrup escapes, so as supply the
" place of that which has been withdrawn, by which means the
" syrups are at all times obtained as free from colour as at
" the commencement of the operation."

[Printed, 4d. No Drawings.]

A.D. 1854, June 17.—N° 1320.

ASPINALL, JOHN.— (*Provisional protection only.*) —" An im-
" proved means of creating a vacuum, or partial vacuum for
" evaporating purposes." This consists " in creating a vacuum or
" partial vacuum in sugar or other like pans by means of steam
" introduced through a blast pipe."

[Printed, 4d. No Drawings.]

A.D. 1854, June 24.—N° 1399.

THOMSON, John.—"Improvements in centrifugal apparatus
" used in the manufacture of sugar." These are, "introducing
" fibrous or other absorbent matters near the centre of such
" apparatus, in such manner that the absorbent matter having
" taken up a quantity of cleansing fluid, that fluid shall, by the
" rotation of the centrifugal apparatus, be gradually separated
" from the absorbent material and applied to cleanse the sugar in
" the centrifugal apparatus." The " vessel containing the porous
" matter is, at its outer periphery or surface, to be made with
" wire cloth or wire work, or it may be made with perforated
" metal."

[Printed, 4d. No Drawings.]

A.D. 1854, June 30.—N° 1433.

SHEARS, Daniel Towers.—(A communication.)—"Improve-
" ments in curing or separating moisture from sugar and other
" substances." These are, constructing centrifugal machines in
such manner that they may open at the circumference and throw
the sugar matters therefrom, delivering them without stopping.
In one arrangement, "where the revolving vessel is in the form of
" a frustum of a cone, open at the top and bottom. The cone
" can be raised on the shaft from the bottom, which is fixed on the
" shaft, so as to leave an open space all round for the escape of
" the sugar" as soon as the vessel is raised by a lever or by any
other suitable means, "and the sugar will escape on the raising."
" But machinery can be made to open at the circumference in a
" variety of ways, differing only in mechanical details."

[Printed, 8d. Drawing.]

A.D. 1854, August 7.—N° 1729.

DUQUESNE, Emmanuel François.—(Provisional protection
only.)—"An improved mode of manufacturing gas for illumina-
" tion." These are substituting "bones and other animal refuse
" of a like nature for coal, resin, and such like carbonaceous
" matters heretofore employed for the manufacture of gas," using
the ordinary apparatus as for coal or resin, " some modifications
" being, however, made in the purifying apparatus." " The re-
" siduum contained in the retort after distillation is animal black

" of superior quality, which is of great utility, and may be
" employed in the manufacture of blacking, the clarification of
" sugar, or as a manure, &c."

[Printed, 4d. No Drawings.]

A.D. 1854, August 31.—N° 1904.

HEATHER, John.—(*Provisional protection not allowed.*)—" Con-
" sisting of sugar nippers combined with sugar tongs, to be used
" for the purpose of cutting or breaking lumps of loaf and crys-
" tallized sugar, and distributing the same at the tea and break-
" fast table, to be called ' Blackwell's' combined sugar nippers and
" ' tongs.' " These consist of sugar tongs or forceps resembling
the sugar tongs now in use, " having blades or wedges formed of
" steel or other hard mineral or metallic substance fixed on the
" inner sides near the tops thereof." The sides of the tongs are
made to expand by means of a joint at the top, so as to receive
lumps of loaf or crystallized sugar between the said blades or
wedges, which may be easily broken by pressing together with the
hand the points or ends of the tongs. " The points or ends of
" the tongs or forceps are kept sufficiently distended for picking
" up the lumps of sugar as with the ordinary sugar tongs, by
" means of a spring placed at or in the aforesaid joint," and
" which yields to the pressure of the hand, so as to take hold of
" and distribute the sugar in the usual manner."

[Printed, 6d. Drawing.]

A.D. 1854, September 2.—N° 1921.

DECOSTER, Pierre André. — " Certain improvements in
" extracting the saccharine parts of the sugar reeds and of other
" sacchariferous substances." These are, cutting the sugar reeds
or cases into " small pieces by means of a rapidly revolving disc
" which carries a series of cutters or knives," introducing these
pieces into " a hopper inclined at an angle of forty-five degrees
" more or less. The sliced or chopped cane falls upon an endless
" web of wire cloth under which is a vessel to receive any juice
" which may exude and drop through the wire cloth. The
" endless web conveys the sliced cane to the sugar mill which is
" constructed with three horizontal rolls and furnished with an
" intermediate guide plate attached to a lever with a weight.
" This guide plate guides the fragments of cane after they have

" been pressed or squeezed between the first and second rolls to
" receive a second pressure between the second and third rolls.
" The fragments of cane then fall into a truck or waggon the
" bottom of which is of wire cloth. Several of these waggons
" are employed and each one when filled is conveyed upon a
" railway to a position directly over a pan in which the cane
" juice is being boiled or concentrated. A cover either closed or
" partially opened at top is placed over the waggon to confine
" the steam, and the fragments of cane thus become saturated
" with the steam." "The fragments of cane are then passed
" through a second mill, the rolls of which may be upon the same
" axes as those of the first mill." "The saccharine liquors are
" concentrated and crystallized or granulated and the sugar thus
" obtained is purified from molasses or uncrystallizable sugar by
" centrifugal apparatus, consisting of a perforated cylindrical
" drum, lined with wire gauze, and attached to a horizontal plate
" or disc which is caused to rotate with great rapidity by means
" of straps and pulleys. Two long conical pulleys placed in
" reverse positions are employed with an endless strap or belt
" capable of being shifted from one end of the pulleys to the other."
" The horizontal disc of the centrifugal machine is driven by a
" strap from the conical pulley passing round a pulley placed
" under the disc, the spindle of which turns in a long socket well
" supplied with oil." "The perforated drum is not permanently
" fixed to the disc but connected so that it may be readily
" removed when the operation is completed and another charge
" drum substituted." "It is advisable to employ self-oiling
" apparatus with the bearings of the spindles which revolve at
" high velocities or are exposed to great pressure." The centrifugal machine may be of various forms, the drum may be in
" the form of a sugar loaf or of any other required form." The
drum is not permanently fixed and "may be lifted off and
" replaced at pleasure."

[Printed, 10d. Drawings.]

A.D. 1854, October 24.—N° 2267.

WELSH, JOHN.—" Improvements in extracting liquids from
" saccharine and other matters." These relate " more particu-
" larly to the extraction of the syrup or treacle and moisture from
" sugar in the process of refining," and consist " in subjecting

" the loaves or cones of sugar which are to be strained to the
" action of an air pump or exhauster." By one modification, the
apparatus is an air-tight chamber or vessel put in communication
with the suction pipe of an air pump, and also with the bottom
of the sugar moulds, which are provided with a pipe and a stop-
cock to regulate the sucking or exhausting action. A mouth-
piece of caoutchouc, gutta percha, or some other flexible material
is fixed on the end of the pipe, the atmospheric pressure causes
the mouth-piece to clasp the pipe, and so form an air-tight or
nearly air-tight joint. " The communication of the air-tight
" receptacle with the air pump is to be contrived so that the
" syrup may not obtain access to the pump." " Two or more
" pipes may be fitted to the air-tight receptacle in order that two
" or more moulds may be operated upon at once."

[Printed, 10d. Drawing.]

A.D. 1854, October 28.—Nᵒ 2301.

BROOMAN, RICHARD ARCHIBALD. — (*A communication.*) —
" Improvements in centrifugal machines and in driving the
" same." These are, first, the adaptation " to the drums of centri-
" fugal machines of rollers or cylinders formed plain, grooved,
" perforated, or with projections, to produce a stamping effect or
" otherwise, which are made to act on the goods under operation so
" that they shall not only have the centrifugal power exerted upon
" them but also that due to the action of the rollers."
Second, " the adaptation to the drums of centrifugal machines
" of apparatuses for moving and displacing the goods while under
" operation." The centrifugal drum is composed of a bottom, a
solid periphery fixed to the bottom, and a moveable and perforated
periphery placed concentrically with the periphery. The perforated periphery is divided into as many parts as there are
compartments between the division plates, and these parts are
fitted into the opening of the compartments so as to be capable
of moving in their like pistons towards the centre of the drum.
" By means of these parts or pistons the matter under operation
" is displaced and pushed towards the centre of the drum and
" the pistons are pressed outwards to their former position by
" means of springs. These machines it is said may also be used
" for forming or moulding the solid parts remaining in the drum
" such as sugar, and in this case the above division plates

" forming the sides of the compartments are to be parallel
" throughout.

Third, in balancing the drums of centrifugal machines "the
" drum is hung or centred loosely on the shaft and the goods are
" loaded equally so as to make it balance; or the drum may
" be made to balance by liquid flowing in a pipe or channel
" round the drum and displaced by air being forced in which will
" drive the liquid to the lighter part or side of the drum and
" thus balance the goods."

Fourth, the combination with centrifugal machines of a crane
" for the purpose of loading and unloading the drum thereof."
The crane is attached by a hook to the top of the drum of the
centrifugal machine by which it is lifted for the purpose of loading
or unloading.

Fifth, in driving centrifugal machines by "imparting the power
" from a main pulley to friction bands or straps which are passed
" round a pulley and the shaft of a centrifugal machine. One
" main pulley may thus be made to drive several machines by
" frictional contact between its periphery and a driving band
" passed over a pulley and the shaft of a centrifugal machine, a
" lever is fitted to each small pulley whereby the band may be
" brought into contact with the periphery of the main pulley or
" removed therefrom."

[Printed, 1s. Drawings.]

A.D. 1854, October 30.—N° 2307. (* *)

WRAY, LEONARD.—(*Provisional protection not allowed.*)—"A
" new manufacture of sugar and other products ordinarily obtained
" from saccharine matters." This consists in "obtaining from
" the plant botanically known under the name or title of holcus
" saccharatus or holcus saccharatum, and on the coast of Africa
" by the name of 'imphee,' a saccharine juice, and in manu-
" facturing therefrom syrup, molasses, treacle, and crystallized
" sugar;" also "vinegar, alcohol, &c."

[Printed, 4d. No Drawings.]

A.D. 1854, November 10.—N° 2387.

LOYSEL, EDWARD.—" Improvements in obtaining infusions or
" extracts from various substances." These are, in reference to

this subject, extracting "the saccharine matter or juices from " beetroot," by " causing water or other liquid to ascend by " hydrostatic pressure through the mass of material to be operated " upon, and, after properly macerating the same, to carry off the " useful extractive matter from the upper part of the macerating " vessel " as follows :—An elevated vessel containing water or other liquid is heated by a fire below it, a pipe from near the bottom of this vessel descends below another vessel in which the material to be operated upon is placed upon a series of perforated plates, so arranged as to divide the layers one from the other. On the descending pipe is a stop-cock to regulate the supply of the liquid. On the top of the pulverized mass is placed another per- forated plate "which will prevent the upper layer of the mass " from mixing with the liquid extract, which will pass through " the plate and rise to the upper part of the (lower) vessel," and from which it may be drawn off when required through a stop- cock placed above the perforated plate last mentioned. When the cock on the descending pipe is opened the heated liquid from the elevated vessel will descend and pass up through the layers of materials in the lower vessel.

[Printed, 8d. Drawing.]

A.D. 1854, December 7.—N° 2570.

FAIRRIE, John.—"Improvements in preparing solutions of " sugar for filtration." These are, dissolving sugar in suitable " proportions of water at a low temperature and heating the " solution to the point suitable for filtration by the application of " a surface or surfaces heated by steam or otherwise, and this " under the ordinary pressure of the atmosphere or in a vacuum " more or less approaching to perfectness." The sugar is dissolved by stirring it in the requisite quantity of water by means of a stirrer, in preference, having four blades in the form of a screw, the solution brought to a proper density is allowed to rest until " the undissolved particles of sugar subside," the clearer part of the solution is then run off, heated by passing it over a heated surface (tubes heated by steam preferred) enclosed in an air-tight - vessel exhausted by an air pump or otherwise when at the proper temperature it is " passed through filters in the ordinary way."

[Printed, 4d. No Drawings.]

S. N

CASTELOT, Eloi Paulin. — (*Provisional protection only.*)—
" Improvements in decolorizing the juices of beet root, sugar cane,
" and raw sugar, and reducing or neutralizing the excess of
" lime contained therein." These are, proposing to use "animal
" black, or chorcoal in powder, blood, milk, &c., in direct com-
" bination with the juices of beet root, or sugar cane, or by
" preparatory ebullition with raw sugar and water, and thus
" defecate such juices, rendering their crystals brighter and
" larger, and the saccharine produce or refining more considerable
" in quantity and finer and whiter in appearance."

[Printed, 4d. No Drawings.]

A.D. 1854, December 28.—N° 2742.

BENSEN, Gerd Jacob.—"An improvement in refining sugar."
This consists in "arranging or combining apparatus for facili-
" tating the dissolving of sugar in water by the application of
" streams of air." Numerous streams of air are "introduced
" below the sugar and water when in a suitable open vessel or
" pan, and heated by pipes having steam or hot fluid within
" them. For this purpose, it is preferred to employ a series of
" perforated pipes near the bottom of the pan or vessel and above
" them to have a series of heated pipes capable of being raised
" out of the fluid; but the arrangement of the air pipes and the
" means of applying heat may be varied."

[Printed, 6d. Drawing.]

A.D. 1854, December 29.—N° 2752.

PILLANS, James.—"Improvements in the preparation of hema-
" tosin, and fibrinous and serous matters." These are,—the blood
of animals is collected in vessels, allowed to coagulate, transferred
into a vessel with a strainer at bottom, and the coagulated clot is
cut and manipulated so as to separate as much as possible the
serum clear and colorless. The clot is finally " put between
" rollers or into a press by which nearly all the moisture left in it
" may be squeezed out. This liquid part as well as the solid part,
" which is mostly febrine, remaining in the press or resulting from
" the rollers is dried on separate trays or shelves at about 110° to

" 115° F., along with any colored serum resulting from the
" former operations." When these two portions of blood, which
I distinguish by the name of hematosin, are dry, they may for
convenience sake be ground to powder, and they are suited in a
greater or less degree, particularly with the exception of the
fibrinous part, for certain manufacturing processes, for instance, in
dyeing and sugar refining." The clear serum evaporated as
above is " in a fit state to be employed under the name of
" albumen."

[Printed, 4d. No Drawings.]

1855.

A.D. 1855, January 11.—N° 74.

OXLAND, ROBERT.—"Improvements in the manufacture and
" revivification of animal charcoal." These are, first, employing
" carbonic acid or the products of combustion of fuel, at
" gradually increasing temperatures, for the destructive distilla-
" tion of bones and manufacture of animal charcoal." In preference, an upright cylindrical retort is employed, about three feet
in diameter and ten feet high, set in brickwork with flues all round,
and the lower part heated to a low red heat, within this upright
retort a cylinder rises from the bottom nearly to the top of the
outer retort. The broken bones are placed between the two, the
retort is covered in at the top. At the bottom of the retort is a
tube or pipe, by which the liquid products may flow away, and
near the bottom of the retort is a pipe to convey away the volatile
products from the retort. The interior cylinder is charged with
anthracite or charcoal or coke, and the same is ignited, and streams
of air are driven in at the bottom, the products of the combustion
of this fuel pass out of the open top of the cylinder, and pass
down amongst the pieces of bone. The animal charcoal is withdrawn through openings near the bottom of the retort, which are
closed during the charring process.

Second, " the employment of carbonic acid and water to sepa-
" rate lime in the revivification of animal charcoal." The charcoal is first well washed and covered with water in a close vessel,

and carbonic acid is forced in till the pressure is about fifteen
pounds to the square inch, after about six hours the water is
drawn off and the charcoal washed until on testing with oxalate
of ammonia no lime or little lime is in the washings.

Third, "drying or re-heating animal charcoal in retorts or
" apparatus subject to streams of heated carbonic acid or products
" of combustion of fuel."

[Printed, 4d. No Drawings.]

A.D. 1855, January 30.—N° 228.

BROOMAN, RICHARD ARCHIBALD. — (A communication.) —
(Provisional protection only.) — "An improved filter." This
" consists of a layer or layers of cotton fibre or cotton waste,
" flax fibre, flax, cotton, wool, or other like vegetable and animal
" fibres, placed and held between two perforated plates, frames,
" or sets of laths, or other suitable contrivance, for confining the
" fibrous material, and keeping it in a proper position to act as a
" filter." The sugar in the form of syrup is poured on to the
top of the filter previously moistened, is caught in a receiver
below. " On leaving the filter, the syrup is in a sufficiently clear
" state to undergo the remaining processes followed in the refining
" or manufacture of sugar." The filters becoming charged with
matters from the sugar, " they may be cleansed or washed by steam
" or water in any convenient manner."

[Printed, 4d. No Drawings.]

A.D. 1855, February 12.—N° 326.

KERR, ROBERT. — " Certain improvements in preparing loaf
" sugar for use and certain apparatus for the same." These are,
in place of breaking up the loaves of sugar "by a chopping knife
" into unsystematic, irregular, and unequal morsels of suitable
" size for use," "dividing it into systematic, regular, and equal
" morsels by means of saws and stamps and otherwise." "The
" sugar may be subjected three times to one series of saws, or
" several series may be employed working in the several directions,
" or a series may be fixed in a frame and used by hand."
Circular saws may be employed accurately fixed, and a good deal
set, or a series of small circular saws, about 6 or 8 inches in
diameter, on one spindle fixed at the required intervals "projecting above a table, to cut one morsel deep, cut the loaf once

" with these, and a second time (or this with a second similar
" series) crosswise, then subject it to a single larger saw to cut off the
" slice of morsels," the saws being worked by ordinary machinery.
Or the sugar is cast or sawn into slabs one morsel thick and these
are subjected to pressure between reticulated edge metal stamps
or gratings, or the slabs are divided by such means into sticks
and then again into morsels; or cast or sawn sticks are divided by
such means at once or by the ordinary fixed chopping knife with
a stop fixed beyond to regulate the size cut off. The lever may
for the above purposes be cast rectangular. The dust produced is
of a novel description. Steam is applied for a moment to remove
any dust.

[Printed, 4d. No Drawings.]

A.D. 1855, February 15.—Nº 346.

DELABARRE, CHRISTOPHE FRANÇOIS.—" Improved apparatus
" to be used in propelling gases and forcing liquids." These
are, " the use and employment of a mixture of steam and air or
" other gases," " projecting a relatively small and more or less
" compressed current of steam or other attracting fluid into a
" large single or multiple recipient pipe or ·channel so as to give
" access to and cause the attraction of a considerable mass of air
" to be carried along with the steam, &c." " It has been usual
" till now to give a bell mouth or funnel shape to the upper part
" of the pipes receiving the jet." " I replace this pipe by several
" successive pipes, each of which has a diameter greater than the
" one preceding it." " If it is desired to propel liquids a current
" of mixed gas is made to act upon them as before stated."
" Although any gaseous fluid and any liquid whatever may be
" employed for producing attraction, air and steam are fittest for
" that purpose." This principle may be applied to a number of
purposes which are named, and in reference to this subject, " large
" and high chimneys may be dispensed with, and the chimney
" may be made to consist of a pipe which is placed horizontally
" so as to render the caloric given off by the smoke available "
for the purpose of " drying articles such as sugar, cane trash,"
&c. It " may also be applied to getting lime and pure carbonic
" acid " for sugar refining. The carbonate of lime is distilled in
" a retort consisting of several discs of metal or fireproof sub-
" stances, superposed and luted together," and the gas collected

" into a gas meter." It may also be applied for raising sugar juices and extracting the vapours from close vessels like retorts containing such juices by a jet of steam into their necks.

[Printed, 1s. 2d. Drawing.]

A.D. 1855, March 14.—N° 565.

RILEY, GEORGE.—(*Provisional protection only.*)—" An improved " process for the manufacture of starch or grape sugar." This consists in boiling the meal of any of the cereals " in water acidu- " lated with sulphuric acid, under a pressure greater than that of " the atmosphere."

[Printed, 4d. No Drawings.]

A.D. 1855, May 11—N° 1065.

STEELE, JAMES.—" Improvements in effecting the drainage of " moulded sugar." These are, first, " the application and use in " moulds used in the draining or refining of sugar of a plug or " stopper, having a spike or projection, which enters the mould " and produces a cavity in the sugar, for the purpose of aiding " the drainage thereof."

Second, "the system or mode of forming the cavity which is " required in the apex of a mould of sugar for aiding the drainage " action, by means of a spike or spiked plug, which is inserted in " the mould prior to the deposition of the sugar therein."

Third, "the system or mode of producing the drainage " cavities in moulds of sugar by moulding the same therein."

In the old process it is said the attendant drives a pin into the apex of the mould, and the above is to supersede the old process. " The pins have broad heads or flange pieces, carrying cloth or " elastic washers to fit up against the mould aperture, hence all " wasteful escape is prevented." When "the drainage is to " commence, the operator withdraws the pin, leaving the drainage " aperture moulded, as it were, in the sugar."

[Printed, 6d. Drawing.]

A.D. 1855, May 26.—N° 1196.

ASPINALL, JOHN.—" Improvements in machinery for extract- " ing moisture from substances, and for separating liquid from " solid bodies, applicable to the refining of sugar, drying of " goods, and to purposes for which centrifugal machines are

" employed." These are, " placing a perforated or wire gauze
" cylinder or drum, open at both ends, horizontally or in an
" inclined position, in causing the same to rotate by frictional
" contact with driving rollers, or by strap pullies, to which motion
" is communicated from a steam engine or other prime mover,
" and in introducing through the centre of this cylinder a perfo-
" rated pipe, which is stationary," and to which, for some purposes,
is fitted a series of inclined directing plates. Or, instead of fixing
the inclined directing plates upon the perforated pipe, " introducing
" a fixed shaft or spindle through the centre of the drum or
" cylinder, and fixing the inclined directing plates thereon."
There is a tube or pipe passing through the cylinder which
supports " a series of perforated nozzles, through which fine
" liquor or water, or steam is directed, if desired, upon the
" materials under treatment in the cylinder." The " mode of
" operating with this machine is as follows :—Supposing sugar,
" in its semi-fluid state, be required to have the liquid separated
" from the granular matter or particles, and at the same time to
" be decoloured or whitened," it is fed into one end of the drum
by a hopper, fine liquor or steam is then introduced through the
pipe, the cylinder rapidly revolved, the sugar gets massed at the
bottom of each directing plate, after which it will be transferred
from plate to plate, until it reaches the opposite end of the
cylinder, from which it will be ejected into a receiver in a dry
granular state. The liquor thrown off from the sugar is received
by the outer cylinder. The process is completed without stopping
the machinery.

[Printed, 6d. Drawing.]

A.D. 1855, June 8.—N° 1313.

CHANTRELL, GEORGE FREDERICK.—" Improvements in appa-
" ratus applicable to the manufacture and revivification of
" animal or vegetable charcoal." These it is said are chiefly
applicable to apparatus described in No. 2388, A.D. 1853, and
consist, first, in " raising the furnaces nine inches or more above
" the bottom of the char or retort chambers," thereby " reducing
" the intensity of the heat at the bottom of the chambers."

Second, " forming the openings in the bed plates at the bottom
" of the char chamber of a hopper shape."

Third, " connecting and jointing the pipes to the under side of
" the bed plates and to the coolers," by " deep socket joints,

" recesses, and projections being formed on the under sides of the
" plates to receive the heads of the pipes, which are made fast
" thereto by screw bolts, which pass upwards through flanges
" cast on the upper ends of the pipes, and I form them with such
" curves or bends as will cause the charge to fall into the centre
" of the coolers."

Fourth, using "three or more coolers to each char chamber,"
" by applying three pipes to each chamber to connect the bottoms
" thereof with the three coolers, one leading to the centre cooler
" and one to each of the side ones," when two coolers only are
used two pipes are employed.

Fifth, " creating of draught or currents of air to act against the
" coolers to facilitate the cooling operation," by "means of a
" fan blast, or draught flue, or flues as may be found most
" convenient."

[Printed, 8d. Drawing.]

A.D. 1855, June 13.—N° 1356.

LODGE, Edwin, and MARSHALL, George. — (*Provisional
protection only.*)—" Certain improvements in the production of
" animal and vegetable naptha, ammonia, and charcoal, and also
" for the evolution of the carburetted and oleficient gases there-
" from." These are, in reference to this subject, " the refuse waste
" of wool and cotton otherwise almost useless " is placed " in a
" retort made of iron or of fire clay, and set over a fire in the
" ordinary way," and heat applied " regularly and moderately "
will " evolve tar and ammoniacal water therefrom, the tar being of
" an oily consistence." The charcoal left in the retort is " of a
" very superior quality," and " is therefore applicable to the
" manufacture of superior substances or compounds in which
" charcoal is required to be employed."

[Printed, 4d. No Drawings.]

A.D. 1855, June 28.—N° 1473.

MOREAU-DARLUC, Charles. — " An improved mode of
" separating substances of different nature or composition by
" means of displacement and substitution." This consists, in
reference to this subject, as follows :—The muscovadoes or impure
crystals of sugar are put on the false bottom in the interior of a
vessel, the lid is put on and hermetically closed and secured, a pipe
under the false bottom is opened, and a pipe into the vessel above

the sugar delivers heated atmospheric air, "which forces the
" syrup to percolate through the crystals, and to be drawn off by
" the lower pipe, after which pure water is introduced above the
crystals, and this is forced likewise to percolate through them,
and finally, heated dry air is forced through until the crystals
become entirely dried. This process it is said is "applicable to
" the treatment of all such substances containing certain parts
" which may be made soluble by any suitable menstruum, and
" are wished to be extracted from them, such, for instance, as the
" refining or purification of sugar, or the extracting of saccharine
" matters from substances containing the same," and a number
of other purposes which are named. There is also claimed the
application of electricity from a battery or apparatus "to the
" dilating of air, gas, or gases for heating the air, gas, or liquids
forced through the material on the false bottom, &c.

[Printed, 4d. No Drawings.]

A.D. 1855, July 18.—N° 1616.

ELLIS, JOHN.—" Certain improvements in the process of manu-
" facturing ammonia, charcoal, animal and vegetable naphtha."
These are, in reference to this subject, using three retorts set in
brickwork, a separate pipe leading from each to a condenser immersed in a cistern of cold water, a pipe having a valve for the
escape of gas extends from the condenser to a vat. The retorts
" are all charged with one or other or a mixture of any two or
" more of the following ingredients, matters, or substances,
" namely :—

All kinds of animal oil mixed with animal refuse				
do.	do.	fats	do.	do.
do.	do.	oils from fish	do.	

" all kinds of cotton and wool refuse, waste cloth, &c., containing
" oil; all kinds of refuse from tallow chandlers and butchers ; all
" kinds of horns, hoofs, bones, and such like substances; all
" kinds of materials which contain or are impregnated with any
" greasy or oleaginous matters or substances." The residue left
in the retorts is animal charcoal.

[Printed, 10d. Drawing.]

A.D. 1855, July 26.—N° 1693.

SCHIELE, CHRISTIAN.—" Certain improvements in obtaining
" and applying motive power." These are, in reference to this

subject, the use of "water as a means of drawing and evacuating "air from the condenser of a steam engine." There is a combined pair of rotatory steam engines for actuating the screw propeller of a steam vessel, and consists of two wheels or runners mounted on a horizontal shaft passing through a steam and air tight circular casing surrounding each engine and fitted with tangential openings formed rectangular and made of thin plates fastened into an adjustable ring to regulate the quantity of steam let into the wheels, the steam acting on vanes is indicated by a dial outside the case. These wheels have fitted on their external surfaces thin plates of metal in the form of two circular arcs meeting in an apex in the middle. In working the engines steam enters the annular chamber of the propelling engine from a regulating valve and passes through the openings into ducts formed on the external periphery of the wheel where the force of the steam is expended. The wheels and propeller are thus put into motion at the same time water is supplied to the condenser and the heated water and air ejected by a rotatory air pump. In reversing the propeller the regulating valve is turned half way round, allowing the steam to act on the other ducts in a contrary direction, and when the engine is to be stopped it is turned half way between the two. In order to transmit rapid motion to the main shaft of engines, wheels are placed in contact therewith so that the pressure required to produce sufficient adhesion is balanced by opposite pressure. The improvement relating to a centrifugal or rotatory, air-pump whereby water is used as a means of drawing and ejecting air from the condenser of a steam engine, a fan is employed, for receiving the water and air ejected in which the water passes through the central tube and out of the fan through a number of small jet pipes arranged in a circle. The air in like manner passes to the fan through the intervals between the jet pipes from the, reservoir which surrounds the water tube, the water and air are therefore momentarily mixed in their passage through the fan. This centrifugal or rotary pump is also applied in certain processes where a more or less perfect vacuum is required, as for instance in the manufacture of sugar, gas, or sulphuric acid, the apparatus being modified accordingly so as to admit of other fluids than water. The same apparatus may be applied for the purpose of producing high pressure blast by supplying air from the atmosphere and conveying it from the separating vessel in pipes in the usual manner.

[Printed, 1s. 2d. Drawings.]

SAVAGE, Alfred.—" Improvements in the means or mechanism
" for treating tea, sugar, coffee, chicory, and such substances as
" require the processes of separation, reduction of size, and
" mixing, or any one or two thereof." There are, in reference to
this subject, " the combination of a lever of variable radius, with
" a knife or knives affixed thereon, and any means of moving the
" said lever by the action of the user's foot or feet for the purpose
" of chopping loaf sugar." In carrying out this, a tray is fixed,
" by bolts or other usual means which afford facilities for its
" removal when required, on the top surface of the case of the
" machine, termed Savage's noiseless machine," for " sifting tea
" and cutting the large leaves at the same time, with apparatus
" for mixing such previously separated portions;" and on the
said tray erecting " the supports of the lever of various radius,
" which lever is counter-balanced by a spring, so that it may
" oscillate freely, as is usual, when a lever having a knife affixed
" to it is employed for chopping loaf sugar;" but this lever is
made " in two parts, one part being connected with the other
" by being received into a socket formed thereon, or otherwise,
" so that the acting length of the said lever may be increased to
" facilitate the chopping of large masses;" and the said lever is
connected " with a treddle, to attach a cord and stirrup thereto,
" whereby the foot of the user may communicate the desired
" motion to it."

[Printed, 4d. No Drawings.]

A.D. 1855, October 13.—N° 2299.

STENHOUSE, John.—" Improvements in the preparation of
" uncolourising materials." These are, employing " vegetable
" charcoal and solutions of phosphate of lime, solutions of alumina,
" and oxide of iron." " The charcoal is boiled in solutions of
" the substances above mentioned till the air contained in its
" pores is expelled and the charcoal saturated with the solution.
" The charcoal is then dried and heated to redness in close vessels."
The charcoal employed is either in " powder or in a granular
" state, similar to the animal charcoal usually employed by sugar
" refiners."

[Printed, 4d. No Drawings.]

A.D. 1855, November 7.—No. 2514.

SIEMENS, Charles William.—(*Provisional protection only.*)
—"Improvements in evaporating brine and other liquids and in
" distillation." These are, in evaporating saccharine or other
liquids, exposing them " in a covered pan with a double bottom
" to the influence of a circulating current of a permanent gas in
" such manner that the said current passes in a zig-zag or circular
" manner along the surface of the evaporating liquid, commencing
" at the end opposite to the source of heat, and advancing towards
" the heated end ; at certain intervals portions of the current
" descend into the double bottom below the liquid, and return
" towards the cooler end of the pan, passing finally through a
" condenser or chamber containing a series of pipes filled with
" the cold liquid ; the current is urged by means of a propeller
" or fan to pass again over the surface of the evaporating liquid
" as before described. The source of heat consists of one or more
" steam boilers and furnaces. The steam generated in the boilers
" passes first of all either partially or wholly through an engine
" (which I prefer to consist simply of a reaction wheel), in order to
" produce the necessary power for working the fan pumps or other
" machinery. The steam next enters one or more steam cham-
" bers or system of tubes in contact with the evaporating fluid at
" the extremity of the pan opposite the entry of the circular cur-
" rent, where it condenses in boiling. The liquid and the water
" of condensation is forced back into the steam boilers. The fire
" after it leaves the steam boilers passes through a system of
" flues made of metal or pottery that cover the entire length of
" the pan above the evaporating liquid, and finally reaches the
" chimney."

[Printed, 4d. No Drawings.]

A.D. 1855, December 3.—N° 2722.

LEITCH, James. — (*Provisional protection only.*) — " Certain
" improvements in melting and blowing up sugars." These are :
—" The sugar is placed in a receiver, through the bottom of
" which a pipe passes to a steam boiler situated underneath the
" receiver. The pipe descends some distance into the boiler, so
" that when there is a pressure of steam in the boiler the hot
" water is forced up the pipe amongst the sugar, until the water
" in the boiler sinks below the end of the pipe, and then the

" steam in place of water passes up and agitates the mixture
" until the sugar is completely dissolved, when it is drawn off by
" a cock at the bottom of the receiver."

[Printed, 4d. No Drawings.]

A.D. 1855, December 7.—N° 2767.

LEITCH, JAMES. — (*Provisional protection only.*) — " Improve-
" ments in melting, blowing up, and filtering sugars and other
" saccharine matters." These are, " two vessels connected by a
" tube carried to within a few inches of the bottom of the lower ;
" it is proposed to put the requisite quantity of water (to
" reduce a given quantity of sugar to the required density) into
" the lower vessel, where it is raised to the boiling point by the
" introduction of either open steam through a perforated pipe
" immersed in it, or jacketed steam retained within a coil pipe, or
" any other heating arrangement most convenient." When the
" water boils pressure is produced, which forces a portion of the
" water into the upper, where it is mixed and kept in agitation till all
" is dissolved ; then the heating medium being quickly withdrawn,
" a vacuum is formed in the lower vessel, which draws the liquor
" (through a sieve placed on the top of the connecting tube) from
" the upper into the under vessel, or can be retained in the upper
" vessel by means of a ·stop valve in tube (if more convenient),
" and drawn off therefrom, when it can be drawn off in a partially
" clarified state at pleasure. A vacuum in the lower vessel may
" be formed by any other means besides the withdrawal of heat,
" such as, injection of cold water, or air pump, if necessary."

[Printed, 4d. No Drawings.]

A.D. 1855, December 8.—N° 2774.

RADCLIFFE, JOHN, and FAVELL, THOMAS VICKERS.—
" Improvements in machinery or apparatus for cutting sugar and
" other substances." These are said to be the general constructions and arrangements of the " machinery or apparatus," and
of " compound indented cutters, having their edges at right
" angles to each other," and of " apparatus or mechanism for
" clearing the cutters of knives after each cut, to prevent them
" from clogging," and for " actuating the knives of machinery,"
and for " presenting the slices of sugar and other substances to
" be cut in a suitable and proper position to be acted upon by the

" knives." Also, "the application and use of indented cutters
" or knives " for the above purpose, and " the system or mode of
" cutting loaf sugar or other similar substances by means of two
" horizontal knives in combination with a central transverse
" fixed vertical knife," as follows :—

The knife edges are made so as to form a series of points or
teeth; "the first set of cutters consist simply of a blade with
" an indented cutting edge; and the second set consists of a
" serrated edged blade, bolted or screwed on to a foundation
" plate, on which plate are also fitted another set of teeth or
" cutters, placed at right angles to the serrated blade, and
" arranged to come opposite either the centres of the notches or
" indentations or the centres of the teeth forming the serrated
" edge of such cutter. The machine in which these cutters are
" placed consists of a rectangular metal framing, in the sides of
" which are fitted the bearings for the actuating shafts, one of
" which carries a fly wheel and a winch handle, for driving the
" machine. The top pair of cutters work opposite each other in
" suitable horizontal guides, being actuated by a cam or tappet
" motion from the main shaft. The top pair of cutters serve to
" cut a slice from off the bottom of the loaf of sugar or other
" substance, which is inserted into the machine from above."
The slice having been cut off, " the pressure of the loaf or other
" substance forces it on to a fixed indented cross knife, which
" severs it into two parts," each part falls into a separate hopper
or conductor, which conveys them " to the second set of cutters,
" which work horizontally immediately below the mouths of the
" hoppers. These latter cutters are similar to the second arrange-
" ment of cutters herein-before referred to, and are carried by a
" vibrating frame, which is actuated by levers and cams and
" suitable gearing from the main shaft, immediately in front of
" the knives on the vibrating frame, which is actuated by levers
" and cams and suitable gearing from the main shaft." Immediately in front of the knives on the vibrating frame are
corresponding knives of a similar construction fixed to the main
framing, and between these fixed and moving knives the sugar
or other substance is cut. " Spring guard-plates or shields
" are employed for clearing the knives after each cut, and to
" prevent them from becoming clogged or filled up. Two spring
" plates are fitted inside the hoppers, for the purpose of presenting
" the large pieces of sugar or other substance in a proper manner

" to the cutters, which cut them into small lumps. By a slight
" modification in the machinery the knives herein-before described
" may be worked by a treadle in place of a winch handle."

[Printed, 1s. 4d. Drawings.]

A.D. 1855, December 15.—N° 2833.

ASPINALL, John.—" Improvements in machinery for curing
" sugar and extracting moisture therefrom, parts of which are
" applicable to separating liquids and moisture from substances
" containing the same." These are, " two hollow truncated
" cones (one placed within the other) are mounted on a central
" vertical shaft, caused to revolve by some suitable motive agent.
The cones may be of the same or different lengths;" " a space is
" left between the inner surface of the outer cone and the outer
" surface of the inner cone; this space may be regulated by
" altering the position of one or other of the cones." " A hopper
" is placed above the cones for the introduction of the sugar or
" other substance to be treated. The cones are mounted in a
" casing, which also supports the bearing in which the shaft is
" stepped. The sides of the cones are of perforated metal, covered
" or not with wire gauze, or of wire gauze, or of other pervious
" material." " When the apparatus is used for curing sugar, a
" centrifugal pump is keyed on the shaft, and communicates
" with a cistern fitted round the shaft and supported on the
" casing of the apparatus. This pump, which is worked by the
" rotary movement of the shaft, is for the purpose of supplying
" liquid to ' whiten ' or ' liquor ' the sugar."

[Printed, 8d. Drawing.]

A.D. 1855, December 17.—N° 2850.

GOLDING, George Gotts. — " Improvements in boilers for
" heating, warming, or raising steam." These consist, of a
boiler the external form of a screw, the thread of which is made
very deep, so as to project considerably from the central cylindrical portion. The thread may be " square, angular, or any other
" suitable form, and may be made with one, two, three, or more
" convolutions. The projecting part of the boiler, forming the
" thread of the screw, is made hollow." In the centre of the
boiler is a vertical aperture from top to bottom, and when set in
brickwork the aperture is over the centre of the furnace. The

aperture forms a reservoir for coal, and is large enough to contain enough fuel to supply the furnace for twelve hours or
upwards.   This boiler is made of cast or wrought iron or
copper, and fitted with induction and eduction pipes, and with
the usual gauges and valves.   This boiler may be applied among
other purposes for the supply of steam "to apparatus heated by
" steam, such as vacuum pans used in the manufacture of sugar,
" and other similar purposes."

[Printed, 1s. Drawings.]

A.D. 1855, December 17.—N° 2852.

LEITCH, JAMES. — (*Provisional protection only.*) — "Improve-
" ments in filtering sugars and other saccharine matters." These
are, using "a series of filter bags, one being placed immediately
" over the other, instead of the single filter bag now generally
" employed." The first bag, receiving the sugar or other saccharine matter, is made of a comparatively open material, and those
placed under it of materials gradually increasing in fineness or
closeness down to the lowest. The whole series of bags are
suspended within a vessel in which an equal temperature is maintained.   " A more effectual filtration is obtained, and the liquor
" runs clear from the lowest bag from the beginning of the
" operation."

[Printed, 4d. No Drawings.]

1856.

A.D. 1856, January 26.—N° 212.

GARDNER, EDWARD VINCENT—"Improvements in heating,
" drying, desicating, and evaporating."   These are, in reference
to this subject, in drying a number of substances, among which
are named " sugar, &c." The following apparatus is described :—
A furnace is supported on brickwork or castings; over the
furnace is arranged a casing or chamber formed of metal or other
suitable material, containing a series of plates which serve to
direct the heated products in a zig-zag course upwards until they
enter the chimney at the top; between these plates are fixed a

series of iron chambers, open at each side, which "are heated by
" the heated products from the furnace passing around them."
At the end of each of the chambers are rollers, "and over each
" pair an endless band or sheet of wire gauze or other suitable
" material is caused to travel, the rollers being actuated by suit-
" able machinery." Over the whole apparatus an outer casing is
placed; inside this are sloping plates or shelves to direct the
materials as they pass from one band to the next. Scrapers are
attached to the sides of the chamber. A hopper is at the top.
" For evaporating fluids the bands and rollers are dispensed with,
" and in lieu thereof evaporating pans or dishes containing the
" fluid are introduced into the chambers." When the vapour or
gas is to be collected, an outlet pipe is attached to the outer
chamber, and connected to a receiver.

[Printed, 10d. Drawings.]

A.D. 1856, February 14.—Nº 387.

BLACKWELL, THOMAS EVANS. — "Improvements in con-
" densing steam, and in cooling and heating fluids." These are,
in reference to this subject, in "apparatus to be applied to a sugar
" pan, the steam or vapour issuing from the pan (and such is
" also the case when applied to other processes)" is conducted
by a pipe into a cylinder or condenser in which are a number of
pipes running through it in which cold water circulates. The
pipes have connections at their two ends with an ascending and
also with a descending pipe. The ascending pipe is provided at
its lower end with a flexible disc valve to prevent the reflux of
water, and the descending pipe is shewn with a pump for drawing
the water through the ascending pipe, the tubes in the condenser,
and the descending pipe to the outlet of the pump. In cases
where the descending pipe can be longer than the ascending pipe
the pump may be dispensed with. The condensed vapor is carried by a pipe from the bottom of the condenser. There is a
small air pump attached near the lower end of the descending
pipe for getting rid of any air in the condensing apparatus so that
it may be filled entirely with water. This pump, it is said "can
" also be used for the purpose of exhausting or partially exhaust-
" ing the cylinder or condenser of air or vapours, and of starting
" the flow of vapour from the sugar pan or other evaporating vessel
" into the said cylinder."

[Printed, 10d. Drawing.]

BROOMAN, RICHARD ARCHIBALD.—(*A communication from Pierre Gèdeon Barry.*)—"Improvements in treating bituminous " shale, boghead mineral, and other like schistous bodies, in order " to obtain various commercial products therefrom." These are, producing a number of substances which are named, among which are, "A 'black,' having the same properties as animal black, and " which may be used in the decolorization of liquids, vinegar, raw " sugar, &c., and also in the manufacture of blacking." In reference to this subject, retorts for decomposing the schistous bodies are made use of, to which are attached condensers.

[Printed, 8*d*. Drawing.]

A.D. 1856, March 3.—N° 542.

ASPINALL, JOHN.—"Improvements in machinery for curing " sugar or extracting moisture therefrom, applicable to separating " liquids from solids." These are, in machinery for the above purposes "made self-feeding and also self-delivering," as follows :—"Through or in a line with the centre of a stationary " annular hopper a shaft passes free to revolve on rotary motion " being communicated thereto ; near to the bottom of the hopper, " and fixed by arms to the shaft so as to revolve with it, is a " conical guide or distributor. The base of the cone is spread " out at bottom, and runs parallel with the bottom of the vessel " in which the 'curing' or extraction of moisture is effected, but " the edge of the base of the cone does not meet or touch the " side of the vessel. The vessel in which the curing or extraction " of moisture is effected is fixed to the central shaft, so as to " revolve with it, and is itself conical ; but instead of the sides of " the vessel extending and widening as they rise in an unbroken " line, these sides consist of gradations or steps, each higher step " in succession being of greater diameter than that immediately " below it." At or near the top of each step or gradation is fitted "a horizontal guide or plate, and on the vessel being made " to revolve the sugar (or other substance under treatment) as it " rises is forced between this plate and the horizontal part of the " step, and so on in succession until the sugar or other substance " rises to the top of the vessel, when the topmost guide plate " directs it out of the machine on to a circular platform or other " recipient. The sides of the vessel are of perforated metal or of

" gauze, as is usual in centrifugal machines; the horizontal parts
" of the steps and guide plates are solid. A liquoring apparatus
" may be made to act on the inside of the vessel if found
" necessary."

[Printed, 10d. Drawing.]

A.D. 1856, March 15.—N° 623.

RICHARD, Louis Joseph.—" Improvements in sugar manu-
" facture." These are, applying " the carbonates of soda, potash,
" or any other chemical agent acting in a similar manner as those
" carbonates in respect to saccharine matters, to the defecation
" purification, clarifying, or refining of saccharine juices, syrups,
" raw or other sugars, whether the same have been obtained from
" cane, beet root, or any other saccharine matters," after the same
" has been defecated in the ordinary manner by means of lime,
then stirring and boiling again the said saccharine liquids for
some moments, skimming, filtering, concentrating, and allowing
the same to crystallize in the ordinary manner.

[Printed, 4d. No Drawings.]

A.D. 1856, March 24.—N° 691.

BRYANT, James, the younger.—" Improvements in machinery
" or apparatus for the reburning of animal charcoal." These
are, constructing and setting retorts for the above purpose as
follows :—The retort " is enclosed in brickwork, and heated by a
" suitable furnace beneath, communicating with flues which
" encircle the retort. A reciprocating or reversing rotatory
" motion on its axis is imparted to the retort during the process
" of reburning, whereby the animal charcoal is kept constantly
" stirred or agitated." " For this purpose the retort is suspended
" by an endless chain or chains passing round an overhead
" pulley or pullies, the back end being further supported by a
" trunnion which must be sufficiently strong to bear the weight
" of the retort when charged. The gearing for giving the reci-
" procating rotatory motion to the retort is so arranged by means
" of an ordinary mangle wheel that the motion will be reversed
" after each partial rotation of the retort, and may be easily
" thrown out of gear by an ordinary clutch, when requisite for
" charging the retort, which may be done through an opening
" in the front and provided with a moveable cover. A suitable

" outlet pipe is passed through the centre of the trunnion to
" carry off the foul air and vapour arising from the charcoal."

[Printed, 1s. 6d. Drawings.]

A.D. 1856, April 19.—No 942.

VARILLAT, WILLIAM JEAN JULES.—(*Provisional protection
only.*)—"Improvements in the apparatus for the extraction of
" colourings, tanning, and saccharine matters from vegetable
" matters." These consist of a number of cylinders which are
" made to communicate with each other by means of cocks placed
" at the top and bottom; the lower ones are placed beneath a
" double bottom with trellice work, and the upper ones above
" other trellice work, which rests on the matter to be treated,
" which is placed in each cylinder in such wise as to allow the
" liquids to filter through. The cylinders are placed on a turning
" platform, so that each time one of the cylinders is emptied or
" filled the cylinder may be placed in a right position." Instead
of a turning platform the cylinders may be placed in a straight line
or square. " The object of this apparatus is to obtain a regular
" displacement with progressive heat." Thus " the water, which
" is passing through three other vats has imbibed the extractable
" principles and lost a part of its caloric, filters into the cylinder
" charged with fresh matters." " The filtration is performed by
" the difference of level of the liquid from one cylinder to the
" other."

[Printed, 4d. No Drawings.]

A.D. 1856, May 8.—No 1083.

FINZEL, CONRAD WILLIAM, NEEDHAM, WILLIAM, and
BARTON, JOHN.—"Improvements in apparatus for filtering
" sugar and saccharine juices." These are said to be "the
" adaptation to the filtering of sugar and saccharine juices"
of certain apparatus described in No. 1669, A.D. 1853, No.
2709, A.D. 1855 and No. 1303, A.D. 1854. "The essential
feature of the aforesaid apparatus" is said to be "the combina-
" tion of straining or filtering cloths or other textile fabrics
" with grooved slabs or planks in such manner that while the
" slabs or planks receive and resist the pressure to which the
" fluids or other semi-fluids are subjected in order to strain or
" filter them, the grooves form passages or channels by which
" the expressed and filtered fluids pass away to suitable receivers

" or otherwise." The cloths are each turned over and folded down around their edges, so that they form for the time bags or chambers into which the fluid or semifluid enters by a pipe. Two grooved slabs or planks and one cloth form one filtering chamber. In adapting this apparatus to the filtration of sugar and saccharine juices, "employing a series of chambers formed as above- " described," and in some cases placing " within each chamber a " loose grooved slab or plank with filtering medium on each side " of the slab, the object of which is to assist in compressing the " substance remaining in the cloth or cloths. This affect is " obtained by shutting off the connection between the supply " pipe and the chamber," and by allowing " the pressure to be " continued in the chamber above or below it or in both, the " affect being to force the juice out of the said chamber and to " press the loose slab or plank against the substance or substances " remaining in the cloth or other fabric on the other side of the " floating slab, whereby the expression of the juice from it to " the greatest possible extent will be obtained." To cleanse the cloths should they become choked during the filtration steam or hot water is passed through them " by means of suitable pipes " before the chamber is opened and the remaining substance or " substances are removed from it."

[Printed, 4d. No Drawings.]

A.D. 1856, May 12.—N° 1119.

NEWTON, WILLIAM EDWARD.—(A communication from George Denison.)—" Certain improvements in machinery for pumping and " forcing water and other fluids." These are, first, employing for the above purposes "an elastic tube, so acted upon externally " by the pressure of rollers, or their equivalents, as to be alter- " nately collapsed or closed and allowed to recover itself, and thus " alternately to form a vacuum in the tube, into which the water " or fluid is drawn, and expel the water or fluid from the tube." A single coil of flexible tube (strong vulcanized india-rubber hose of a cylindrical form) "is arranged within a stationary circular " frame," " and having arranged within it a central shaft " carrying a roller, which, revolved by the shaft, presses the tube against the interior of the frame so as to produce the above effect.

Second, " equalizing the resistance of the tube, by a truncated " cone or a gradually diminishing thickness at that part where

" the roller or rollers, or their equivalent, leave the tube, in com-
" bination with a lift below or at the back."

Third, " relieving the tube from the pressure of the roller, or its
" equivalent, by means of cams and sliding journal boxes or
" bearings."

Fourth, "the mode of attaching the tube to the circular
" framing," " by forming the latter in two parts, and providing a
" lip or bead" on " the former."

This pump, it is said, may be applied to a number of purposes
which are named, "and to raising and moving substances of
" partial fluidity such as molasses and syrups of sugar during the
" process of manufacture."

[Printed, 8d. Drawing.]

A.D. 1856, May 31.—No 1288.

NEEDHAM, WILLIAM, and KITE, JAMES (secundus.)—
" Improvements in machinery or apparatus for expressing liquids
" or moisture from substances." These are a frame work, in
which are placed a number of slabs or floats which are grooved,
between each slab or float, supported by blocks, one, two, or more
cloths or other filtering medium are placed, which cloths are
attached to a series of short pipes with cock at the sides, through
which the material to be pressed passes into the press, filling the
whole vacant space or spaces thereof, " the pump continuing to
" work till the liquid or moisture yielding to the pressure finds
" vent through the cloth, runs along the channels (formed by the
" grooves) and makes its escape, leaving the more solid particles
" between the slabs or floats until the press refuses to receive any
" more, which condition will be easily ascertained by the slower
" speed or stoppage of the drainage from the press." When this
stage is attained, each alternate cock is closed or shut off, when
by the pressure from below the float boards will be raised. The
blocks under the ends of the slabs are removed; " the material
" still being forced into the lower chambers, that in the upper
" becomes from the pressure more solid." At this stage, after
shutting off all the cocks, the side rods of the frame are removed,
the pipes are unscrewed at their joints, and the solid matter
removed from the upper chambers; the side rods " are then re-
" placed, and the lower cocks are then turned off, and the upper
" ones turned on; the material operated upon then fills the upper
" chambers, whereby the material is in like manner in its turn

" made solid and ready for removal, and the operation reversed,
" and so on alternately." Among the applications named to
which the foregoing apparatus may be applied is the filtration of
sugar.

[Printed, 10d. Drawing.]

A.D. 1856, May 31.—N° 1289.

ALLMAN, FENNELL, and BETHUNE, DONALD.—(*Provisional
protection only*.)—". Certain improvements in apparatus for sepa-
" rating fluids from solids, or for separating the more fluid
" particles from the more solid of various bodies." These are,
" the employment of centrifugal force, exhibited in vacuum, by
" means of suitable mechanism actuated by water, wind, steam,
" or other power, and the forcing of hot air into the chamber or
" vessel in which the vacuum or partial vacuum is formed by
" means of an air force-pump." A strong chamber is put in
communication with an air-pump or exhausted receiver. In this
chamber is placed an inner revolving receiver of perforated
material, in which is put the substance to be operated upon. " A
" vacuum is produced in the outer chamber, and on motion being
" given to the inner receiver, all the smaller and more fluid par-
" ticles are projected with great force into the vacuum." " The
" air-pump cock is then closed and heated air is forced into the
" chamber through a pipe," and removed by the exhaust pipe.
Sometimes heat is applied to the apparatus. Among the applications named to which the foregoing apparatus may be applied is
the filtration of sugar.

[Printed, 4d. No Drawings.]

A.D. 1856, June 6.—N° 1350.

GARDISSAL, CHARLES DURAND.—(*A communication*.)—" Im-
" provements in machinery for extracting fibrous and other
" products from vegetable substances." These are, " separating
" the fibre of plants from the matters which bind them together
" preparatory to their further manufacture for which it is well
" adapted, the fibres being preserved of their entire length or
" nearly so." In these machines green plants or other substances
among which are named " sugar cane, grapes, and other fruits
" and vegetable substances generally in a green or dried state
" for the purpose of expressing or extracting the juice or syrups
" from such fruit or substances," are " submitted to friction and
" pressure, but without at any time exceeding the limits of

" resistance of the fibres or parts so treated, and from which
" pressure the plant or substances can at any time escape when
" the fibres are on the point of being broken or otherwise injured.
" They are composed of plane or curved surfaces furnished with
" flats or grooves; one of these grooved rubbing surfaces is
" formed of a series of moveable parts linked together, and acting
" on the fixed rubbing surface which is also grooved, or the
" moveable rubbing surface may be rigid while the stationary
" surface is formed in pieces linked together." " The rubbing or
" crushing cylinders or sections of cylinders and other rubbing
" surfaces may be of any suitable dimensions so grooved with any
" suitable form of groove, and the grooves may either be parallel
" with the axis of such cylinders or sections of cylinders, or
" produced in a helical direction on the cylinder, or in any other
" suitable form or direction, and such surface may have a con-
" tinuous or alternating motion imparted to them for effecting
" the desired operations."

[Printed, 2s. 4d. Drawings.]

A.D. 1856, June 12.—N° 1395.

STENHOUSE, JOHN.—" Improvements in the preparation of a
" decoloring material suitable for the treatment of acid, alkaline,
" and neutral solutions." These are, a substance to " be used
" as a substitute in neutral and alkaline solutions for common
" bone black, and in acid solutions for what is called purified
" animal charcoal." This " porous vegetable charcoal is pro-
" duced chiefly as follows :—I form a very intimate mixture of
" either hydrate of lime, unslacked lime in the state of the finest
" powder, calcined magnesia, or the light subcarbonate of mag-
" nesia of the shops, with certain vegetable substances, such as
" maize, wheat, and other kinds of flour, common resin, or colo-
" phonium, pitch, wood tar, asphalt, or bitumen, coal tar, and
" coal tar pitch." This mixture is then heated to redness in
" covered crucibles or cast iron retorts until the vegetable matter
" is entirely carbonized. The mixture when cold is then digested
" with hydrochloric or sulphuric acids, according as lime or
" magnesia has been employed, and repeatedly edulcorated with
" water on a filter until everything soluble has been removed.
" The porous charcoal remaining on the filter is the decolorizing
" agent."

[Printed, 4d. No Drawings.]

LEWSEY, Charles John. — (*Provisional protection only.*)—
" Improvements in sugar cane mills." These are in place " of the
" large and massive ' side frames ' or standards of cast iron here-
" tofore used (whether with or without wrought-iron tie rods),"
employing " wrought-iron standards composed of numerous plates
so placed and connected together as to present far greater strength
than those heretofore used, at the same time the strain is more
equally distributed. "These plates are arranged in four sets at
" each end of the mill," "and they are at their lower ends fixed
" to horizontal plates which are secured to the base plate of the
" mill." "The upper ends of the plates of each set are connected
" together, and they are keyed so as to press on the sets of plates
" which they pass between." The result is a framing consisting
of an arrangement or combination of several plates of wrought
iron arranged in diagonal lines, and where cane juice pumps
are required, a self-adjusting pump is applied, whereby a continuous flow of cane juice is maintained without the use of
clacks.

[Printed, 4d. No Drawings.]

A.D. 1856, June 27.—N° 1516.

BETHUNE, Donald.—(*Provisional Protection only.*)—" Certain
" improvements in apparatus for separating the more fluid particles
" from the more solid of various bodies." These are, employing
" centrifugal force in a vessel or chamber, either entirely cr
" partially closed by means of a suitable mechanism, actuated by
" water, steam, animal, or other power, and the employment of
" heated air in such chamber." In a strong chamber is mounted
a revolving perforated cylinder, in which are placed the material
desired to be operated upon, among which sugar is named. On
the axis or spindle of this cylinder " is a small pulley or rigger,
" which obtains a rotary motion from any suitable source of
" power," which motion it communicates to the cylinder. Hot
air is admitted by a pipe through the cover of the outer chamber.

[Printed, 4d. No Drawings.]

A.D. 1856, July 29.—N° 1797.

ANDERSON, Alexander Williams.—(*Provisional protection
only.*)—"Improvements in refining sugar." These are dissolving

3 lbs. of sulphate of copper in 1 gallon of boiling water, adding to
the solution about $1\frac{1}{2}$ lbs. of granulated tin, "when all the copper
" is precipitated, filter; the filtered liquid will be a colorless solu-
" tion of sulphate of tin to be used in sugar refining." To the
solution of sugar and water in the blow-up cistern, at a density
not higher than 20° Beaume, three ounces of hydrate of lime are
added to every 100 lbs. of sugar in solution, and an imperial pint
of the above sulphate of tin to every three ounces of the lime added.
The proportions of lime and sulphate of tin may be slightly
diminished or increased according to the quality of the sugar;
when the liquid has been brought to the usual degree of heat it
is passed through bag filters and beds of granulated charcoal."
When no lime is used the same quantity of sulphate of tin is
added and the sugar solution proceeded with as above, but the
solution filtered from the animal charcoal should be tested in a
wine glass with sulphuretted hydrogen for tin and tincture of
litmus for acid. If tin in solution be indicated the sugar solution
is allowed to remain in contact with the animal charcoal till all
trace of tin disappears. If the time for this disappearance exceeds
ten minutes, the charcoal should be re-burned. If the liquor
indicates acidity, by the liquor being reddened by the litmus, the
charcoal must be cleansed and re-burned. It is advantageous once
a week or once in two weeks to allow the charcoal to remain for
about half an hour in contact with water, with from two to five
per cent. of hydrochloric acid, and afterwards cleansing it with
steam or water. The tin solution may be evaporated to dryness,
and five ounces of the residue used to every one hundred pounds
of sugar in place of the solution.

[Printed, 4d. No Drawings.]

A.D. 1856, October 9.—N° 2364.

KING, THOMAS.—"An improved continuous compressing ma-
" chine." This consists "in arranging two endless perforated
" bands or webs of articulated bars or links in such a manner
" that whilst each works around two suitably-formed drums,
" placed at such a distance apart as corresponds with its length,
" their faces shall not throughout their entire length be parallel
" to each other in the direction of their length, but the axes and
" diameters of the drums at one end of each of the bands shall
" approach one another so that there is but a smaller space or

" distance between the faces of the band at one end than at the
" other end; thus the space between the faces of the two con-
" tinuous webs, bands, or belts is somewhat wedge-like, and any
" materials placed on or fitted in between them will, upon the
" webs, bands, or belts being caused to move from the wide end
" toward the narrow end be gradually compressed until they
" reach a point in a line with the centres of the drums in closer
" proximity." "The two drums may be geared together, and
" driven at any suitable speed by any suitable means, and the
" four drums and their connections may be mounted on a suitable
" framing or standards." A shoot or hopper at the wide end of
the machine conveys the articles to be compressed into it. Whilst
the compressed matters on leaving the narrow end of the machine
fall on to a shoot or hopper. This apparatus among a number of
appliances named may be used for expressing the juice from " beet
" root, sugar cane, and other vegetable substances."

[Printed, 10d. Drawing.]

A.D. 1856, November 5.—Nº 2603.

SIEVIER, ROBERT WILLIAM.—(*Provisional protection only.*)—
"An improvement in the mode of treating saccharine juices in
" in the manufacture of sugar." This consists in treating juices
containing sugar with sulphurous acid gas alone. As the beet
root (all other juices being treated in a similar manner) is being
ground the sulphurous gas is introduced into it in a close vessel
by means of a tube. The juice and pulp pass into " the usual
" receptacle, and should it not have taken up enough of the sul-
" phurous acid gas so as to prevent its losing its white or natural
" color, more gas must be introduced " into it. In cases where the
mass of juice and pulp is very thick, a solution of sulphurous acid in
water is employed. The juice is then filtered through a filter made
of a close textile fabric or sand. "A small quantity of chalk or
" other alkaline body must be added to take out any acid," as
is now the practice. The solution is then evaporated for crystallization. Blood, eggs, or any other albuminous substance are
used, if necessary, in the usual manner. These improvements
render the use of charcoal unnecessary. In preference the juice is
filtered twice, once before and once after " the chalk or alkaline
" body is used."

[Printed, 4d. No Drawings.]

A.D. 1856, November 24.—N° 2785.

LEWSEY, CHARLES JOHN.—" Improvements in sugar cane
" mills." These are, " the combination of the framing and means
" of applying and adjusting the brasses or brazings therein;" also
the combining with sugar mills pumps without clacks or valves
to the suction pipes. In place of the " massive ' side frames '
" or standards of cast iron" having wrought iron standards,
composed of numerous plates. The base plate of the press is of
cast iron, to which are bolted the sets of angle pieces made of
strong boiler plate, which carry the bearings of the two lower
rollers. There are wedges by which the rollers are set up, as may
be required. To the angle pieces are bolted the tie plates, which
are arranged alternately with the angle pieces. At the points
where the different sets of tin plates intersect each other, the
plates of one set pass between the plates of the other set, and
they are connected together by bolts. The bearings of the top
roller are carried in the diamond-formed spaces enclosed by the
meeting of the different sets of plates, and the lower brass of the
bearings is kept set up by wedges. The pump referred to above
has two buckets on two separate pump rods, each has a valve
opening in the same direction, and they respectively open as they
move towards the suction end of the cylinder, and close as they
move towards the delivery end. There is a suction pipe which
has no clack or valve. There is a delivery pipe which conveys the
juice away, attached to which is a pipe, and in which pipe is a
regulating valve, to which is attached a float, this valve remains
closed so long as there is the requisite quantity of juice in the
cistern to rise above the lower end of the suction pipe, but should
the level of the juice fall, then the float will descend and open
the valve, which will cause part of the juice to return into the
cistern, so as to maintain the juice above the suction pipe, and so
prevent air being drawn into the pump.

[Printed, 10d. Drawing.]

A.D. 1856, December 2.—N° 2849.

LONGBOTTOM, JOHN. — " Improvements in apparatus for
" drying, roasting, carbonizing, and calcining vegetable, mineral,
" and animal substances." These are, " the application and use
" of hot water circulating pipes for the purpose of drying,
" roasting, carbonizing, and calcining vegetable, mineral, and

" animal substances " as follows : — A round " vacuum car-
" bonizing vessel, pan, or boiler " is described, fitted with a
series of shelves for laying the substances upon; above the
furnace is a chamber in which are a series of hot water heating
tubes " for surcharging steam. These pipes are continued also
" in the form of coils " at the top of the round vacuum carbonizing vessel, " for the purpose of surcharging the steam
" emitted from the substances under treatment." Inside the
vacuum vessel is a perforated metallic cylinder for containing the
substances to be dried or carbonized. There are two taps for
shutting off the vapour arising from the vegetable substances,
and also two taps to draw the water off therefrom. The furnace
is fitted with hot water tubes to which is attached a chimney.
In using the apparatus the round vessel above is charged with
" peat or waste tar, sawdust, chips of wood, or other vegetable
" substance," and made tight, and the air and water is pumped
out " by any known process until a partial vacuum is formed in
" the carbonizing vessel. In this state sulphuric acid is pumped
" in, which fills all the pores of the vegetable substance, there-
" by rendering it fit for carbonizing, which is effected by steam
" passing from a common boiler mounted or not on wheels."
This steam passes into the chamber where the pipes are heated
by the circulation of hot water from pipes in the furnace, and
thus superheats the steam before it enters the carbonizing vessel.
Or air may be used by passing it over pipes in the furnace.

[Printed, 8d. Drawing.]

A.D. 1856, December 31.—N° 3104.

TERRY, ALEXANDER ROBERT.—(*Provisional protection only.*)—
" Improvements in machinery for cutting sugar and other sub-
" stances." These are, connecting and moving " certain cutting
" edges in such a manner that the sugar or other substance "
may be reduced by them " into lumps of any desired size." This
is effected by working two different cutters at right angles with
each other, and attaching to the last a series of small cutters.
The first or cross cutter causing the loaf to be cut into slices or
slabs, is worked by a fly wheel shaft, by means of proper connections, the second acts upon the slab cutting it into strips, by
means of suitable connection with the fly wheel shaft. The
third operation, that of reducing the strips into lumps, about
cubical, is performed by a series of cutters attached to the second
cutter, and acting simultaneously with it, any binding of the

material in this last operation being prevented by suitable apparatus. The same object is attained " by using a set of cutters " fixed at right angles with each other, and so arranged that they " will reduce the slabs into lumps by a single stroke, instead of a " succession of strokes as required by the former plan."

[Printed, 4d. No Drawings.]

1857.

A.D. 1857, January 24.—Nº 220.

McONIE, ANDREW. — (*Provisional protection only.*) — " Im-" provements in the construction of centrifugal machines or " 'hydroextractors' used for the manufacture of sugar and other " purposes, and in the arrangement of appliances to give motion " to such machines by steam power." These are, a drum of cast iron or of other cast metal, turned inside and out, after mounting on its spindle, has its sides or periphery perforated all over with holes countersunk in the inside. The spindle of the drum tapers towards each end; the lower end is stepped into a cup-formed bearing, into which it descends more and more as the brass wears, this adjustment being effected by a screw passing through the bottom of the brass. The upper end of the spindle passes through a bush screwed into the bridge of the machine, and as it wears it is forced down by an arrangement with a key or wedge. To stop the drum when necessary " a friction strap is brought to " bear on the lower part or the periphery of the drum, by means " of an eccentric mounted on an axis passing through the top of " the case, and having a handle at its upper end." To drive this machine, employing " a small horizontal steam engine, mounted " on a frame in connection with the frame which carries the drum. " This steam engine, by means of a connecting rod and crank, " gives motion to a vertical axis carrying a large pulley, from " which a band passes to a small pulley on the spindle of the " drum."

[Printed, 4d. No Drawings.]

A.D. 1857, February 10.—Nº 387.

PARTZ, AUGUST FREDERICK WILLIAM. — " An improved " method of evaporating fluids, condensing and absorbing vapors,

" gases, and fumes, arresting and precipitating floeculent, me-
" tallic, or other particles, and transferring heat from air or steam
" to fluids and pulverulent substances." This consists in " the
" use of perforated disks, plates, or sheets," as follows :—The liquid
" to be evaporated, as cane juice," is placed in a vessel through
which a shaft passes, supported in journals and driven at any
desired speed, by a pulley attached to it, or in any convenient
manner, a series of " disks of woven wire perforated metal, fibrous
" substances, or other equivalent material, attached to the said
" shaft." " A semi-cylindrical hood " encloses the upper halves
of the discs. As these discs rotate they carry a film of the liquid
and present the same to the vapors or gases that are led into the
hood by means of a pipe or otherwise. If cane juice is under
operation " a current of dry or heated air is forced through the
" apparatus by means of a blast generator " and a pipe carries
the vapors away.

[Printed, 8d. Drawing.]

A.D. 1857, February 26.—N° 567.

EDWARDS, JOSEPH SLATTERIE.—" The preparation and novel
" application of a certain foreign fruit or vegatable as an article
" of food, confectionary, or to be used is brewing or distilling, or
" for the manufacture of sugar and gum." " The foreign fruit
" or vegetable which it is proposed to prepare and apply to the
" various useful purposes above named, is the Ceratonia Siliqua,
" commonly known as the carob or locust pod, and sometimes
" called St. John's bread." In reference to this subject, it is
applied as follows, the pods are ground " under edge runners, or
" other suitable mill," and the stones, skin, and husk are sifted
or not, as desired, from the meal, of which a thick compost is
made with hot or cold water, and after allowing this mixture to
stand a time, it is placed " in flannels or other cloths, which are
" placed on frames or boxes," &c., and subjected to pressure, the
strong syrup thus obtained, in preference, is evaporated in a vacuum
pan " at as a low temperature as convenient, until a strong syrup
" or liquid sugar is formed, and which may be subsequently
" crystallized."

[Printed, 4d. No Drawings.]

A.D. 1857, February 27.—N° 577.

MUCKLOW, EDWARD.— " Certain improvements in appara-
" tus to be employed for the purposes of cooling and evaporating."

These are, " cooling such liquids as require to be evaporated,
" whilst at or below a boiling temperature, particularly sugar, dye
" extracts, and other liquids, requiring to be evaporated at a low
" temperature," by the " employment, application, and use of
" an endless cloth, or other suitable equivalent, waving ribs,
" folds, lags, or buckets formed upon its external surface for the
" purpose of taking up liquid to be cooled and evaporated," revolving on " a wince or roller or rollers " above and into the vat or
cistern containing the liquor to be evaporated. " A similar effect
will be obtained by causing a wheel or wheels having buckets
" attached to their peripheries to revolve in the vat."

[Printed, 6d. Drawing.]

A.D. 1857, March 5.—N° 646.

ANSENS, ANTHONY. — (*Provisional protection only.*) — " Im-
" provements in moulds or forms for loaves of sugar." These
are, making " moulds (whether of iron, copper, or other metal or
" material) square or oblong square, with a cornice at their bottom
" part, from thence gradually tapering (but still with flat or
" square sides) till within a small space from the top, then the
" taper is changed to an angle till it reaches the top, when it
" becomes a diamond point, with a hole in the centre thereof,
" to allow the syrup which may still remain in the sugar after it
" is placed in the mould to drain off."

[Printed, 4d. No Drawings.]

A.D. 1857, March 11.—N° 703.

MOUNTFORD, GEORGE. — " Improvements in machinery or
" apparatus for cutting or chopping loaf sugar, roots, and other
" substances." These are, apparatus consisting " of a semi-cir-
" cular hopper or trough, or of an oblong box in which the sugar
" loaf, titler, or lump, or other substance to be operated upon is
" placed, and propelled towards one end thereof by means of a
" plate or piston in connexion with studs which run in grooves in
" the sides of the hopper, trough, or box with traverse chains,
" and four chain pullies and ratchet wheel, the last of which is
" put in motion by a connecting rod with a lever and weight fixed
" on the two uprights herein-after mentioned. To enable the
" piston propelling the loaf or lump to return to its original posi-
" tion, it is let loose by a pin, tappet, or stud in the ratchet
" wheel, which throws the ratchet or pall on each ratchet wheel
" out of gear; the ratchet is thrown in gear by a pin or stud on

" the ratchet wheels or framework, connected with a shaft having
" studs and shaft, and a horizontal shaft, chain pullies, and
" weights." When the article " operated upon arrives at the far
" end of the hopper, trough, or box in which it is placed, it is sliced
" by the knife " first herein-after mentioned. Two uprights are
fixed with grooves, in which a knife block slides, having a suitable knife either straight or curved in the cutting edge fixed
thereto. This knife descends in a vertical direction upon the
sugar or other substance to be operated upon, whereby a slice
of the thickness required is cut or chopped off the bulk. The
knife block is worked by connecting rods fitted on stud pins
fastened in it at the upper end with the other end of connecting
rods placed on crank pins fixed into toothed gear wheels. The
slice of sugar or other substance operated upon " when cut falls
" upon a horizontal revolving table having numerous holes, aper-
" tures, or perforations of sufficient size, through which the slice
" of sugar or other substance operated upon passes after being
" cut into the required cubes, nobs, or pieces." Above the horizontal revolving table is a knife block, wherein a series of knives
are placed which descend in a vertical direction upon the slice of
sugar or other substance and cut it as required; to assist this a
stationary knife or knives is or are fixed in a horizontal position.
" The revolving table is driven by a ratchet wheel on a horizontal
" shaft with a toothed pinion gearing into teeth under it. The
" knife block is worked up and down by a crank shaft and con-
" necting rods, the shaft working in bearings and driven by
" wheels from the main shaft, and having a connecting rod
" actuating the ratchet work of the revolving table. The main
" shaft with fly wheel or wheels may be driven by hand, treadle,
" or any other power, and is connected with wheels to work the
" whole of the machinery fixed in upright brackets and rails.
" The whole of the apparatus or machinery (except the fly wheel
" or wheels and winch handles, pullies, and treadles for driving
" the machinery) is enclosed in framework of rectangular and
" suitable form for the purpose."

[Printed, 1s. 4d. Drawings.]

A.D. 1857, May 5.—No 1266.

SIEVIER, ROBERT WILLIAM.—" An improvement in the mode
" of treating saccharine juices in the manufacture of sugar."
This consists, first, "the use of sulphurous acid gas in the com-

s.   P

" mencement of making sugar applied to the pulp and juices of
" beet root or cane before boiling, and then adding lime to excess
" to defecate well the juice and neutralize that excess."

Second, " the use of sulphurous acid gas as above stated in the
" manufacture of molasses."

The sulphurous acid gas made in a close vessel is washed by
passing it through water into a gas holder, and from thence to
the juice. The materials recommended for making it are sulphur
and peroxide of manganese or copper and sulphuric acid. " In
" its application to the manufacture of beet-root sugar; all other
" saccharine juice being treated in a similar manner. As the
" beet root is being rasped or ground by the usual process,"
" introduce through a tube amongst it sulphurous acid gas," in
a close vessel, mix the whole together and if not sufficient gas
has been used "to prevent its losing its white or natural colour,
" more must be introduced." In some cases where the mass of
pulp is very thick, the juice and pulp are sprinkled with sulphurous acid water. The juice is separated from the pulp by
means of a textile or sand filter, a small quantity of chalk or lime
or other alkaline matter is then added, it is then boiled, the scum
removed, and the solution boiled for crystallization, or lime is
added in excess and the excess neutralized by sulphurous acid,
boiled down to ten degrees and filtered. The sliced roots may be
treated with the gas, &c.

Second, the use of vessels made of wood, stone, slate, glazed
metal, or earthenware, for boiling saccharine solutions from which
sugar or molasses are made.

[Printed, 8d. Drawing.]

A.D. 1857, May 5.—Nº 1267.

KEDDY, THOMAS—(*Provisional protection only.*)—" New or
" improved machinery for cutting sugar and other substances."
This consists as follows :—The loaf or cone of sugar is divided
into a series of discs " by a machine in which a series of cutters
" situated in the same plane approach simultaneously and con-
" verge upon the sugar loaf." These cutters enter " the sugar
" loaf at several points in the same plane, and cut or split off a
" disc of the desired thickness. After the cutters have opened,
" the sugar loaf is advanced to the proper distance, and the
" cutters again close and cut off another disc or slice, and so on
" until the whole of the loaf has been cut up. The cutters may

" approach or close upon the loaf either in the direction of radii
" or by an oblique motion. The slices or discs of sugar are next
" put into a hopper, the sides of which collapse so as to suit discs
" of different thicknesses. The sugar descends in the said hopper
" until it is arrested by a stop. A pair of knives one 'on either
" side by an alternating motion divide the disc of sugar into a
" series of small bars, which as they are cut off one allowed to
" pass, by the withdrawal of the stop, to be acted upon by another
" set of knives. The last-named knives act in a plane at right
" angles to that in which the first-named act, and divide the bars
" of sugar transversely, thus cutting it into small square pieces.
" The advancing motion of the knives is effected by cams on a
" rotating shaft, and the return motion by springs."

[Printed, 4d. No Drawings.]

A.D. 1857, May 7.—No 1287.

ZIEGLER, ERNST. — "A substitute for animal charcoal,
" applicable also as a coloring matter." This consists in " manu-
" facturing a substitute for animal charcoal from clay " as
follows :—Clay as free from extraneous matter as possible is
mixed with any substance which " burnt in hermetically sealed
" retorts gives a pure carbonaceous substance; it should, how-
" ever, be fluid or capable of being dissolved, such as glue, oils,
" rosin, gum, &c., treacle," but these substances it is said are
" expensive, and tar of coal or wood, by preference, is substituted,
the oil from the tar may be previously separated from it by distillation. The tar is added to water, and this is intimately mixed
with the clay, "by means of apparatus, such as is commonly
" used in large potteries." This substance so mixed is "made
" into small cakes, say, in hollow cylinders of from 4 to 5 inches
" long, and 1 inch in diameter," about $\frac{1}{8}$th of an inch thick,
dried and carbonized in the same ovens as for carbonizing bones
in air-tight pots or retorts; "a temperature of from 25 to 32
" degrees Wedgwood " has been found to answer well, and clay
retorts are preferred to iron. The grinding and sifting is the
same as for bone black, and the material washed with water to
remove the fine dust which it contains may then be employed in
sugar refining manufactories. "After its absorbing power becomes
" lost, it may be revived by the same process as is at present
" employed for re-animating animal charcoal." A more effective
material "may be obtained by mixing with the tar or other material

" employed (before its amalgamation with the clay) from 10 to 20
" per cent. of salt, potash, soda, or lime dissolved in water," and
proceeding as above.   After burning, grinding, &c., these ingredients must be extracted by water and acids.   A carbon similar
to carbonized blood is obtained by adding tar to a fixed alkali,
and proceeding as above, " but the carbon so obtained, as is the
" case with carbonized blood, can only be used in a powdered form
" for refining sugar, and is therefore not capable of being
" revivified."   Another substitute for animal charcoal is made
by adding to the clay " ¼ to ½ part of tar, coal, or other combus-
" tible substance that has been impregnated or mixed with a
" solution of alum or other salt ;" when it is dried, burned in
closed retorts or in an open fire, then granulated and ground.
" I have found clay, mixed with an acid solution and burned to
possess some effectiveness."

[Printed, 4d.   No Drawings.]

A.D. 1857, May 15.—N° 1382.

BROOMAN, Richard Archibald.—(*A communication from
Pierre André de Coster.*) — " Improvements in machinery to be
" employed in the refining of sugar."   These are, first, " the
" construction of centrifugal machinery for purifying sugar and
" for separating the granular from the liquid portions thereof,"
which acts as follows :—The charge may be placed in the perforated metal drum by removing the outer cover and the cover
of the drum, but in preference, it is charged through a funnel and
tube at the top and spreads itself over the inside of the hollow
shaft issues through the apertures in the perforated tube and
thus finds its way into the interior of the drum.   By the rotary
motion communicated to the drum through the driving gear,
the matters are projected against the sides, the liquid is driven
through the perforations and apertures in the sides of the drum
while the solid matters remain inside the drum ; from the rapidity
with which the drum is driven, say about two thousand revolutions a minute, in about three minutes " the grains and crystals
" remaining inside the drum are perfectly cleansed and dry.   The
" liquid issues from a spout connected to the outer casing."
    Second, constructing " machinery for breaking up lumps and
" cakes of sugar, and for mixing it with liquor prior to being
" treated in a centrifugal machine," consisting of a pair of rollers
with rings or flutes formed thereon, the rings on the rollers

coming between each other. "These rollers are geared together
" and caused to rotate in opposite directions, at the bottom of a
" hopper fed with the lumps to be broken and with water or
" other purifying liquor. After issuing from the rollers the sugar
" and liquor fall into a vessel, in which there is a revolving
" agitator, by which they become regularly mixed, and from the
" said vessel they are drawn off to be introduced into a centri-
" fugal machine."

[Printed, 10d. No Drawing.]

A.D. 1857, May 23.—N° 1456.

TRAVIS, EDWIN, and CASARTELLI, JOSEPH LOUIS.—" An
" improved apparatus for regulating the supply and discharge of
" steam, air, water, and other fluids." This consists of a "box
" of brass or other suitable anticorosive metal or material, with
" a slide tube at one end, an adjusting plug, piston, or valve
" at the other, and an escape tap or hole at the side;" this is
named "a 'steam trap.' We fix this apparatus to the end
" of an additional pipe, which is attached to the ordinary steam
" pipe or other vessel on which it is intended to act, forming one
" continuous tube. We give the said additional pipe a com-
" pensating arrangement with itself by bend, coil, disc, or other
" similar agency, to act when the pressure of the steam exceeds
" the point at which the apparatus was adjusted. Or, we pro-
" vide a compensating arrangement with the apparatus, by spring
" or other mechanical contrivance. We also make the said appa-
" ratus with a swivel or universal centre to enable it to accommo-
" date itself to any vibration or divergence of the connexion to
" the sliding tube, or we simply screw the apparatus to a wall
" or other convenient place, so that when the steam is first raised,
" it will force the air or fluid through the ordinary pipe into the
" additional pipe, which will then escape through the escape tap
" or hole. As soon as the steam begins to escape through the said
" hole or tap, and the additional pipe becomes heated by the
" steam and expanded, we adjust the plug, piston, or valve,
" so as to stop the escape of steam, and when it cools the con-
" traction of the metal pipe will draw the sliding tube from the
" piston, plug, or valve, and cause it to open, thus allowing
" any condensed steam or water to escape, or it may be applied
" directly to any range of piping, and thus dispense with the
" additional pipe. We also chamber the inside of the trap to
" allow the free discharge of the air or fluid from the tube and

" outlet, and to an adjusting piston, plug, or valve, we apply a
" spring, if found necessary for the purpose of compensating for
" any additional expansion after it has been adjusted; but the
" trap will act efficiently, either with or without the spring. We
" make the trap or apparatus in such manner that by trans-
" posing the sliding tube and the plug, piston, or valve it may
" be used for right hand or left hand." This apparatus it is
said is applicable to a number of purposes which are named,
and among which, are " to pipes and vessels used in boiling and
" refining sugar, brewing, distilling, cooking, evaporating pans,
" and other similar purposes."

[Printed, 1s. Drawings.]

A.D. 1857, June 4.—Nº 1573.

MILLER, WILLIAM.—" Improvements in the manufacture of
" sugar." These are, conducting " the progressive processes of
" the manufacture of sugar " in metal vessels which shall never
be heated materially above 212° F. and the heat being kept at or
as nearly as may be at 212° F. In the process called " blowing up "
a vessel is employed surrounded by a jacket, in the lower part of
which is some water. A steam pipe supplying steam from a boiler
enters the jacket under the water, and there is an overflow pipe,
by which the water in the jacket is kept at the same level. In
the upper part of the jacket is an escape pipe for the steam. The
sugar and water in the vessel are kept stirred till the sugar is
melted, when the solution is run into bag filters, the filtered solution then heated from 180° to 190° F. in a similar pan, after which
it is run into charcoal filters. When evaporating cane juice or
other saccharine solutions a similar vessel to that just described
is employed, but in addition, there are three series of flat tubular
passages through the pan or vessel in which the saccharine
matters are placed, in order to present a greater extent of heated
metal surface to the saccharine solution in the pan or vessel, and
two octagon barrels to the axes of which rotary motion is given
are fixed a long way above the pan or vessel, over these are suspended two endless bands, in preference, of strong sail cloth,
while there are two smaller rotary rollers in the pan under the
solution, under which the two endless bands pass, on these
rollers are stirrers or agitators, the sides of the apparatus are
closed in, to above the upper rollers but the top is left open.

[Printed, 10d. Drawing.]

A.D. 1857, July 31.—Nº 2085.

GALY-CAZALET, Antoine, and HUILLARD, Adolphe.— "An improved apparatus for and mode of manufacturing sul- "phuret of carbon, animal charcoal, and carbonic acid." This consists, in reference to this subject, as follows :—"We construct "a furnace or cupola of brick or stone, inclosed in an air-tight "case of sheet iron or of other suitable material;" this is divided into an upper and lower compartment by a fire-clay grate. "The "lower compartment has two lateral openings with air-tight "doors for feeding it with fuel and for withdrawing the ashes;" a pipe is connected to the upper part of the fire-place. Above the upper compartment or oven is a chamber, which also serves as a chimney, and is closed by a lid, which can be raised and lowered by a chain passing over a pulley. At the bottom of this chamber are two doors turning on vertical spindles passing through the casing, and capable of being moved by levers on the outside. When the grate has become hot enough the two doors at the bottom of the chamber are closed, and the chamber is filled with bones, the lid is closed, and the doors at the bottom of the chamber are opened, so as to allow the bones to fall in small quantities at a time upon the grate in the oven below. The oily or greasy vapours are conducted by the pipe connected to the upper part of the fire-place, to a vessel of cold water where they are condensed. The residue or animal charcoal is withdrawn at a door at the side of the oven.

[Printed, 6d. Drawing.]

A.D. 1857, August 5.—Nº 2116.

BOTTURI, Sebastien.—"An apparatus and oven for the "carbonization and distillation of all animal and vegetable "matters." Some of the matters specified are "textile, vege- "table, and animal, viz., bullocks' blood, night soil, turf used as "fuel, refuse of any kind, and, in a word, all those matters which "may be carbonized." An oven is made of fire-bricks, in the interior of which is a safety valve. The flame pipes are con- structed so "that when the flame and smoke has passed through "the base of the oven under the platform, they divide and pene- "trate into the two chimneys which are on the right and left "sides of the oven; after having passed through each side of "the two chimneys," one conveys it to the front of the oven, the

other to the back, there the flame unites to be divided again to
the right and left into two other chimnies, from whence " the
" flame having lost its strength escapes by the discharging
" chimney." At each extremity of these chimneys is a stopper
removable at will. In the interior of the oven is a box to receive
the matters to be carbonized; the box is filled by a tube reaching
nearly to the bottom of the box which has a funnel and stop-
cock. Glass tubes are at each side of the oven to take off the
less volatile fluids, which are conducted into bottles. Upon the
oven is a glass dome, from which is a pipe to convey off the
more volatile fluids to a series of vases or bottles with a series
of syphons "made in such a manner that neither gas or fluid
" can evaporate," all of which arrangements are described and
claimed.

[Printed, 1s. Drawings.]

A.D. 1857, August 20.—N° 2212.

BROOMAN, RICHARD ARCHIBALD.—(A communication.)—" A
" new method of defecating sugar and other saccharine matters,
" and of refining or rectifying alcohol." These are, defecating
saccharine matters, whether proceeding from the cane or other
vegetable produce by first treating them with lime or its equivalent in excess, and then with a saponifiable matter so as to form
an insoluble compound or insoluble soap with the substance
itself, or with one of the constituents of the defecating substance.
The right of employing all soaps is claimed, but the preference is
given to olive oil as a saponifiable matter and amongst alkalies to
soda. Three soaps are formed and used according to circumstances what is termed a neutral soap with 10 per cent. of alkali
called No. 1. No. 2, soap with 7 parts of alkali, and No. 3, an
alkaline soap, by adding alkali to No. 1. Ammonia is made use
of in cases where saccharine liquor is not intended to present any
alkaline quality.

Alcohols are refined or rectified in a similar manner.

[Printed, 6d. No Drawings.]

A.D. 1857, September 15.—N° 2391.

BENSEN, GERD JACOB.—"An improvement in drying sugar."
This consists as follows :—" Where the sugar has been crystallized
" in a vacuum pan, it is run into pneumatic pans having false
" bottoms perforated," covered with wire gauze, the sugar, in

preference, is filled into about the depth of three feet, the bottoms of such pneumatic pans are connected by suitable pipes with iron tanks which are kept vacuous by an air-pump or otherwise. " The " sugar having been thus treated is then conveyed into a hot " chamber," in preference, heated from 120° to 140° F. and the sugar is placed in similar pneumatic pans as above, the spaces under the false bottoms are kept " vacuous when at work, by " which, the heated air of the chamber will rush through and " amongst the crystals of sugar, and dry the same, so as to " render the sugar fit for the market."

[Printed, 6*d*. Drawing.]

A.D. 1857, October 6.—N° 2561.

FINZEL, WILLIAM CONRAD, and BRYANT, JAMES.—" Im-" provements in cleansing animal charcoal, and in removing iron " and other impurities therefrom." These are, " the cleansing " of animal charcoal, and the removing of iron and other im-" purities therefrom by driving hot water, steam, hot air, or " other cleansing and purifying agent or agents through the " same by centrifugal force." " The charcoal may or may not " afterwards require reburning in order completely to revivify " it." In the Provisional Specification one part of the invention consists " in exposing granulated animal charcoal to magnets in " order to extract any particles of iron which may be found " among the grains."

[Printed, 4*d*. No Drawings.]

A.D. 1857, November 21.—N° 2925.

BENSEN, GERD JACOB.—" An improvement in the manufacture " of moulded sugar." This consists " in removing the syrups or " liquors in which the crystals of sugar have been formed from " the crystallized sugar and then substituting or mixing there-" with clear liquor or syrup, and at once finishing in moulds," as follows :—Sugar, in preference, crystallized as in No. 2391, A.D. 1857, and the syrup separated, " a quantity of fine clear " liquor or syrup produced from refined sugar is to be added " and well mixed with the crystals of sugar, and the proportion " of liquor or syrup is that, the mixture will just run from a " spoon," and this may be done in a vacuum or open pan or a pan in which streams of air are blown in order to mix the whole intimately together, and heated to about 170° to 180° F., and then filled into moulds, the lower ends of which are stopped. When the sugar is set the plugs or stoppers are removed, the liquor draining from the sugar in the moulds is collected in a receiver, and filtered through charcoal may be used again for a similar purpose. "The sugar in the moulds is then stoved in the ordi-"nary manner."

[Printed, 4d. No Drawings.]

A.D. 1857, December 26.—N° 3167.

PARSONS, CHARLES FREDERICK.—(*Provisiuual protection only.*) —" Cleansing and reburning animal charcoal." This is effected as follows :—A vertical metal plate is made to revolve in an air-tight furnace with a double fire under it. "At the end of each "fire there is a bridge at right angles with the bars of the "furnace for the purpose of retaining the heat. The flame "travels twice round the revolving plate; the plate travels on an "upright shaft, and is driven by either bevil or meter gearing. "The upright shaft is protected from the fire by means of a "hollow tube." The revolving plate is fed with charcoal by a hopper at top, the mouth of which is so constructed that it can spread the charcoal over the plate any thickness. There are two elevators that stand at right angles in the furnace to assist in spreading the charcoal to any given thickness. There is also a scraper, which stands at an angle of 45°, which is lowered down by means of a double-acting screw and bevil gearing, which scrapes the charcoal when burned off the plate into an internal tube with an oblique bottom, which discharges it outside the furnace into any place assigned for it. "The plate described "above may be worked so as not to revolve."

[Printed, 4d. No Drawings.]

1858.

A.D. 1858, March 5.—N° 445.

PARSONS, CHARLES FREDERICK.—"Improvements in ma-"chinery for producing and revifying animal charcoal." These are obtaining an extended surface for the above purposes, "from

" which the atmosphere is excluded in order to prevent com-
" bustion of the bones or charcoal." This consists of a circular
frame in the lower part of which are one or more furnaces, the
heat passes by flues round the frame, and into a chimney or shaft
common to both or all of the flues. " Over the furnace and flues
" there is a metal disc or platform with a central aperture, and
" the rings or flanges in which the edges of the plates forming
" the disc or platform are connected or terminate, are free to
" revolve in sand troughs. The central aperture opens into a
" fixed spout leading to the side of the frame, the outlet being
" closed by a tight fitting door or valve. There is a cover fitted
" perfectly tight over the revolving platform ; spreaders and
" spreading rollers are fixed between the cover and the upper
" surface of the platform for the purpose of distributing the
" material to be operated upon evenly thereon." . The feeding
hopper with a valve or slide to prevent the access of air passes
through the cover nearly to the surface of the platform. Two or
more ventilating pipes with valves opening outwards are fitted
between the cover and the furnace shaft to convey away matters
given off by the material under treatment. The platform is
connected by arms to a central shaft, to which rotary motion is
given by any well-known means. When the matters are sufficiently carbonized " on the platform, an angular guide is let down
" upon the surface thereof, whereby the matters are guided off
" the platform down into the central spout to be withdrawn
" from the machine. On commencing operations a gas or other
" flame or agent may be introduced to consume or draw off all
" the oxygen which may be inside the machine."

[Printed, 10d. Drawing.]

A.D. 1858, March 24.—No. 621.

BRINJES, JOHN FREDERICK, the younger, and COLLINS,
HENRY JOSEPH.—" Improvements in the manufacture and re-
" burning of animal charcoal." These are, for the above purposes the retorts are set so as to have a flue space around them
and placed horizontally or nearly so over the heating furnace,
and are each supplied or charged with the materials to be burnt
or reburnt by a vertical pipe leading downwards from a hopper,
and opening into one end of the retort. If desired an archimedean screw may be worked in this pipe by a pinion and worm
wheel for regulating the feed. The materials are slowly and

continuously worked through the retorts by another archimedean
screw rotated by worms or worm wheels or other means. This
latter screw has projections for moving the contents of the retort.
At the discharge end is a chamber with a pipe for carrying off
the gases and vapors to be condensed. "The bottom side of the
" chamber above mentioned is furnished with a receiver for the
" charcoal, which receiver is fitted with slides, so as to divide it
" into separate compartments. The upper portion of the receiver
" is surrounded by flue space, whilst the lower portion thereof
" is beyond the brickwork, so as to allow the charcoal contained
" therein to cool before it is finally discharged. In place of the
" archimedean screws for traversing and agitating the contents
" of the retorts, the retorts may be made to revolve, and a screw
" may be formed on their interior surface, or travelling tables
" inside the retorts may be used for the same purpose."

[Printed, 10d. Drawing.]

A.D. 1858, March 27.—Nº 655.

GILBEE, WILLIAM ARMAND.—(A communication.)—"Improve-
" ments in treating saccharine fluids." These are, " the applica-
" tion to the treatment of saccharine fluids of alcohol and agents
" capable of effecting in conjunction with alcohol the elimination
" of organic matters which are mixed with the sugar in the juices
" of sacchariferous plants," as follows :—The fluid is " concen-
" trated to from 18° to 30° Beaumé's areometer, either after defeca-
" tion in the ordinary manner or direct, care being taken to
" neutralize the acidity of the syrup as soon as it appears, by
" lime or other base." If too alkaline, they "they are first neu-
" tralised by carbonic or sulphuric acid, and then conveyed into
" a closed sheet-iron vessel provided with an agitator," and alcohol
" added till the mixture marks 60° to 70° by Gay Lussac's alco-
" holimeter. Three volumes of alcohol at 93°, and one volume
" of syrup at 20° Beaumé, are the suitable proportions for this
" mixture. After agitating for a few minutes the whole is allowed
" to settle, and the clear liquid decanted from the impurities which
" contain a small quantity of sugar. This deposit may be either
" washed several times with alcohol (which will serve for a sub-
" sequent precipitation), or after being freed from alcohol by
" passing through it a jet of steam, it may be used like molasses
" in distilleries. The clear decanted liquid is treated with "the
" purifying agents; an acid, or an acid salt, such as sulphuric,

" oxalic, or tartaric acid, and sulphate of alumina, forming with
" potass and soda insoluble compounds in alcohol, will separate
" these alkalies." In preference, using sulphuric acid diluted
in alcohol, agitating in the cold, neutralizing the acids by a base
of lime, barytes, strontian, oxide of lead, or other suitable base, or
one of their basic salts. On distilling, the ammonia is evolved
and condensed, the deposit is removed, and the alcohol condensed,
while the saccharine matter is concentrated, clarified, filtered, &c.

[Printed, 4d. No Drawings.]

A.D. 1858, April 1.—N° 697.

WARD, HENRY.—"Improved machinery for expressing liquids
" from organic substances." This "relates chiefly to the ex-
" traction by a pressing process of the saccharine juices from
" beet-root pulp," and consists in "the use of a reciprocating
" plunger in combination with a compressing chamber," as
follows :—Standards are bolted to a bed plate which carry at their
upper ends a crank shaft, to which the power for working the
machine is given in any convenient manner. Pendant from this
crank shaft are two crank rods, to the lower ends of which is
secured the crosshead of a plunger or piston. The ends of this
crosshead work in guides carried by the side standards to ensure the
parallel action of the plunger. Immediately below the plunger
and supported on an iron frame, to which it is securely bolted, is
the compressing chamber "the form of which may be somewhat
" varied. I prefer, however, that it should be narrow in cross
" section but broad laterally, so that it may present an extended
" surface to the pulp to be placed therein. The capacity of the
" chamber will also contract towards its lower end. The sides of
" this chamber being intended to act as strainers or filters, and
" permit the juices to exude while the solid portions of the pulp
" are prevented escaping with the liquid."

[Printed, 10d. Drawing.]

A.D. 1858, April 8.—N° 753.

RICHMOND, EDWARD. — (*A communication from Thomas
Blanchard.*)—"Certain new and useful mechanism for reducing,
" or reducing and crushing, and in various other respects treating
" grain, sugar cane, tobacco, or other substance or substances."
These are, first, "in the employment in combination with each

" other of two series of circular discs, those of each series being
" arranged on a separate shaft side by side with each other, and
" with washers of less diameter and of slightly greater thickness
" interposed between them, the two shafts being placed parallel to
" each other and at such distance apart that a portion of the peri-
" phery of each of the discs on each shaft shall pass into the space
" between the two discs on the other shaft, the said shaft being
" geared to rotate in opposite directions and with equal velocity."
Second, " a combination of the two series of discs combined
" as above," and " one series of clearers or stationary eccentric
" plates for each series of discs, such clearers being interposed
" between the discs " for " the purpose of forcing the particles or
" pieces of the reduced substance from between any two of the
" discs, and discharging such at their peripheries."
Third, " a combination of such rotary shears, or the same and
" their clearers with crushing rollers or mechanism, or such, and
" an endless apron, or other suitable device for the removal of
" the reduced, or reduced and crushed material from the reducing
" mechanism in such direction as circumstances may require."
Fourth, " a combination of an endless apron or conveyor, one
" or more sets of shearing or reducing mechanism and a bolting
" or sifting mechanism arranged together."

[Printed, 10d. Drawing.]

A.D. 1858, April 23.—N° 898.

SILLEM, HERMAN JAMES.—" Improvements in the machinery
" for the manufacture of sugar." These are, " an apparatus
" formed of suitably shaped scrapers, plates, and brushes, or other
" like appliances, which may be attached to or brought into
" proximity with the centrifugal machine in such way that their
" distance from the inner perforated surface of the revolving
" drum may be regulated at pleasure, for the purpose of equally
" distributing the charge at its introduction into the machine,
" and also for the purpose of removing the charge from the
" rotating surface after it has been converted into ' cured ' sugar
" without the necessity of stopping the machine." The before-
named scrapers, plates, and brushes, or other like appliances
are likewise used " to remove the sugar from the rotating surface
" of the machine, and which, as they do not rotate with the
" revolving drum, will, through the means given for regulating
" their distance from the perforated surface, discharge the sugar

" through a hopper, spout, or other conveyance, into suitable
" receptacles, by means of the pressure received from the re-
" volving cylinder acting upon the charge when the plates and
" so forth are brought up to or towards the inner circular surface
" of the revolving cylinder."

[Printed, 10d. Drawing.]

A.D. 1858, May 7.—N° 1024.

FIELD, James John.—" Improvements in evaporating or in
" extracting moisture from liquids, and from substances in a liquid
" state and in apparatus to be employed therein." These are,
employing for the above purposes gases or vapours more or less
perfectly dried, of such nature as to prevent when desirable the
oxidation of the matters acted upon by forcing them through the
liquids or substances divided into numerous small streams, or
sheets, or jets, or over the liquid or substance. The moist gases
or vapours are then passed " through or over chlor de of calcium
" or other chemical agent having a great affinity for water or
" liquid," and thus dried they may be used again, or " they
" may be passed into a mechanical drying chamber in addition
" to or in substitution for the chemical drying." The apparatus
for subdividing the liquids as above " consists of a pipe leading
" from the chemical or mechanical drying chamber into a reservoir,
" having one of its sides perforated with numerous small holes or
" narrow slits, through which perforations the dried gases or
" vapours pass into a chamber wherein the liquid or other sub-
" stances to be operated upon (divided into numerous small
" streams, or sheets, or jets by perforated spreaders or other like
" mechanical agents) is presented to their action;" the gases
or vapours then pass through another series of perforations into
the chemical or mechanical drying chambers above mentioned.
" After the fluids fall in the liquid chamber, I raise them by a
" pump, or otherwise, to be again divided," and so on as desired.
This process, it is said, is particularly applicable to among other
purposes which are named " the evaporation of sugar."

[Printed, 8d. Drawing.]

A.D. 1858, June 5.—N° 1271.

MANBRÉ, Alexandre.—"An improved method of preparing
" malt and other grain, and in extracting the saccharine matter

" therefrom, whether for the purposes of brewing, distilling, or
" otherwise." This consists, in reference to this subject, first,
in "separating the bran or husk from the grain by grinding or
" otherwise," and using only the flour. Second, "extracting
" the saccharine matter from the flour," as follows :—A vessel
containing water is heated by a naked fire " from 50° to
" 100° Fahrenheit, according to the temperature of the exter-
" nal air," the flour is gradually stirred in, and the whole
heated up to about 185° F., and thorougly mixed, "is allowed to
" settle when infusion and maceration only are required. Where
" deemed requisite, the whole is made to boil for from one hour
" upwards, and then allowed to settle for from 1½ to 2½ hours,
" and the clear liquor is drawn off into another vessel to
" undergo a last boiling." " Another method of extracting
" saccharine matter from grain or flour :"—The flour is put on
felt or flannel, resting on a perforated false bottom, in a hermetically closed vessel, water at from 50 to 100 F. is admitted
through a cock to moisten the goods, after which " water at about
" 175° to 185° F. is added, and the whole is left for from 10
" minutes to one hour. Then air or steam is forced in over the
" mass, when the clear liquid, carrying with it all the saccharine
" matter, will be forced through the filter and perforated false
" bottom, to be drawn off and used for the manufacture " of,
among other things which are named, sugar. " Or, instead of
" introducing any agent from above to force the clear liquid
" through the filter, a pump or vacuum may be employed to
" draw it through it."

[Printed, 4d. No Drawings.]

A.D. 1858, June 7.—N° 1283.

LOMBARD, Jean Baptiste Adolphe, and ESQUIRON,
Xavier Tristan.—" A new or improved means of obtaining
" saccharine substances from cereal and vegetable matters, and
" applying the products obtained to various useful purposes."
These are, in reference to this subject, as follows :—" The sac-
" charification of all cereals, such as barley, wheat, rye, oats,
" maize, rice, &c., and of all vegetable matters, such as ligneous
" fibres, parenchyma, pulps, roots, the rind of fruits, the peels of
" potatoes, and all feculous fruits in general," as follows :—" All
" substances containing starch must first be reduced to flour, or
" coarsely ground " and " sifted, in order to have only the flour

" to treat; they are mixed for some hours before use in once,
" twice, three, four, five, or six times, and more, their weight of
" water," but this may be dispensed with ; " the mixture is then
" placed in a close boiler, furnished with a safety valve and high-
" pressure guage," and acid is added, the proportions of which
vary with the substance operated upon. The temperature of the
boiler is raised " either by a bath of fatty bodies, metal infusion,
" or other substances capable of developing pressure . in their
" apparatus, and consequently heat above 212° F." A current of
steam may also be employed, and a current of carbonic acid may
be injected into the boiler. When saccharification has reached the
degree desired, the apparatus may be cooled by tubes of cold
water into it, and the liquid drawn off, and the acid neutralized by
lime, baryta, strontia, or their carbonates. If sulphuric acid is
used, adding a little oxalic acid, or oxalate of ammonia dissolved
in water, which precipitates the calcareous salts. The liquid is
drawn off, and if not " clear, it is clarified by albumen, white of
" eggs, blood, &c., or other known means."   " For the manufac-
" ture of syrups the operation is the same, the length of the
" action only is increased, we stop when the alcohol no longer
" precipitates dextrine, the syrups thus obtained are then
" evaporated, in order to bring them to the desired degree of
" concentration."

[Printed, 8d. Drawing.]

A.D. 1858, June 15.—N° 1355.

WARNER, HAMILTON SHIRLEY.—(*Letters Patent void for want
of Final Specification.*)—" Improvements in the manufacture of
" decolorizing and purifying charcoal." These are, making such
charcoal as follows :—First, " pitch from the pitch lake at La
" Brea in Trinidad " is melted and " the porous stone òr gravel
" of the hill of San Fernando in Trinidad (which is I believe a
" silicate of alumina and magnesia)," or any other substance not
fusible or decomposible by heat, " such as pumicestone, broken
" brick, spent animal charcoal, calcined bones, &c., would be
" suitable," ground or otherwise, in a dry state, is added with
stirring, raising, and continuing the heat until all volatile matter
is given off at a red heat ; any dust is removed by sifting.
Second, the materials, pitch. clay, &c. may be ground to powder
and mixed, and heated in small iron pots, or heated while passing
through an iron or other tube, heated to a strong red heat in a
furnace, or the powders may be fed into the upper end of a re-

s.   Q

volving cylinder, open at both ends, heated by a furnace or
carbonized upon a plate with a fire underneath. " Powdered
" pitch may also be mixed with spent animal charcoal in
" reburning."

Third, " the pitch and clay may both be reduced to fine powder
" mixed, then slightly moistened with water, and granulated by
" appropriate machinery." " The grains are then to be dried
" and burned in any of the ways above mentioned."

[Prnted, 4*d*. No Drawings.]

A.D. 1858, June 23.—N° 1415.

SPENCER, THOMAS.—" Improvements in the treatment of iron
" ores and ferruginous sands, and certain applications arising
" therefrom." These are, oxide of iron is mixed with carbonaceous matter, as " charcoal, coke, coal, wood, or peat, or the dust
" of either of these, or that of any other matter containing
" carbon," and heated in a retort such as for gas, having a small
aperture fitted with a stop cock, to permit the escape of any gas
or moisture, for two to six hours, at a dull or cherry red heat ;
when gas ceases to come off the stop cock is shut, and the mass
allowed to cool, or it is cooled under water. If carbonate of iron
is used in place of the oxide above, it is first calcined, to drive off
the carbonic acid before it is mixed with the carbonaceous
matters. The result of the carbonization as above, is a compound
named " ferrosoferric carbide, or indifferently ' oxy-carbide,' or
" ' magnetic-carbide,' each term implying the same substance."
The application of this substance to the purification of several
substances is given, among which is " the purification of saccharine
" fluids, such as those derived from beet-root or the sugar cane,"
by substituting " this ferruginous compound in lieu of charcoal,
" used in the refinement of sugar." " Where the powers of the
" magnetic carbide are expended in the course of the process of
" refinement, they are to be revived by reheating in a retort or
" oven." " A similar process of reheating is now adopted for the
" revivification of the animal charcoal used in the manufacture
" of sugar."

[Printed, 6*d*. No Drawings.]

A.D. 1858, July 13.—N° 1574.

BUCHANAN, GEORGE.—" Improvements in sugar cane mills."
These are, first, " the general arrangement and construction of
" sugar cane mills," as follows :—

Second, " greatly increasing the strength of the frames; depen-
" dence being placed upon the cohesive strength of wrought-iron
" instead of cast-iron.   The latter material is only used as a con-
" venient medium for keeping the whole together and securing it
" to the sole plate."   Reducing the weight of the side frames
one-half and having sheet-iron plates at sides.   Greatly reducing
the height of the frame, which lessens the vibration and makes
the supplying or feeding with canes easier, and reduces the height
of the feed board.

Third, " the mode of drawing out the lower rollers horizontally "
without "raising or lifting the said rollers or moving the top
" roller, as in the ordinary mill, by sliding back the rollers in a
" horizontal direction, an opening being left in the standard
" behind the rollers on each side," in which a wrought-iron tie-
bar is placed, "which can with facility be removed when it is
" wished to draw the rollers out."

Fourth, "the wrought-iron tie-bars at sides and their peculiar
". construction, and the dispensing thereby with the use of nuts
" for adjusting screws of lower rollers."   The adjusting screws
" pass through the wrought-iron tie-bars which are tapped to
" receive them."

Fifth, "dispensing with the use of cast-iron flanges to lower
" rollers " by "placing a sheet-iron plate opposite to ends of top
" rollers."

[Printed, 6d. Drawing.]

A.D. 1858, August 17.—N° 1878.

LICHTENSTADT, DAVID, and DUFF, CHARLES.—(*A communication*.)—(*Provisional protection only*.)—" Improvements in
" treating tan and tanning refuse to obtain valuable products
" therefrom."   These are, in reference to this subject, as follows :
—" After the tan has been dried by the air, I distil it in close iron
" cylinders, generally used for the distillation of coal, and which
" cylinders are in communication with iron pipes so constructed
" that the ligneous acid and tar runs into a lower cistern, and
" the hydrocarbonic gas from thence is conveyed under the fire-
" place of the above-mentioned distilling apparatus to be con-
" sumed."   The charcoal remaining in the retort is prepared
" for filtering, sugar refining, deodorizing, bleaching of spirits
" and other liquids, &c.," by taking the charcoal when hot from
the cylinder, stirring it in a liquid composed of 100 pounds of

water and 5 pounds of muriatic acid, for about half an hour, taking it out, drying it " in a warm place, and it is fit for use " ground fine or in small pieces."

[Printed, 4d. No Drawings.]

A.D. 1858, August 31.—N° 1981.

MARGESSON, PHILIP DAVIES.—"Improvements in treating " sugar canes and other canes containing saccharine matter in " the preparation of food for animals, also in manufacturing " sugar and worts or wash for brewing, distilling, and vinegar- " making, and applying the resulting fibre in the manufacture of " paper." These are, in reference to this subject, the canes as soon as ripe are cut into thin slices or cuttings, and then dried, by preference, in the sun in the open air, and reduced to a fine powder in a suitable mill, and made into cakes by compressing. " The " saccharine matter contained in the compressed powder of cane " may, when about to be dissolved out of the powder by water, " and the solution of syrups thus obtained may be used for " crystallizing into sugar," &c.

[Printed, 4d. No Drawings.]

A.D. 1858, November 16.—N° 2578.

BRUÈRE, ALEXANDRE MARTIN.—(Provisional protection only.) —"The novel application of hydrogen gas to various purposes in " the arts." These are, in reference to this subject, " in operating " upon wood for the production of charcoal; charcoal of better " quality will be produced by the introduction of hydrogen gas " into the carbonizing chamber, as the hydrogen will unite with " and carry off the matters foreign to charcoal contained in the " wood. The same remark will also apply to bones when operated " upon for the purpose of obtaining animal charcoal, and to " lignites and peat."

[Printed, 4d. No Drawings.]

A.D. 1858, November 19.—N° 2613.

HOWE, GEORGE, and NORTON, JOHN. — " An improved " method of boiling water or worts for brewerys, distillerys, &c., " by steam, or for heating rooms, public buildings, churches, " chapels, factorys, &c." An oblong steam boiler is shown, from the upper part of one end of which is an outlet pipe with a tap, which pipe "conducts the water or steam into a coil pipe,

" which may be made of any metal, and which coil pipe is intro-
" duced into a boiling pan or vat, thereby causing the liquid in
" the pan or vat to boil quickly, the pipe is then carried from the
" pan or vat into the boiler " at the top of the end furthest
from the exit pipe, " thereby returning any condensed or waste
" steam, at the same time causing a great saving of coal, and
" preventing loss of water, with much less chance of explosion.
" This plan is adopted for drying houses of every description,
" bleach yards, sugar houses, soap works, and similar purposes;
" also for heating buildings, churches, chapels, factorys, &c."
Many variations may be made in the above arrangements,
without, it is said, " deviating from the principles of the said
" invention."

[Printed, 8d. Drawings.]

A.D. 1858, November 29.—N° 2722.

BENSEN, GERD JACOB.—(*Provisional protection only.*)—" Im-
" provements in cleansing or purifying animal charcoal after it
" has been employed by sugar refiners." The charcoal is laid
uniformly on a strainer or false bottom in a vessel, and " the water
or liquid with which the washing is to be performed is run on to
the surface of the charcoal; a partial vacuum is then produced by
pumps or otherwise in the space between the bottom of the vessel
and the under surface of the strainer or false bottom, and this
causes the washing liquid to pass rapidly through the layer of
charcoal, which is thus quickly cleansed; in a similar manner air
may be drawn through the charcoal, and steam may be blown
through it by leading steam by a pipe into the space between the
two bottoms of the vessel. The charcoal may be afterwards
" revivified by burning in the usual manner."

[Printed, 4d. Drawing.]

A.D. 1858, December 2.—N° 2755.

MAC KIRDY, LAUCHLAN.—" Improvements in the manufac-
" ture of sugar." These are, " the saving of time and labour in
" certain processes " " together with other advantages," as follows :
First, " the filling of sugar moulds direct from the heaters,
" such moulds being arranged in sets in frames " which either
have wheels or they are placed on " carriages for conveyance to
" and from the heaters."

Second, "placing the mould frames or carriages at a level
" under or near the heaters" sufficiently low to run the liquid
sugar direct into the moulds, the heaters having outlets "formed
" with two or more mouths or spouts so 'as to fill two or more
" moulds simultaneously."

Third, "the fixing of the spikes or studs for closing the mould
" apices on fixed or adjustable frames" which close the apex
apertures of the moulds whilst the sugar remains liquid.

Fourth, "providing such mould with two or more apex aper-
" tures and corresponding spikes or studs," when the time arrives
for draining the sugar.

Fifth, "doing away with the old system of draining pots," the
sugar being drained off "into a pan adjusted or fitted into the
" bottom of the frame or carriage," removable when filled or
discharged by an outlet; or "the moulds may be originally
" placed in sockets in the frame or carriage, the pointed studs
" being carried by an adjustable frame, by means of which they
" are lifted into the moulds, and lowered thence when the
" drainage is to commence."

[Printed, 8d. Drawing.]

A.D. 1858, December 3.—N° 2761.

HENRY, MICHAEL.—(A communication from J. A. Lagard.)—
" Improvements in manufacturing and revivifying bone black or
" animal charcoal, and in kilns and apparatus employed therein."
These are, first, carrying out the above "in kilns, receivers, and
" apparatus or means," afterwards described.

Second, "constructing the receivers (or apparatus for holding
" the bones or charcoal) with feet for raising them above the sole "
and with an escape "valve or cap free to slide up and down under
" the pressure of the gases liberated from within the receiver, its
" travel being regulated, preferably, by a bent wire held in the
" orifice."

Third, "constructing and arranging the kiln with a fire-place at
" each angle of the sole" having "rows of holes or passages
" communicating with transverse flues beneath, leading through
" up-flues into an exit flue surrounding the upper part of the
" kiln," the products of combustion finally enter into chimneys
at each end of the kiln. The raised receivers are completely
enveloped in flames, &c.

Fourth, the arrangement of kilns as above together with the arrangements necessary to the carrying on of the operation continuously. The kilns sufficiently heated, the receivers are introduced therein by means of an instrument by which they are clutched, and which has a shank running on rollers and terminating in a handle ; " one receiver is set over each hole. The doors " through which they are introduced are arranged alternately so " that each may front one of the vertical flues corresponding to " each transverse row of holes, by the time the last batch of " receivers is in the first will be ready for removal and may be " replaced by fresh ones."

Fifth, "the mode of constructing, combining, or providing kilns " with columns or tubes arranged thus :—an inner within an " outer, concentric to it or nearly so, and so contrived and com- " bined that the bone black undergoing revivification may pass " or be contained between two heated surfaces, the inner of which " provides for the escape of the vaporous productsof dessiccation," and the arrangement and combination of the same.

[Printed, 10*d*. Drawings.]

A.D. 1858, December 4.—N° 2771.

CAMERON, JOHN. — "Improvements in apparatus for the " manufacture of sugar." In the Provisional Specification these are said to be, first, " constructing a battery for boiling or evapo- " rating cane juice or other saccharine liquids of a series of open " pans, which are fitted and supported by their conical edges or " flanges in cast iron plates, which form the top of the building " or furnace; these cast iron plates are bolted together and " formed with a gutter along each side to receive and convey the " cane juice to the grand or large pan, and also to carry off the " skimmings. The cast iron plates may be inclined or conical, " so as to conduct back into the pans any liquid which may boil " over.

Second, cleansing the evaporated or granulated sugar from molasses by " submitting it to the action of a vacuum in a pan, " or in a series of pans or boxes with false bottoms formed of " thin bars of galvanized iron or copper or other suitable ma- " terial fitted or placed very close together. The space under " the false bottom communicates with a close vessel or monte- " juice, which is exhausted by an air pump worked by a steam " engine. The molasses is then drawn through into the monte-

" juice." The molasses is forced up a pipe by means of steam into the monte-juice into a cistern, and from thence into what is called " a wetzel pan, having a revolving cylinder or assemblage of " tubes traversed and heated by the waste steam of the engine " which drives the air pump," and placed in the vacuum boxes or pans for crystallizable sugar. " The cane juice or other saccharine " liquid evaporated or concentrated in the battery may be further " concentrated and granulated in the witzel pan, and run into the " vacuum boxes and cleansed as above mentioned." The sugar emptied, the boxes are cleansed by blowing steam into the under part of each box.

In the Final Specification which is filed by order of the Lord Chancellor by Samuel Cameron ; it is said, the sugar may be kept warm in the vacuum boxes by surrounding the apparatus with a steam chamber or jacket, and also " a shower of water or syrup " may, if desired, be allowed to fall upon the sugar during " the operation, or one portion of the operation," so as to wash off the molasses.

[Printed, 8d. Drawing.]

A.D. 1858, December 8.—Nº 2810.

CHANTRELL, GEORGE FREDERIC.—" Improvements in appa- " ratus applicable to the manufacture and revivification of " animal or vegetable charcoal." These improvements are said to be chiefly applicable to apparatus described in No. 1313, A.D. 1858, and are, first, " the construction of the retort or " char chambers with corrugated or undulated sides," the corru- gations, in preference, running horizontally, so as to cause agitation in the char in passing down.

Second, " forming the openings in the bed plates and the upper " portion of the junction pipes between the bottom of the retort " and char chamber with as great a horizontal sectional areal " opening as the retort chambers themselves, and dividing the " lower part of the junction pipes into two lateral branches to which the coolers are attached, " so that one-half of the char " runs into the right, and the other half into the left hand " cooler."

Third, constructing " metal charcoal coolers with corrugated " or undulating sides;" in preference the corrugations run vertically.

Fourth, "the construction and application of the iron lumps to
" the under side of the bed plate, upon which the kiln is built,
" having oblique passages formed therethrough, and provided
" with sliding covers and rods for working the same." Each
lump has six hopper-mouthed holes, which are carried down
obliquely on alternate sides, for the purpose of dividing the
charcoal that it may fall into two coolers.

[Printed, 10d. Drawing.]

A.D. 1858, December 24.—N° 2940.

ELERS, WILLIAM, and FINK, LUDWIG.—(*Provisional protection not allowed.*)—"The filtration of dissolved sugar & other
" liquors." This consists of "a vessel with two bottoms, the
" upper one perforated with one or more holes, to which are
" attached the bag or filter. The two bottoms are placed apart
" from each other, thereby forming a chamber into which the
" liquor to be filtered is introduced by a pipe from a higher level,
" which liquor rises through the hole or holes into the bag or
" filter, & when filtered escapes by a passage or pipe placed above
" the chamber, as before described, between the two bottoms,
" while the scum, sullage, and refuse remains in the said cham-
" ber, or may fall into another chamber placed below it, from
" whence, through a door or aperture constructed for that
" purpose, the scum, sullage, or refuse can be removed, and the
" chamber cleansed without disturbing the bags or filter."

[Printed, 4d. No Drawings.]

1859.

A.D. 1859, January 1.—N° 15.

PRINCE, ALEXANDER.—(*A communication from Prince Sergius
Dolgoruki.*)—(*Provisional protection only.*)—" Improvements in
" the construction of cylindrical presses." These are, in reference to this subject, an arrangement of apparatus adapted " to
" the extraction of sugar from beet roots or sugar cane, and
" other similar purposes." The " sugar cane or other substance
" is first ground or grated into a pulp by any convenient means,
" and is then passed into the space or opening formed by upright

" metallic plates or planks, and between two revolving cylinders
" which give the required pressure. These cylinders are formed
" of metallic discs at their ends, having cross wooden pieces
" extending from one disc to the other, which form the cylinders,
" they are then covered with an elastic or flexible substance, such
" as vulcanized caoutchouc, for the purpose of squeezing and
" extracting the liquid from the solid parts, and at the same time
" delivering the solid matter, leaving the liquid behind. The
" ends of the cylinders are also covered to prevent the liquid
" escaping between them and the outer planks of the machine
" The further cylinder rests on moveable bags, which are held in
" the desired position by a lever, to which is attached a weight
" according to the pressure required. A case or bowed box is
" made to pass to and fro or up and down, as may be found best,
" at the same time dislodging the solid parts from the cylinders,
" leaving the juice or liquid behind." Spiral springs are placed
at each end of the cylinders to ease the pressure on their surfaces.
The pulp rises between the cylinders, " at the same time retaining
" the juice which the case or box draws to throw behind into a
" reservoir to receive it."

[Printed, 4d. No Drawings.]

A.D. 1859, January 7.—N° 58.

REYNOLDS, HENRY.—(*Provisional protection only.*)—" Im-
" provements in refining and decolorizing saccharine substances."
These are, decolorizing " sugar, molasses, treacle, &c., by other
" means than, or in combination with animal charcoal;" by
mixing " meta-stannate of alumina," or " stannate and meta-
" stannate of lime as well as free meta-stannic and stannic
" acids, and several other chemical compounds of tin, for ex-
" ample, the protochloride, and perchloride, the sulphate, the
" bichloride, and nitro-muriate, also the stannates of soda and
" potash," in " about the proportion of 1 part of the base to
" about 1000 or 2000 parts of sugar according to the intensity
" of its color," boiling " the whole; and as the scum rises to the
" surface, it brings with it the whole of the base, with more or
" less of the coloring matter; this scum is removed in the
" ordinary manner. The solution is then passed through the
" filtering bags as usual, and also through a bed of animal
" charcoal if the color is not sufficiently discharged; the animal
" charcoal will also remove any remaining trace of the discoloring

" agent employed." To overcome the odour of blood when used
in the clarification of such solutions, adding to such solutions a
minute quantity of chloride of zinc.

[Printed, 4d. No Drawings.]

A.D. 1859, January 26.—N° 235.

ALEXANDER, WILLIAM RICHMOND. — " Improvements in
" furnaces and apparatus for the manufacture of sugar and for
" the consumption or prevention of smoke." These consist " of
" a long ' fire mouth ' or drying chamber, fitted up immediately
" in front of the furnace," formed " with a double bottom, per-
" forated on the upper side for the discharge into the chamber
" of heated air. The ' megass ' or waste sugar cane is deposited
" in this chamber, and pushed into the furnace from time to time,
" for use through a perforated door, which forms the communi-
" cation between the drying chamber and the furnace. The
" furnace bars are tubular, opening at one end into a hollow
" fire-bridge at the back end of the furnace, and at the other
" into the false bottom space of the front drying chamber. The
" grate bars are also minutely perforated on their upper side, for
" the discharge of air amongst the burning fuel upon the bars.
" Air is forced into the furnace by a blowing fan or other me-
" chanical force at the extreme back end of the structure of the
" furnace. The pipe from the fan conducts the air into and
" through a set of heating cylinders or chambers studded with
" lateral spikes or projecting pieces of metal inside and out, to
" enable the air to take up a superior amount of heat. After
" leaving the heater, the air is forced along a conducting pipe
" laid along the bottom of the main flue, and beneath the steam
" boiler, usually fitted up at the after end of these structures, and
" beneath the ' teaches ' or evaporating pans at the front. The
" air thus takes up heat as it passes along until it is finally
" discharged into the hollow furnace bridge, whence it passes
" partly into the furnace directly amongst the burning fuel, and
" partly through the bars into the space of the false bottom of
" the drying chamber; the latter portion dries the deposited
" megass, and then finds its way from the chamber through the
" perforated division door into the furnace also; in this way the
" full and efficient supply of highly heated air, subdivided into
" numerous minute jets, effects the perfect combustion of the
" fuel, and prevents all or nearly all discharge of smoke. The

" same arrangements or modifications of it will answer for
" burning coal and other fuel." Air alone has hitherto been
used in the above description of the apparatus, but it is proposed "when the nature of the fuel shall render it expedient to
" convey a jet of steam into the air pipe, and so cause the mingled
" air and steam to enter the furnace."

[Printed, 1s. 4d. Drawings.]

A.D. 1859, February 7.—N° 338.

CHANTRELL, GEORGE FREDERIC. — " Improvements in the
" treatment of charcoal after its manufacture or revivification."
It is stated that newly prepared or reburned charcoal "is not
" so powerful in decolorizing as it is after having been exposed
" for some time to the action of the atmosphere," and the spreading out and exposing, also causing a current of air to act on
such charcoal, likewise the apparatus employed, are claimed.
The charcoal is fed to the upper apron of a series of endless
bands from a hopper. " The charcoal falling upon the first
" endless band is carried forward until it falls off on to a sieve or
" riddle placed in an oblique direction ; the dust falling through
" the meshes of the riddle or sieve is received in a closeted box,
" from which it is removed from time to time as required. The
" charcoal slides on to the next endless band, which travels in an
" opposite direction to the first;" from this band it passes on to
a sieve as before, and from the sieve on to a third band and sieve ;
if the charcoal is then free from dust, endless bands are now only
employed. Any required number of these horizontal endless
bands are used, their surface being kept " flat or nearly so by
" applying one or more rollers to the under surface of the lower
" portion of the web. These rollers may be mounted upon the
" ends of lever beams, which are weighted so as to press the
" rollers against the under surface of the web ; or the stretching
" and supporting rollers may be banded together by means of
" narrow endless straps or bands." In preference, the band
material is permeable to air, and a natural or artificial draught is
applied to the charcoal in its passage through the machine.

[Printed, 8d. Drawing.]

A.D. 1859, February 8.—N° 355.

ASPINALL, JOHN. — " An improvement in the refining of
" sugar." This is in the " blowing up " or melting of raw

sugar, and is for the purpose of preventing the sugar, when mixed with water and blood, the sugar being of greater specific gravity than the water, falling to the bottom, and coming in contact with the steam becoming carbonized, and thus much treacle being produced ; by " causing the melting of the sugar to take place before " coming in contact with the steam, steam pipes, or heated " bottom or heated sides of the pan, whereby no carbonization " will take place, and consequently there will be less treacle than " is now produced in the ordinary method of melting." To effect this, the sugar is supported upon " a perforated false bottom " or by other suitable means," and the water level kept " up to, " or just above the level at which the bottom or lowest stratum " of the sugar is maintained on the perforated false bottom or " other support. The water melts the sugar, and the sugar " enters the lower part of the pan in a state of solution or ' liquor,' " in which state steam or steam pipes or other heated medium " exercise no carbonizing effects." It is obvious that pans may be variously constructed to carry out the above operation, which may be termed " surface melting."

[Printed, 6d. Drawing.]

A.D. 1859, Febuary 9.—No. 370.

NEWTON, WILLIAM EDWARD.—(*A communication from Emile Rousseau.*)—" An improved mode of bleaching and purifying or " refining sugar and vegetable juices or extracts." This consists in employing " for the purpose of bleaching and purifying or " refining sugar and vegetable juices and extracts of some highly " oxidized substances, such as the peroxides of manganese and " iron, either artificial or natural, or the carbonates of these " bases." The peroxides are hydrated. To give an approximate proportion, the quantity of hydrated peroxide of iron " employed " in the state of a firm paste is generally from three to five per " cent. by weight to the weight of sugar." If the sugar be acid, necessarily a quantity of chalk is first introduced into the boiler, " sufficient to neutralize the acid. If on the contrary the sugar " be highly charged with lime, a small quantity of carbonate of " iron must be added to the peroxide," and the carbonic acid from same will precipitate the lime as carbonate. " When from " any cause, the sugar has dissolved a small quantity of oxide of " iron, it may readily be removed by adding to the syrup a small

" quantity of chalk (carbonate of lime) before filtration. The
" addition of this substance will in all cases produce a beneficial
" effect."

[Printed, 4d. Drawings.]

A.D. 1859, February 23.—N° 492.

DAVIES, George.—(*A communication from Augustus Jouan.*)—
(*Provisional protection only.*)—" Improved apparatus applicable
" to the evaporation of saccharine liquids, and for the concen-
" tration of heat for other purposes." " This apparatus is
" applicable to the ordinary pans employed for boiling sugar,"
and consists in applying a floating cover thereto, the exterior or
outer form of which " is convex, or nearly hemispherical, with a
" tube projecting upwards therefrom, and the lower or interior
" surface is of a concave, conical, or funnel shaped form. The
" lower edges of these two parts are united or joined together,
" and they are each furnished at the upper part with a hole, in
" which is inserted a tube, the two tubes being situated one
" within the other, and having an internal or space between
" them. The outer tube is made slightly conical at the top, so
" as to reduce its diameter to about that of the inner tube, and
" the tubes are there united or joined together." There is a
small hole near the top of the outer tube to allow of the expansion of the air between the tubes and inside the hollow cover
when heated. The base of the cover is made circular or of other
form, so as nearly to fit the interior of the pan. At the top of
the tubes (which is open) is placed a moveable sieve, designed for
detaining the grosser parts of the liquid at the earlier part of the
operation, and beneath this on the neck of the outer tube is a
larger sieve of an annular form. The cover is furnished with
screw rods passing through screwed bushes, for the purpose of
plunging and holding the cover below the surface of the liquid."

[Printed, 4d. No Drawings.]

A.D. 1859, February 24.—N° 505.

WAGNER, Jean Henri Guillaume Daniel.—" Apparatus
" for cleaning water and removing all matters in suspension and
" dissolution contained in it, water intended to feed generators of
" all sorts (applicable also to other purposes), which, besides
" previous to its getting into generators, is heated to the highest
" degree without almost any expense." These are, in reference

to this subject, a vessel made of iron divided into a receiver above and a reservoir below; in the upper part of the receiver is an inlet pipe, and about the same level is an outlet pipe for the escape of any excess of liquid; a cock regulates the flow of liquid from the receiver into the reservoir; this cock is worked from the outside by means of hand gear, with an index hand moving over a divided scale on the cover. The reservoir has a series of partitions, so as to form a continuous passage. Near the bottom of the reservoir is a pipe for injecting a current of air or other gas, and near the top of the same is an outlet pipe for the air or gas injected. The outlet pipe is on the other side at the bottom of the reservoir, and a cast-iron cage is interposed upon it having a copper sieve covered with a disc of filtering felt. " If " it is desired to concentrate a syrup, I run the liquid divided " in thin continuous sheets or films throughout the surface of " all the partitions, injecting always inversely to the run of the " liquid," "a current of air or other gas heated or not by means " of an exhausting or blowing ventilator."

[Printed, 2s. 8d. Drawings.]

A.D. 1859, March 7.—No 595.

ASPINALL, JOHN.—" Improvements in evaporating in vacuo." These are, "evaporating in vacuo by means of steam or hot air " pipes, whether free to rise and fall or stationary, maintained at " the upper part of the vacuum pan, or at or about the surface " of the liquor to be evaporated," the object being to avoid the formation of treacle, which is said is produced by the direct con- tact of the saccharine fluid with hot surfaces. The bottom of the pan is surrounded with "a jacket for containing hot water, air, or " steam."

[Printed, 10d. Drawings.]

A.D. 1859, March 16.—No 656.

SEYMOUR, GEORGE.—(A communication from Hypolite Leplay.) —(Complete Specification but no Letters Patent.)—" Improvements " in making refined sugar, and making potash and soda from the " residues." These are, " employing caustic baryta or sulphuret " of barium " in " making refined sugar from raw sugar or syrup " from beet root or other saccharine vegetables, or from molasses " made during the fabrication of sugar from beetroot and other

" saccharine vegetables ;" also " in the fabrication of potash and
" soda," as follows :—Caustic baryta is dissolved in a series of
vessels containing hot water, and steam is introduced into each by
pipes. The solution of baryta should be 30° Beaumé. The syrup,
however obtained, is heated in another vessel by means of a steam
pipe, and both being boiling about two parts of the baryta solution
are added to one of molasses when the molasses is of the specific
gravity of 48°. About ten gallons of molasses are operated upon
at once, and are thrown into a series of vessels furnished with false
bottoms, the mother liquors filter through the wire gauze; the
saccharate of baryta is washed with a weak solution of caustic
baryta, and is finally decomposed by suspending it in water
and passing carbonic acid gas through it. Carbonate of baryta is
formed and precipitates, and the sugar is in solution with a little
bicarbonate of baryta and traces of carbonate ; this is boiled,
treated with a soluble sulphate (of lime or alumina), clarified by
albumen, and " filtered by a Taylor's filter," and is ready to be made
into loaves in the usual manner. The mother liquors containing
caustic baryta, &c. are treated with carbonic acid gas, or crystals of
hydrate of baryta, are separated by evaporation or cooling. The water
is evaporated and the residue calcined in a reverberatory furnace ;
" the potash, and salts of potash, and soda remain, having a con-
" siderable commercial value." When sulphuret of barium is used
the solution should be 32° Beaumé and the quantity at least one-
third greater than the solution of caustic baryta.

[Printed, 1s. 4d. Drawings.]

A.D. 1859, March 17.—N° 675.

HUGHES, EDWARD THOMAS.—(A communication from Saint
Jean Thérése.)—(Provisional protection only.)—" Improvements
" in machinery or apparatus for crushing sugar canes and other
" materials." These are, fixing " upon a platform or foundation
" a strong framework of wood or metal, having bearers for the
" shafts of the crushing cylinders, and also for the main driving
" shaft, which latter is divided in the centre and furnished with
" two cranks. Upon the shafts of the two lower cylinders,
" toothed wheels are fixed in the usual manner, geared into
" another wheel on the top cylinder shaft, at the end of which is
" a large toothed wheel geared into a pinion on the driving shaft,
" the cranks of which are connected by rods to a double set of

" compound levers, which are again connected to long levers at
" each side or end of the apparatus. The aforesaid long levers
" are acted upon alternately either by hand or power, giving
" motion to the intermediate levers and cranks, the effect of
" which is transmitted by means of the gearing to the crushing
" cylinders, which act effectually either by the power of two men
" or by any other equivalent obtained by water or any other
" motive power."

[Printed, 4d. No Drawings.]

A.D. 1859, March 21.—N° 708.

BAUCQ, ARNOULD.—(*Provisional protection only*.) — " Main-
" taining graters mechanically." This consists as follows, in the
manufacture of sugar to feed or maintain the graters, by placing
a " moveable bottom on a carriage in the feeding trough, where it
" is supported from above, it is moved backward and forward by
" excentrics and pullies motived by hand or other power; pieces
" of wood are used to retain the matters during the backward
" movement, and triangular knives cut and distribute them. A
" knife is also placed at each end of the moveable bottom, rollers
" on the balance bring back the pushers, a catch releases the
" moveable bottom, and a sheet iron guard spreads the matter of
" each movement, or if the motion be given from beneath, an
" arch board may be substituted for the guard. It is found
" advantageous to couple two graters and feed or maintain them
" by one apparatus; but for one grater the machinery is less
" complicated and one knife will suffice. The mode of com-
" municating motion may be varied, and be from above or below,
" and this improvement may be found applicable to other
" purposes than sugar making."

[Printed, 4d. No Drawings.]

A.D. 1859, April 23.—N° 1025.

MARSHALL, JOHN, junior.—" Improvements in filtering and
" depurating fluids." These are, in reference to this subject:—
A vessel or reservoir constructed with an aperture at top, through
which it is filled with the liquid to be operated on. Inside this
vessel is a bag of waterproof material, somewhat larger than the
interior capacity of this vessel; at the bottom of this vessel is an
opening for forcing water or other fluid under the flexible bag,

and so force the liquid to be filtered out through a tube in the top of the vessel or reservoir into filters constructed of a series of hollow frames or boxes, covered on both sides with filtering material held down by wire gauze or other convenient substance; both the filtering material and the wire gauze overlap the edges of the box, so that when the boxes are brought together alternately they will make a secure joint. Sometimes the liquid to be filtered is put into the flexible bag and forced out of it by the introduction of water into the vessel outside of the bag, the bag being connected to the pipe leading to the filters. "When it is " desirable to depurate or exert any chemical action on the fluid " to be filtered," the boxes with the filters " are charged with the " material necessary to effect the object, or the same may be " previously incorporated with the fluid prior to its passing to " the filter boxes." Instead of making the discharge boxes of iron, a block of wood cut to the length and breadth of the filtering boxes may be employed. These blocks are to be grooved or corrugated on each side, avoiding only those parts which come in contact with the edge of the filter boxes, and which parts must be planed true to make a perfect joint; over the grooves is wire gauze.

[Printed, 10d. Drawing.]

A.D. 1859, April 30.—N° 1085.

FRANCIS, EDWARD.—" Improved apparatus applicable to the " treatment of tea and other useful purposes." This consists of " an horizontal flat sieve or riddle with a vertical circular side " which may be formed solid or perforated," and " provided with " horizontal radial arms which spring from a revolving centre " vertical shaft," "formed flat and placed at an angle (of say " about 60°) to the bottom of the sieve" but they " may be " formed of any convenient shape and placed in any convenient " position" and so fitted that their distance can be regulated from the bottom at pleasure "and that they will yield upwards " upon any undue amount of pressure" being applied to them, and "they may have a tongue of leather, india-rubber, or other " suitable material fitted to their lower edges." In some cases they may be entirely or partially covered "with soft leather, " woollen cloth, or other suitable material." If preferred "an " horizontal revolving wheel having a suitable number of slots or " openings (which radiate from the centre)" may be substituted

in lieu of the radial arms. The whole is covered and at the top
is a hopper with a slide in the throat for regulating the supply of
the material. Below is a receptacle for the material after it has
been acted upon. In some cases the hopper is placed a short
distance from the mill and connected by an enclosed shoot to
which a vibrating motion may be given if required, and the under
side is a riddle or sieve through which the smaller particles fall
into a receptacle the larger portions only passing into the mill.
The shaft carrying the radial arms or wheel is rotated " by power
" applied through a wheel or pinion a band or other suitable
" mechanical contrivance." "This apparatus is adapted to
" breaking and sifting tea, dressing seeds, disintegrating chicory,
" raw sugar, and a variety of articles."

[Printed, 8d. Drawing.]

A.D. 1859, May 5.—N° 1131.

REYNOLDS, HENRY.—" Improvements in refining sugar and
" other saccharine substances." These are, first, " the dissolution
" of sugar and other saccharine substances by means of an
" ascending or descending current of fluid propelled by mechanical
" means." In carrying out this, employing " a vessel constructed
" as an ordinary 'blow up pan' or other suitable vessel fitted
" with a perforated platform or false bottom," through the central
portion of which is driven " a current of the dissolving fluid by
" means of a pump, or screw, or fan, or any other convenient
" method, which driving apparatus may be fixed in the melting
" pan itself or the same apparatus may be made to serve several
" pans." In melting sugar by free steam " causing an external
" circumferential descending current of fluid to pass over the
" steam pipes in its passage to the inferior central portion of the
" false bottom upon which the unmelted sugar rests, " causing
" the sugar to be suspended in and thus to be always surrounded
" by the dissolving fluid."

Second, " the employment of the salts of tungsten either as the
" oxide or in combination with a base," for example, lime or
soda, potash, or a metallic oxide. The agent above referred to is
mixed in the proportion of about one part to one thousandth
parts of sugar according to the intensity of its colour, boil the
whole, and if the solution is acid to litmus paper add lime free
or combined, remove the scum and "the partially decolourized

" liquor is run through the bag filters in the usual way and also
" through a bed of animal charcoal if the colour is not satis-
" factory."

[Printed, 4d. No Drawings.]

A.D. 1859, May 11.—N° 1178.

MANBRÉ, ALEXANDRE.—(*Provisional protection only.*)—" An
" improved method of extracting and purifying sugar called
" glucose and 'sirop de fecule,' from potatoes or fecula, or
" starch, or dextrine, for the purposes and uses of brewers, dis-
" tillers, vinegar makers, colouring makers, or otherwise." These
are, first, "the saccharification of a mixture of nine parts of
" potatoes or fecula, or starch, or dextrine, with one part of flour
" from rice or maize, together with forty parts of water, the said
" mixture to be converted into saccharine matter by the re-action
" of the disastasis, or by the re-action of germinated barley,
" called malt, or by the re-action of sulphuric or any other acid,
" or by the action of heat."

Second, in filtering apparatus, having a hermetically closed
vessel with a perforated false bottom. " Supposing the apparatus
" charged with saccharine liquid, then air or steam or any other
" agent is introduced in over the mass through a cock provided
" for the purpose ;" when the liquid "is forced through, the filter
" and false bottom is completely purified," is evaporated to obtain
" the sugar either in a state of syrup or hardness, or granulation,
" or instead of introducing any agent from above to force the
" saccharified liquid through the filter, a pump or vacuum may be
" employed to draw it through it."

[Printed, 4d. No Drawings.]

A.D. 1859, May 18.—N° 1230.

TERRY, ALEXANDER ROBERT.—(*Provisional protection only.*)—
" Improvements in apparatus for sawing and cutting up loaf
" sugar." These are, first, "the use of a circular saw to reduce
" the common sugar loaf into long square pieces called sticks."

Second, "the use of a cross-cut knife in connection with a
" feeding apparatus to reduce the sticks into lumps of the sizes
" required."

Third, " the use of a sieve to separate the imperfect lumps."

"The parts performing these various operations are connected
" together in the usual manner and embodied in one machine,"

[Printed, 4d. No Drawings.]

A.D. 1859, May 30.—N° 1332.

GREEN, WILLIAM.—" Improvements in washing or purifying
" and treating sugar." These are, first, "removing of color from
" the drippings from ' bastards ' or ' low pieces,' commonly termed
" molasses or treacle, by means of animal charcoal or other car-
" bonaceous matters, by preference, in a finely divided state, or
" arranged in thin layers of different degrees of fineness, and
" kept in motion by the force of gravity or by mechanical
" means." Also " the precipitation of the carbonaceous matters
" employed for such decoloration or other matters in treacle, by
" means of aluminous, siliceous, or argilaceous earths, then pass-
" ing the treacle through a filter of granulous charcoal, the said
" matters being kept cold or at a low temperature during the
" process."

Second, " the use of purified or decolorized treacle for dis-
" placing dark or inferior bastards or other low quality sugars,
" and at the same time recoating such cleansed crystals."

Third, " the use of fluids of reduced temperature for dissolving
" or moistening the treacle or dark coloring matter contained
" upon ' bastards ' or other low quality sugar, also the use of ice
" for such purpose."

Fourth, "the manufacture of small blocks of sugar of regular,
" irregular, or ornamental shape," irregular preferred, by forcing
the same into dies or moulds.

Fifth, " bleaching or decolorizing of sugar by means of sul-
" phur, sulphurous acid gas, or other bleaching or decolorizing
" agent," and also in connection with centrifugal force, for
forcing these agents over the crystals of sugar, &c.

Sixth, " dissolving sugar to be manufactured into refined sugar
" by means of fluids, by preference, at a low temperature, and of
" centrifugal force, also the dissolving of sugar and filtering the
" syrups so obtained by synchronous operations."

Seventh, " imparting a yellow tinge to sugar, by artificially
" coating the same with saffron or other suitable coloring matter,
" as also the improving the appearance of crystals of sugar which
" have been washed nearly to whiteness by imparting to them a
" blue tinge as commonly practised upon refined sugar."

Eighth, "removing syrups and sugar from centrifugal ma-
" chines " by employing " a scoop which is free to slide up and
" down a stout shaft, and which scoop is capable of elongation or
" divergence when lowered into the basket."

Ninth, " revivifying animal charcal or similar matters by the
" combined action of centrifugal force, chemical agents, and of
" high pressure or superheated steam, also cleansing and revivify-
" ing animal charcoal or similar matters while in the ordinary
" filtering cylinder or vessel."

Tenth, " coating of crystals of sugar, by mechanically agitating
" the same while in contact with the syrups or other matters with
" which they are coated."

[Printed, 6d. No Drawings.]

A.D. 1859, June 2.—N° 1355.

SMITH, ALEXANDER and SMITH, WILLIAM.—(*Provisional
protection only.*)—" Improvements in machinery for curing sugar
" and for separating solid and liquid substances by centrifugal
" force." These are, in place of a wire gauze cylinder revolving
on its axis with great rapidity " a series of narrow bars or rings,
" placed edgeways are employed, these are packed as close
" together as may be desired, and form a very strong and com-
" paratively smooth surface, which can be cleaned with great
" facility. The bars or rings may be tapered or thicker at one
" edge than the other, if desired, to facilitate the escape of the
" liquid."

[Printed, 4d. No Drawings.]

A.D. 1859, July 28.—N° 1752.

ASPINALL, JOHN.—" Improvements in evaporating and in
" apparatus for the same, especially applicable to the evaporation
" of syrup. These are, first, " evaporating substances by means
" of pipes being situated at or near the surface of the substances
" to be evaporated, or in other words such pipes being wholly or
" partially (by preference wholly) immersed in the substances but
" not allowed to reach the bottom of the same.

Second, " constructing apparatus for evaporating with movable
" steam pipes, hot air pipes, or other pipes so arranged that they
" the said pipes may be caused to rise and fall as may be required
" according to changes of level in the substances evaporated in
" order that the said pipes may be kept always at or near the
" surface of the said substances."

[Printed, 6d. Drawing.]

A.D. 1859, August 6.—N° 1816.

DE LISLE, ALFRED THEODODE.—" Improvements in clarifying
" and decolorizing solutions of sugar and other liquids." These
are, in reference to this subject, as follows:—" Mix with the
" solution of sugar or other liquid powdered animal charcoal in a
" heated state, and afterwards separate the charcoal by filtration.
" I have found that the more finely the charcoal is powdered, and
" the higher its heat when mixed with the solution or liquid the
" better is the effect produced."

[Printed, 4d. No Drawings.]

A.D. 1859, August 12.—N° 1861.

POSSOZ, LOUIS ANTOINE. — "Improvements in the manu-
" facture and baking of sugars. These are, " purifying the
" saccharine juice produced from beet root, sugar cane, maple,
" sorgho, and all other saccharine vegetables, as well as the
" syrups or solutions of sugar, raw or impure, by processes which
" dispense with the employment of animal charcoal or blood,
" both in the manufacture and refining of sugar. These processes
" have for basis the employment of successive additions of lime
" and carbonic acid in fixed proportions," and such processes are
carried into practice " by the aid of certain particular arrange-
" ments of machinery and apparatus. The proportions of lime
to be used vary.' " For the first addition of lime with beet root
" there should be used (on an average) ten parts in weight of
" real lime for one thousand of raw juice to be purified. The
mode of mixing the lime is given. It is preferred to add the
lime to the juice at a temperature between 50° and 86° F. The
clear juice is drawn or not from the precipitate and carbonated
by carbonic acid avoiding an excess of gas so as to leave the
juice alkaline. After standing or by centrifugal force the clear
liquor is separated from the precipitate. The juice is treated
as above three times with lime and the lime finally completely saturated by carbonic acid and the whole carried up to
boiling to drive off "excess of carbonic acid and precipitate the
" carbonate of lime which might be held in solution;" after
which the juice is "clarified by the centrifugal machine, filtered,
" evaporated, and immediately baked without interruption or
" any other filtration being necessary." This process may be
varied.

[Printed, 2s. Drawings.]

A.D. 1859, August 31.—N° 1980.   (* *)

VON KANIG, WILHELM ADOLF.—"Improvements in the manu-
" facture of starch and compounds of starch, and in extracting
" gum dextrine and grape sugar therefrom." These are, first,
" I take the sago of commerce, either whole, in which case
" it must be ground fine, or I use flour of sago and proceed to
" wash and cleanse it from impurities in cold water, and then mix
" therewith a purified solution of chloride of lime made by dis-
" solving one pound of chloride of lime in three gallons of water,
" and I stir the whole together until well mixed, and leave it to
" bleach and settle for three hours more or less. The liquid is
" then separated from the solid portion, and the solid residue
" or starch well washed with cold water until all traces of lime
" are removed therefrom. This residue is then left to dry, and
" is fit for use. By this invention it will be seen that I convert
" sago into a soluble starch, and can obtain therefrom as com-
" monly practised in the manufacture of starch, gum dextrine
" and grape sugar."
     Second, in making compounds of starch by mixing the
" improved starch with ordinary starch during its manufacture, " by which it is refined and improved and rendered
" soluble. I also propose to combine the improved starch with
" gum, by which it is rendered useful for printing and for other
" purposes for which gum is now used;" and 3, in the use of a
solution of chloride of lime in the manufacture of ordinary starch
for the purpose of separating the gluten therefrom and for
making soluble starch, and for converting insoluble into soluble
starch.

[Printed, 4d. No Drawings.]

A.D. 1859, September 3.—N° 2013.

SCHRAMM, HENRY ROBERT LOUIS.—(*Provisional protection
only.*)—" A new process for pressing and separating simultaneously
" the fibres and pellicles contained in the constituent matters of
" the beet-root sugar, beer, grains, alcohol, potatoes, beets, and
" other similar substances." This consists of a frame having
two pairs of rollers, "one pair being above the other, and each
" pair having a pair of knives so arranged as to scrape the sur-
" faces clear of the adhering substances. A pipe leading from a
" tube or vat, within which the substance to be pressed is kept in

" a partly liquid state, transmits it between the first pair of
" rollers, which thoroughly press it, separating the greater part
" of the insoluble matters. The liquid then passes through a
" sieve or sieves, and falling upon an oblique plate drops between
" the next pair of rollers, and is a second time pressed by them;
" and after passing through another sieve is delivered perfectly
" pure, yet containing all the soluble matters. In this state it is
" ready for any future operations. Each pair of rollers has a
" pair of adjusting screws to regulate the distance between each."
" Where beds are made use of to press upon," they are formed
" of red cast iron, a composition of iron and copper where the
" copper predominates."

[Printed, 4d. No Drawings.]

A.D. 1859, September 7.—N° 2037.

LYONS, JAMES JOHN.—(*Provisional protection only.*)—" Im-
" provements in the manufacture of sugar." These are, " apply-
" ing to the upper portions of the pans used in the manufacture
" of sugar an open pipe or flue, into which an up current of steam
" is introduced through a pipe connected with a boiler in which
" the steam is generated." " In some cases, in any convenient
" position on the steam pipe, a vessel may be carried to receive
" any condensed water that may arise in the steam pipe, from
" which it may be re-conveyed to the steam boiler."

[Printed, 4d. No Drawings.]

A.D. 1859, September 10.—N° 2065.

ROBINSON, HENRY OLIVER.—" Improvements in machinery
" or apparatus for the manufacture of sugar." These are, " a
" new construction and combination of the gearing of wheels and
" pinions for communicating the motive power of steam engines
" to sugar cane mills, the mill and engine being fitted on to a
" base plate or frame of iron. The motive power is communi-
" cated to the mill from the main shaft of the engine by means
" of two pairs of internal toothed wheels and spur pinions,
" arranged in such a form or manner that the cranked or main
" shaft of the engine may be placed at a low elevation with respect
" to the plane of the bed plate or bed frame, in lieu of having
" it elevated, as it necessarily is " in a former patent, No. 10,345,
Old Law. It is stated that " by means of this new combination,".

the patentee is "enabled to use certain kinds of steam engines,
" which were practically inapplicable " in his " former combina-
" tion ; the centres of effort and the strains of the machinery are
" brought nearer to the plane of the base line, the fly wheel is
" brought down to a convenient level, and the injurious vibrations
" incident to the former elevated positions of these parts are
" obviated. The descriptions of steam engines to which this
" invention is more particularly applicable are the beam steam
" engine, the table steam engine, the inverted cylinder steam
" engine, and the horizontal steam engine."

[Printed, 1s. Drawing.]

A.D. 1859, September 20.—N° 2138.

MANBRÉ, ALEXANDRE.—" An improved method of extracting
" and purifying sugar, called glucose and ' syrup de Fecule ' from
" potatoes, or fecula or starch or dextrine." This consists in
" extracting and purifying glucose," as follows :—In a vessel or
apparatus capable of being heated by a naked fire, or otherwise,
is placed 20 parts of water, into this is put about " one part of
" diastasis for one thousand parts of the mixture," about to be
described, " or from 20 to 25 parts of malt for 100 parts of the
" mixture, or from 1 to 2 parts of sulphuric acid for 100 parts of
" the mixture, but these proportions may be varied." The water
is heated to about 155° F., and 9 parts of potatoes or fecula, or
starch or dextrine, with 1 part of flour from rice or maize are
mixed with 20 parts of water at 60° F., and the whole gradually
added, stirring being maintained as well as the heat at 185° F.,
until the mixture is completely converted into sugar. The acid is
then neutralized by carbonate of lime or quicklime, or any other
agent, the liquid is allowed to settle, filtered through a woollen
filter, and received in a clarifier filled with lime water, also burnt
bones, bullocks blood, &c., and finally conveyed into a close
vessel with a perforated false bottom on which is a filter, and the
solution is driven through it by steam or air forced in over the
top, or a pump or vacuum may be employed to draw it through
the filter.

[Printed, 4d. No Drawings.]

A.D. 1859, September 29.—N° 2210.

OXLAND, ROBERT.—" Improvements in the treatment of saccha-
" rine matters." These are, " the use of sulphurous acid as a

" bleaching agent on the saccharine matters," either alone or in aid of the decolorizing power of animal charcoal, "for this pur-
" pose mixing gaseous or liquid sulphurous acid with the saccha-
" rine solution at from 25° to 35° Beaumé, and then evaporate
" in the usual manner to the desired strength. The process is
" more especially applicable to solutions such as those of treacle
" or patent syrup which are not required to be converted into the
" crystalline form, although with care it may be advantageously
" employed in the treatment of solutions of sugar intended for
" crystallization." The sulphurous acid from an ordinary sulphur burner is passed through water, and from thence into a column, in preference, of lead or of glazed stoneware pipes, the top of which is filled with large rough pieces of coke or pumice stone. This column stands in a receiver. "If a very pale syrup is required,
" the whole of the syrup to be operated upon may be passed
" through the column, but a very good syrup may be produced
" by the admixture of two to four per cent. of the syrup saturated
" with sulphurous acid, with the syrup which has been passed
" over animal charcoal."

[Printed, 6d. Drawing.]

A.D. 1859, October 11.—N° 2314.

NEWTON, ALFRED VINCENT.—(*A communication from Horatio Nelson Fryatt.*)—"An improved mode of clarifying and defecating
" saccharine solutions and juices." These are, adding to a saccha-
rine solution alcohol, heat being also applied, to precipitate the impurities " albuminous and other nitrogenous, saline, and other
" impurities," while the sugar is in solution in the alcoholic liquid. Sometimes alcohol is added to the saccharine liquid until the liquid is of a certain strength, sometimes the alcohol is combined with water before using it. "The combination of alcohol and
" water found to work best consists of sixty per cent. of alcohol
" and forty per cent. of water or thereabouts, the per-centage by
" volume according to alcohometry. Variations from this per-
" centage will act in degree, but large variations cannot be practi-
" cally employed. The quantity of this combined liquor to be
" used is from fifty to sixty per cent. or thereabouts of the weight
" of the sugar operated upon ;" the mixture is gradually heated until nearly boiling, and kept so for about half an hour, with continual stirring, neutralizing any acidity by small quantities of

milk of lime. The solution is decanted and gradually cooled. In about eight or ten hours the purified solution is drawn off from the sediment, and boiled down and crystallized and purified from the slight colouring matter "in any manner the operator may " prefer, including the washing of their surfaces by alcohol." Close vessels are employed for recovering the alcohol.

[Printed, 4*d*. No Drawings.]

A.D. 1859, November 25.—N° 2668.

CARR, THOMAS.—" Improvements in arrangements and mecha-
" nism for drying glue, moulded clay, sugar, white lead, and
" various other substances and articles of manufacture." These
are, "the combination of a fan or fans" "with a chamber or
" chambers for drying purposes," also " the system or method
" of drying articles or substances in chambers, by causing the
" major part of a current of air artificially heated and set in
" motion by mechanical means to keep continuously circulating
" through the said chambers," as follows :—Two square chambers communicate with each other and with chambers at both
ends ; in one of these chambers the stove or boiler or other heating
apparatus is placed, with pipes or flues from it along one of the
chambers. A shaft is placed in suitable bearings carrying a fan
at each end and one in the middle. These fans may consist of a
boss, with three arms secured in holes or sockets formed in the
boss, to which the vanes or blades are secured, which may be of
thin sheet metal or wood. The rotation of the shaft carrying the
fans will cause a current of air to blow down one chamber and up
the other, one part of the air passing out through an aperture at
one end of the chamber, while an equal quantity of air will flow
in through an aperture at the other end of the chamber.

[Printed, 8*d*. Drawing.]

A.D. 1859, November 28.—N° 2689.

BENTALL, EDWARD HAMMOND.—" Improvements in machi-
" nery for cutting vegetable substances." These are, first, in
apparatus for the above purpose, making the top part of the
framing in one piece it is mounted on legs or vertical supports.
The pedestals or bearings of the several shafts are not fixtures, but
cast separate, and the upper surface of the framing is planed at all

parts where the pedestals or bearings of the several shafts are fitted. It has been found "convenient in practice that the places " for some of the fittings should be planed in one direction, then " the frame should be turned round in the planing machine " and the surfaces for the other fittings planed in a transverse " direction."

Second, arranging the driving gear so that "auxiliary hand " power may be applied thereto. To this end, one of the shafts " which carries a part of the differential gearing is prolonged " beyond the framing, so that a winch handle may be adapted " thereto if required." The differential gearing consists of a series of wheels of different diameters. The wheels are "so arranged that " one set of wheels must be completely ungeared before either of " the others can be put into gear. In order to cut two, three, " or more lengths of chaff as required, two, three, or more sets of " wheels of different diameters may be mounted on the differential " gearing shafts, so that by pushing the one or the other into " gear, the required length of chaff may be produced."

Third, this also relates to gearing, and consists "in the employ-" ment of a toothed wheel provided with internal teeth, whereby " a nut or pinion connected with the driving gear is to be " actuated." This wheel is named "an annular wheel," and by using it the size of some of the toothed wheels are advantageously reduced, and also the machine made "more compact than hereto-" fore. Several parts of the gearing are also covered and con-" cealed from view by means of this annular wheel, while the " other parts are concealed beneath a semicircular or curved " wrought iron cover, which covers all the differential gearing."

Fourth, "mounting the upper pressing rollers in bearings in " side levers or arms, which admit of being raised or lowered " simultaneously." Making "use of two pairs of pressing or " feeding rollers, and the upper rollers of both pairs are mounted " in bearings in the levers, so that when the latter are raised, the " upper rollers are raised with them. By thus raising the upper " feeding rollers, the distance between them and the lower rollers " is increased, and the substance to be operated upon is drawn " between the upper and lower rollers and fed forward to the " knives or cutting instruments." The box and mouth-piece of the machine are cast in one piece, which "adds materially to the " strength and solidity of the machine, and reduces its weight." This machine, it is said, may be employed among the other

purposes which are named for "cutting up sugar canes and other
" vegetable substances."

[Printed, 10*d.* Drawing.]

A.D. 1859, November 29.—N° 2698.

ROBINSON, HENRY OLIVER.—"Improvements in machinery
" and appartus for the manufacture of sugar." These are, first,
" the pan or bottom formed of a rectangular intersection of two
" semi-cylinders, or semi-ellipses, with rectangular mouths or top
" edges adapted to receive upper works or top sides and dividing
" pieces formed of straight pieces, and whether made of cast iron
" or wrought iron."

Second, "the modification of a spheroidal pan or bottom of cast
" iron, partially flattened by cutting off four segments, and
" adapted to receive upper works or top sides, and dividing pieces
" formed of straight pieces."

Third, "the said upper works and top sides and dividing
" plates formed of pieces straight lengthways, adapted to con-
" nect with square-mouthed pans of any form, and to connect at
" their ends with the ends of similar pieces belonging to adjoin-
" ing pans, and whether made of cast iron or wrought iron."

Fourth, "the moveable piece or door in the dividing plate
" between the pans."

Fifth, "the employment of a pump or machine to transfer or
" raise the juice or syrup from one sugar pan to another of a
" series." The pumps are of a centrifugal construction.

[Printed, 10*d.* Drawing.]

A.D. 1859, December 19.—N° 2889.

COWAN, JOHN, and COWAN, PHINEAS.—(*Provisional protection only.*)—"Improvements in revivifying or restoring animal
" charcoal, and in the apparatus employed therein." These are,
in preference, a pipe "is placed in an upright or nearly upright
" position in the furnace or heating apparatus," provided with
" plates or deflectors so arranged that the charcoal in descending "
is retarded in its gravitation or its " downward progress." Or a
twisted pipe or vessel may be employed; or a pipe or vessel may
be used with a number of plates or steps arranged within it at
suitable angles, forming " a sort of spiral staircase for the pro-
" gressive descent of the charcoal from step to step." " Instead
" of only using the hot gases and flames, &c. evolved from fuel

" for revivifying animal charcoal, superheated steam may be
" employed for the purpose, either alone or in combination with
" the heat obtained from other sources."

[Printed, 4d. No Drawings.]

1860.

A.D. 1860, January 5.—N° 36.

ROBINSON, RICHARD ALEXANDER.—" Improvements in sugar
" mills." . These relate to " that class of sugar mills where the
" steam engine, the crushing rollers, and the gearing are all on
" the same bed or base plate, and consists in combining a peculiar
" arrangement of gearing with an extension of the bed or base-
" plate at one side, in order to receive a beam steam engine, and
" thus to bring down the gearing and axes or shafts nearer to the
" base or base plate. For these purposes the bed or base plate has
" an extended or projecting surface formed on one side, on which
" the cylinder of a beam engine is fixed, the further end of the beam,
" by a connecting rod, giving motion to a crank on the main or
" driving axis or shaft, on which axis or shaft are fixed a pinion,
" and also a fly wheel. The pinion gives motion to a cog wheel
" on an axis which passes under the beam, such axis having at its
" other end a pinion, which takes into and drives a cog wheel,
" which it is preferred should have internal teeth, though it may
' be arranged to have external teeth. This cog wheel is fixed
" on an axis or shaft, which, by suitable wheels, gives motion to
" the crushing rollers of the mill, as is well understood."

[Printed, 10d. Drawing.]

A.D. 1860, January 25.—N° 193.

HUGGINS, HORATIO JAMES.—" Improvements in filtering and
" decolorizing cane juice, solutions of sugar, and other liquids,
" and in the manufacture of sugar." These are, a frame in which
is mounted a shaft on suitable bearings driven by a strap, cord, or
pulley; at the upper end of the shaft is a plate or disc, there is a
feed pipe for conducting the cane juice or other liquid into the
casing, which terminates just above the surface of the disc and by
which the liquid to be filtered is equally distributed all round the

axis of the disc. There is a ring of wire gauze or perforated metal lined on the inner side, in preference, with fine wire gauze, although filtering cloth may be used, and at the back it may be lined with cloth or filtering fabric, which may be removed when required; this ring is carried by arms fixed on an axis, against this ring the liquid is thrown on leaving the surface of the disc. Several modifications of this apparatus are described. In some cases employing a perforated drum mounted in a cistern containing the liquid to be filtered, revolving slowly, the liquid percolates from the exterior to the interior of the drum from which it is removed by suitable passages. The liquor, if desired, may pass through an interior perforated ring or rings after filtering through the side of the drum. In order to decolorize sugar and other liquids an apparatus somewhat like the first apparatus is described, but in the cases of wire gauze on the disc is placed animal charcoal or other decolorizing agent. To separate molasses from sugar and for washing sugar, a trough is mounted on a disc, the sugar is run into the trough, sheet metal covers prevent the sugar from running through the perforated sides of the trough, while the crystalliza-tion proceeds the machine is turned slowly and the sugar stirred, the sheet metal covers are removed, the machine moved rapidly, molasses pass out, and the sugar is washed by adding syrup.

[Printed, 1s. 10d. Drawings.]

A.D. 1860, January 27.—N° 212.

DUNCAN, JAMES, SCOTT, ALEXANDER, and DAWSON, JAMES.—" Improvements in re-burning animal charcoal, and in " the application of the products arising therefrom, and in the " apparatus employed therein." These are, first, " the system or " mode of treating or re-burning animal charcoal with the aid of " the mechanical arrangement for withdrawing the gaseous " matters," as follows :—" If the natural draught of the chimney " is not sufficient to draw off the gaseous products with the " requisite velocity, a fan wheel or air pump is arranged " to communicate with a chamber in which are the mouths of the retorts, a gauze wire diaphragm or other suitable medium is in the neck of each retort to prevent any portion of solid particles being withdrawn by the exhausting action.

Second, " the application to, and use in apparatus for re-burn-" ing animal charcoal, of a pipe or pipes for carrying away the

" gaseous products evolved from the used charcoal." The pipe or pipes are perforated with a number of holes, and pass down the whole length of the retort from the neck.

Third, "the system or mode of collecting the gaseous matters " evolved in re-burning animal charcoal, and utilizing the same " either by causing them to assist in heating the retorts, or " applying them to aid in the production of chemical products." The chemical products named are "sulphate of ammonia and " other generally similar chemical products." The charcoal thus re-burnt is superior in quality, and is peculiarly serviceable in deodorizing and decolorizing beet root sugar.

[Printed, 4d. No Drawings.]

A.D. 1860, February 9.—N° 351.

GILBEE, WILLIAM ARMAND.—(A communication from Edmond Pesier.)—"Improvements in treating saccharine fluids." These are, first, "the employment of alcohol for purifying or refining " saccharine syrups," as afterwards described.

Second, "the process of treating saccharine juices for producing " and purifying syrups," as follows :—

" The saccharine juice is first suitably defecated by lime, the clear " part is drawn off and boiled for half an hour; it is afterwards " saturated with carbonic acid or other suitable acid; it is then " concentrated to 18° Beaumè at the lowest, and as high as 25° " Beaumè, in the presence of animal black in fine powder, or " without any addition." The syrup thus obtained has alcohol added to it until it marks "from 60° to 70° by Gay Lussac's " alcoholometer," when an abundant precipitate of mineral and organic matters takes place, while the solution retains the sugar. The alcoholic liquor is distilled and the alcohol condensed for using again, and in the still is " a syrup which can be boiled direct."

[Printed, 4d. No Drawings.]

A.D. 1860, March 16.—N° 695.

WHITE, GEORGE. — (A communication from Johann Georg Leuchs.—"Applying as a substitute for the animal albumen " hitherto obtained from birds' eggs or blood, certain parts of " reptiles, fish, molusca, and articulated or radiated animals." Making use of the roes, spawn, or eggs of all sorts of fish, frogs, " or reptiles, the albuminous parts of the bodies of the animals

s.   s

" called by naturalists medusa, physalia, parpita, velella, rizostom,
" aurellia, cyanea, solen, cardium, or other analogous species."
To obtain the albumen, "the bloody, fleshy, or cuticular parts
" adhering thereto are cleared away therefrom, and the remain-
" der " triturated and crushed, pressed, or strained. " Or a small
" quantity of water or suitable menstruum may be added, in
" order that the albumen may easier be pressed or strained out."
After standing, the clear liquid is decanted, and may be made
use of in this state, or the same may be evaporated partly, or
entirely brought to dryness. It is applicable to a number of purposes named, among which is for fining or clarifying " saccharine
" or solutions or liquids."

[Printed, 4d. No Drawings.]

A.D. 1860, June 7.—N° 1404.

CLARK, WILLIAM.—(A communication from Jules Réné Lion.)—
" Improvements in the preservation of animal and vegetable
" matters." These are, in reference to this subject, "the
" preservation of beetroot for sugar mills and distilleries, which
" to the present time are dried, and thereby incur considerable
" expenses for desiccation and manipulation." "These means
" and processes of preservation consist, 1st, in producing a
" vacuum in any capacity, hermetically closed, in which the
" matters to be preserved are placed; 2ndly, in preserving the
" substances by means of a vacuum in combination with gas or
" hot air previously passed over chloride of calcium; 3rdly, the
" preservation of substances by means of a vacuum produced by
" withdrawing the air by high-pressure steam by engines and
" machines, in a manner which, before this application, were
" never adopted for alimentary substances. 4thly, in using for
" the desiccation of vegetables or grain, also for the preservation
" of alimentary substances, a distilling apparatus." The "vacuum
" produced in any capacity hermetically closed in which are placed
" the substances to be preserved can be obtained, 1st, by an air
" pump; 2ndly, by a barometic (barometric?) column; 3rdly, by
" the condensation of steam, air, ether, chloroform, or other
" volatile liquid or matter. The vacuum produced by the air
" pump simply consists in connecting it with the vessel containing
" the substances, and in extracting the air from the capacity."
" The barometric column " consists " of a tube of earthenware,

" gutta percha, or other matter. The lower part is situated at
" the base of the house, in a cellar, for example ; the upper part
" communicates with the preserving capacity."

[Printed, 8d. Drawing.]

A.D. 1860, June 15.—Nº 1463.

BROOMAN, RICHARD ARCHIBALD.—(*A communication from
Messieurs Légé and Danguy.*)—" Improvements in dessicating
" substances, and in neutralizing or retaining any fœtid gases
" which may be evolved in the process." This invention it is
said "is also applicable to the drying of sugar, chemical products,
" and other substances more or less solid," and "consists in
" forcing currents of hot air, or in some seasons dry air," through
passages in the bottom of a " closed vessel opening into a shaft, in
" which vessel the matters to be dessicated are agitated, tossed
" about, and divided by means of screws or shafts with arms
" caused to devolve rapidly. In the dessication of substances
" giving off fœtid or poisonous gases the shaft is closed by
" raising of a valve which opens communication with a pipe
" leading into a vessel containing some agent which will absorb,
" neutralize, or thow down the particular gas to be treated.
" The vessel contains a pump, which is caused to throw the agent
" from a rose spreader from the top of the vessel. After the
" neutralization, absorption, or deposit of the fœtid and
" poisonous gases, the air and non-injurious gases pass through
" the only outlet at the upper part of the vessel into a tube, and
" back into the first named shaft above the point at which the
" valve is placed. When the gases are not required to be puri-
" fied, the valve is lowered, when it closes the communicaton with
" the purifying vessel, and opens free passage through the shaft.
" When the air is to be employed hot, a furnace is placed under the
" dessicating vessel, with pipes, through which air is driven by a
" fan with sufficient force to drive in the heated air through the
" valves in the bottom of the vessel."

[Printed, 10d. Drawing.]

A.D. 1860, June 19.—Nº 1486.

WALKER, JOHN.—" Improvements in mills or machinery for
" expressing juice from the cane and other like vegetable
" substances." These are, first, " constructing and arrangeing

" rolls in such manner that on cane being fed to them from a
" feeding table, hereafter described, they first split the cane and
" cause the inside thereof to be turned downwards before passing
" between the pressing rolls, whereby the juice is extracted with-
" out being made to pass through the bark or rind of the cane."
One of the bottom rolls is grooved, and a ridge or projection is
left, more or less angular, between every two grooves, and upon a
feed table a rack is fixed, containing as many compartments for
the reception of canes as there are projections on the roll, and the
compartments are arranged " in such manner, that, the centre of
" each cane shall be delivered directly over the ridge to which the
" feed compartment corresponds."

Second, employing " a wrought-iron strap on each side of the
" mill for supporting, surrounding, and adjusting the bearings
" of the rolls " by means of wedges and screws. The straps are
drawn tight by means of keys passing through the bottom part
of the said straps.

Third, " casting the beds in the side frames for the two under
" rolls at right angles to the thrust or pressure at the time the
" cane is passing through the mill."

[Printed, 6*d*. Drawing.]

A.D. 1860, June 26.—N° 1554.

FLETCHER, JOHN.—"Improvements in the apparatus for treat-
" ing saccharine and saline solutions." These are, first, in
evaporating apparatus " composed of a cylindrical metallic vessel
" having an opening at each end, and fitted with trunnions or
" axes running on rollors. It is fitted internally, near its cir-
" cumference, with a number of longitudinal metallic tubes,
" communicating with a second or steam cylinder enclosing the
" first, but having no communication with the interior of the
" evaporating cylinder. Steam is introduced into the second
" cylinder from any conveniently placed steam boiler, through a
" pipe in a stuffing box turning with the cylinders. Another
" pipe attached to the steam cylinder carries off the water formed
" by the condensation of the steam therein ; this pipe also
" revolves in a stuffing box, and is connected with a condensed
" steam box, which regulates its discharge." After the inner
cylinder is about one-third filled " with saccharine or saline
" solution, steam is admitted into the steam cylinder, which

" passes thence into the tubes. The apparatus is then caused to
" rotate slowly on its trunnions, and in its rotation the liquid
" is taken up by the inner surface of the evaporating vessel, and
" by the outer surface of the tubes, and thus it becomes exposed
" to the action of a current of air, which passes through the
" apparatus, and which is increased by means of a jet of steam
" placed in a chimney in communication with one of the aper-
" tures in the inner cylinder; the opposite aperture being for the
" admission of air. The steam cylinder may be fitted with a third
" casing or cylinder, perforated at one end at intervals round its
" circumference for the admission of air, the other end being
" furnished with an aperture having a cover, there being in this
" case no admission of air into the apparatus except by the per-
" forations in the end of the said air cylinder." "The apparatus
" may be constructed without the outer cylinder or current of
" heated air, the jet of steam in the chimney being still employed,
" and a current of cold air admitted into the apparatus by re-
" moving the cover" from the aperture "for allowing access to
" the interior of the evaporating pan for cleaning the same when
" necessary;" or the cover may remain on and the pan, exhausted
by the jet of steam in the chimney. "The operation however
" is not attended with the same advantage as when a current of
" heated air is used."

Second, the cooling apparatus, consisting "of a series of
" metallic pans or troughs placed or encased in one common
" receptacle of metal, so as to have a space for freely circulating
" cold water or cold air, or the two combined between them,
" round their sides and bottom."

Third, " the employment of an improved chain and bucket pump
" consisting of a frame-work with a toothed wheel at each end,
" over which two endless chains are placed, so that by turning one
" pair of wheels the chains are set in motion. Attached to the
" chains is a series of tilting buckets. The bucket frame is lowered
" down to the bottom of the vessel to be emptied by suitable
" gearing. The top wheels being set in motion, the chain of buckets
" begins its circuit, by which each bucket in succession scoops up
" a quantity of the liquid in the vessel, which it carries to the top
" of the frame, where it comes in contact with a stop, which tilts
" the bucket over and causes it to discharge its contents."

[Printed, 1s, 4d. Drawings.]

A.D. 1860, July 12.—N° 1680.

BREARLEY, Thomas. — "Improvements in machinery for "producing and revivifying animal charcoal." These are, combining a rotating grate or fire-place, and rotating spreader and scraper with apparatus afterwards described for "producing "and revivifying animal charcoal." These, it is said, are applicable to machinery described in No. 445, A.D. 1858. In place of a fixed and stationary fire-place or places under a rotating platform or disc," "a fire grating is arranged to have a continuous "or interrupted motion communicated to it." In preference, "the fire grating should be of an annular form, to apply heat "principally to the outer circumference of the disc or platform "on which the materials under process are spread." In preference, also, the dome or cover should be rotated at intervals, and the platform or disc stationary. "The feed hopper and the "spreading apparatus are applied to the cover as before, and "the lower edges of the cover enter a circular groove or recess "in which suitable materials are placed for preventing the "evolved products from the matters under process passing into "the outer atmosphere, and also for preventing atmospheric air "passing into the apparatus. The animal charcoal when completed is removed from the platform through an opening in "its centre, and the charcoal falls through a descending pipe "into a cooler consisting of a series of upright pipes or tubes "surrounded by water. The lower ends of the pipes or tubes "are fixed in a plate which has below it a conical hopper having "a closing valve at its lower end." The waste heat "from the "machinery is conducted to, and in some cases through, a steam "boiler in order to generate steam by such waste heat."

[Printed, 6d. Drawing.]

A.D. 1860, July 25.—N° 1807.

NEWTON, Alfred Vincent.—(A communication from Horatio N. Fryatt.)—"An improvement in the process of concentrating "and crystallizing sugar." This, it is said, does not consist "in "the use of alcohol for the defecation or clarification of sugar or "sugar juices, as that process is known," but is a process "for "concentrating and crystallizing sugar by the use of alcohol "applied to a concentrated solution of sugar after it has parted

" with a large amount of its water, so as to render the mass
" mixable, and thus permit its easy exit from the pan or
" evaporating vessel, and also to be capable of arranging its
" crystals and parting with its coloring matter and fluid portions,
" whereby a larger per-centage of crystals are obtained than by
" the methods heretofore employed." " Instead of stopping the
" boiling when the amount crystallized is from 40 to 60 per
" cent." it is much more concentrated, and while stirring " high-
" proof commercial alcohol " is mixed with it in the vacuum pan,
" a fluid will be furnished capable of holding the crystals in
" division without dissolving them." " When the matter is in
" any purging vessel the alcohol can be permitted to run off,
" leaving the crystals in a pure state." " If desired a condensing
" apparatus may intervene between the vacuum pan and the air
" pump, or between the air pump and the final exit of the vapor,
" by which all the alcohol may be saved." The alcoholic solution
run from the crystals of sugar is distilled and the alcohol " sepa-
" rated and saved, and the sugar passed back into the vacuum
" pan for further evaporation and eventual recovery."

[Printed, 4d. No Drawings.]

A.D. 1860, August 15.—No 1971.

COURTOT, HIPPOLYTE. — "An improved machine, with a
" drawer and moveable knife, used to part and break sugar."
This consists " of a cast-iron framing, supporting a pair of rag
" wheels, chains, and pullies, a couple of steel springs to keep the
" chains asunder with two screws to regulate the distance, and
" for this purpose the wheels will be moveable, and will, by
" suitable gearing, communicate movement to a knife sliding in a
" drawer, which knife will strike twice for each turn of the crank.
" A plate or table (of metal) will occupy the space between the
" rag wheels, chains, and pullies, and on this table the sugar
" (previously cut into long strips) is placed. The crank will be
" then turned, and a fly wheel will, by its excentric weight, carry
" with it a rod, which will set in motion the rag wheel and
" chains " by means of a ratchet wheel, " the chains having hooks
" on their inner side will drag the sugar forward, and the sliding
" knife by the action of an excentric will descend in its drawer
" and depress one end of the table, and divide the sugar between
" itself and a fixed blade beneath the table; it will then rise (as

" will the end of the table by (means of) a spring beneath it) and
" fall again, cutting twice for each revolution of the crank, which
" may easily be turned by a female, who (using the machine above
" described will cut without difficulty upwards of a ton of sugar
" a day."

[Printed, 6d. Drawing.]

A.D. 1860, August 15.—Nº 1981.

FRYER, ALFRED. — " Improvements in centrifugal machines."
These are, first, " cleansing and freeing the meshes or open
" surfaces of which the periphery of the drums of centrifugal
" machines is formed by causing the said drums to revolve in an
" atmosphere moistened and warmed to a temperature of from
" 130° to 140° Fahr. by means of steam."

Second, creating a moist warm atmosphere for the drums of
centrifugal machines to revolve in, and maintaining the temperature thereof at from 130° to 140° Fahr. by means of air and steam
combined and applied as afterwards described.

Third, " constructing the cases of centrifugal machines with a
" lining of wood or other suitable non-conducting material."

Between the top of the case surrounding the drum and the
drum itself, placing a cover to prevent the egress of the air
between the case and drum at the top, and preventing the egress
of air at the outlet for the syrup or other liquid by means of a
trap, lining the inside of the case with wood or other suitable
non-conductor, fitting a pipe opening from near the top of the
case, carrying it to a point below the bottom of the drum, and
bending it inwards to near the axis thereof. Within this pipe
about the part where it turns to the axis fixing a small jet and
connecting it to a steam pipe.

[Printed, 6d. Drawing.]

A.D. 1860, August 31.—Nº 2104.

BELTON, PATRICK MICHAEL.—" The manufacture of a com-
" pound to be used as a substitute for animal charcoal in refining
" sugar and otherwise." This consists in " the use or employ-
" ment of bog peat, treated substantially " as follows :—" As a
" substitute or equivalent for animal charcoal in refining sugar and
" otherwise." To 100 lbs. of the peat, taken as it comes from the
bog is added " 6 lbs. of chalk, or its equivalent " in powder, and

mixed with water to bring the whole in "to a pasty consistence,
" thereby effecting the perfect admixture of the two." It is then
pressed and cut, or formed into flat cakes, which are subjected to
the action of a powerful press, "each cake being covered with a
" cloth or with wire gauze, and placing them between iron plates.
" The cakes are then air or stove dried, until all moisture is
" expelled. They are then ground to a coarse mealy powder, and
" charred much after the manner of charring bones for the same
" purpose." On cooling the material is ready for use.
[Printed, 4d. No Drawings.]

A.D. 1860, September 5.—N° 2139.

VAUVILLÉ, ERNEST. — (*Provisional protection only.*) — " A
" system of extracting juice from beet-root and other plants."
This consists as follows :—" On leaving the rasp or grater the
" pulp falls into a vat, where it meets with an iron plate
" (perforated with holes of a tolerably large size), which arrests
" any filaments while permitting the pulp to pass through, and
" meet a partition which spreads it equally on the cylinders, and
" the juice obtained by their pressure passes through a metal web
" or textile fabric, and is conducted to a funnel, by which it
" escapes. The axles of the cylinders are moveable, and by
" means of an endless screw for one and a lever for the other the
" pressure is increased or diminished by fixing or giving play to
" the shaft of the cylinder in its slide. An aglomeration of pulp
" at any point would only separate the cylinders for an instant ;
" if their strength was not sufficient to reduce it a spring bringing
" the cylinder back again immediately it has passed to the next
" cylinders, which exercise a stronger pressure, and if after this
" operation it is believed that any juice remains in the pulp it is
" submitted to the action of a hydro-steam press, and every
" portion of the liquid is driven from the pulp. The cylinders
" are set in motion by any power, suitable gearing being used,
" and a rag wheel and chains to give an opposite movement to
" the cylinders." In operating as above "the use of sacks or
" bags will be entirely dispensed with."
[Printed, 4d. No Drawings.]

A.D. 1860, September 6.—N° 2153.

WRIGHT, RICHARD.—" Improvements in the manufacture and
" refining of sugar, and in apparatus employed therein." These

consist "in the combined use in apparatus of revolving discs and
" the vapour of water below 212° Fahrenheit." In carrying out
this, the juice or other solution of sugar to be evaporated, is placed
in an open evaporating vessel, in which several discs or moving
surfaces, in preference, of copper, are caused to dip. This evaporating vessel is placed in an outer vessel of such dimensions as
to admit of its containing a quantity of water below the bottom
of the evaporating vessel, an outlet or overflow pipe from the
outer vessel below the level of the bottom of the evaporating
vessel carries off water and maintains the level of the water in the
outer vessel. There is an outlet pipe for escape of steam near
the upper part of the outer vessel. The water in the bottom of
the outer vessel may be heated in any convenient manner, but, in
preference, free steam from a separate vessel is introduced near
the bottom of the outer vessel and below the surface of the water
therein.

[Printed, 6d. Drawing.]

A.D. 1860, September 7.—N° 2163.

STEVENS, CHARLES.—(*A communication from Edouard Four-
meaux.*)—(*Provisional protection only.*)—"Improved bags to be
" used in the manufacture of sugar." This relates to "the
" manufacture of bags for containing the pulp of sugar."
These bags, it is stated, "were originally of spun flax and hemp,
" which were afterwards replaced by others made of wool spun
" and twisted. These, although possessing advantages over
" those first in use, were more expensive." To obviate this, and
to unite "a moderate cost with a durability at least double that
" of the bags hitherto used, the improved bags are made of
" cotton, which is spun double and twisted."

[Printed, 4d. No Drawings.]

A.D. 1860, September 15.—N° 2247.

NAPIER, JAMES MURDOCH.—"Improvements in machinery
" for the manufacture of sugar." These are, in centrifugal
machines "which will discharge not only the liquid, but also the
" dried sugar from the rotating container by the centrifugal
" force; and as the matter may in all cases be supplied to the
" rotating container while it is in motion, the rotating action of
" these machines will be continuous." In preference, the axis

of the rotating container is placed vertically. The rotating
chamber is cylindrical, with apertures in its circumference which
are commanded by valves or slides covered with wire gauze, or
otherwise perforated, which are withdrawn either by the attendant
or by the action of the machine " when the dried sugar is being
" discharged through them out of the rotating container, and
" they may be shut in the same way." The surfaces of the
rotating chambers are corrugated and covered with wire gauze.
Machines are described " in which the rotating container is wholly
" charged and wholly discharged at intervals, the liquoring being
" effected at corresponding intermediate intervals," also machines
" in which the rotating container is constantly receiving fresh
" matter, at the same time it is discharging the liquid as well as
" the finished sugar; the liquoring is also continuous." The outer
fixed casings of the machines may be furnished with a chamber
for steam or hot water. The valves above referred to may be
arranged so that the centrifugal force may either have a tendency
to open them or to shut them. " Crushing rollers may be used,
" if necessary, to break the lumps before the matter passes into
" the machines."

[Printed, 3s. Drawings.]

A.D. 1860, September 20.—N° 2296.

RICHARDSON, THOMAS, and PRENTICE, MANNING.—" Im-
" provements in treating phosphatic matters, and in obtaining
" products therefrom." These are, in reference to this subject,—
In making artificial charcoal a strong solution in muriatic acid of
phosphate of lime is employed; when bones are employed they
are first crushed between rollers or edge stones, and " 20 cwt. of
" bones" are treated with " 20 cwt. of muriatic acid of 1·130
" specific gravity in a suitable vessel; when the phosphates are
" all or nearly all dissolved," the solution is drawn off, and if
needful it is evaporated to 1·300 sp. gr., and dried up with 7 or
8 cwt. of dry sawdust, and the whole is carefully incorporated
with 8 to 9 cwt. of finely ground pitch, and then placed " in a
" clay retort or other suitable close furnace, and exposed to
" a long continued red heat until the phosphates are rendered
" insoluble in water." The charcoal is then withdrawn, and, if
necessary, any muriate of lime is removed by washing it in water;
when dried it may be employed as a substitute for animal charcoal obtained from bones. The process is the same " when other

" phosphatic substance is employed in place of bones, and we
" can substitute charcoal powder or any other organic dryer
" instead of sawdust."

[Printed, 4d. No Drawings.]

A.D. 1860, September 29.—N° 2359.

GREEN, WILLIAM.— "Improvements in refining or treating
" sugar and molasses." These are, first, in " decolorizing treacle
" or molasses and low syrups, and employing the same for
" ' liquoring ' or whitening sugar." The low syrups are obtained
" by concentrating the waters used for washing animal charcoal
" after it has been employed for decolorizing syrups." Also
economizing the process of refining by the use of inferior sugars
or 'syrups to those now employed for the manufacture of fine
liquor.

Second, decolorizing " treacle or other saccharine fluids by
" keeping the same for a much longer period in contact with the
" animal charcoal." " If very low sugars be used, it is in some
" cases necessary to keep the syrup in contact with about its
" own weight of animal charcoal for several days, and then pass
" them through a series of cisterns," five in number, each
capable of holding two tons of animal charcoal.

Third, instead of animal charcoal sometimes employing " che-
" mical bleaching agents for decolorizing treacle, those preferred
" being Nash's finings, or the phosphate of soda." The mixture
is allowed to " stand for some time in shallow vessels kept in a
" warm room." After about a week, more or less, the clear
decolorized treacle is run off from the scum and precipitate.
" When the treacle or low syrups are decolorized, if not suffi-
" ciently thick they are to be concentrated either in vacuo at a
" low temperature, or by exposure to warm air, either in shallow
" vessels or other suitable apparatus." Several arrangements are
described suitable for this purpose, and for concentrating syrups
or liquors. An apparatus is described, " in which the evaporation
" is carried on in a large shallow vessel fitted on to and forming
" part of a tubular boiler." Another apparatus consists of a pan
in which there are a number of tubes near to the bottom for the
circulation of steam. Another apparatus consists of a pan in
which a drum revolves in the fluid contained in it; but " instead
" of forming the drum of a series of bars or rods as usual,
" I form them with one or more series of moveable floats "

" around their circumference. These floats I prefer to have
" made in the form of combs, so that as the drum revolves in
" the fluid contained in the pan they take up on their surface
" thin films thereof, and expose the same to the action of a
" current of hot air." Another apparatus " consists of a series
" of inclined planes, by preference with corrugated bottoms, and
" provided with jackets for the circulation of steam or other
" heating media." The inclined planes may be arranged "as a
" screw, and the syrup caused to pass slowly over the same until
" sufficiently concentrated." Another arrangement is a shallow
vessel "with a flat coil or series of coils, of much greater depth
" than width, for the passage of steam, boiling water, or other
" heating media." It is now proposed "to employ decolorized
" treacle or low syrups decolorized as herein-before described for
" liquoring superior kinds of sugar." "Instead of the gutter
" commonly used for conveying the ' skip' from the pan into the
" ' heater ' or other receiving vessel, I propose to make them with
" a series of narrow compartments, between the alternate spaces
" of which steam is free to circulate."

[Printed, 1s. Drawing.]

A.D. 1860, October 22.—N° 2571.

BROOMAN, RICHARD ARCHIBALD. — (*A communication from
Jean Baptiste Dureau.*)—" Improvements in apparatus for evapo-
" rating and concentrating, specially applicable to the manufacture
" of sugar." These are :—"A cylindrical or other shaped column
" or tube " of a suitable metal is "placed on a boiler or pan, and
" containing double plates or shelves, one above the other, the
" whole or nearly the whole height of the column; communication
" is maintained between the double plates by pipes extending
" somewhat above the plates; pipes are provided for leading
" steam to each of the double plates. A tube or pipe leading
" from the pan runs through the centre of the column. The top
" of the column is dome-shaped, and is placed in communication
" with a reservoir, from which liquors to be evaporated and con-
" centrated are led to the plates in the columns. The dome is
" provided with water and acid valves and draw-off taps, and the
" boiler with the requisite appliances for evaporating and con-
" centrating apparatuses. Liquor from the reservoir falls or is
" forced on to the first plate, when it rises to a level with the top

" of the overflow pipes, by which it descends to the second plate,
" and so on from stage to stage until it arrives at the bottom of
" the column, where pipes lead it to the bottom of the pan. The
" steam escaping from the plates and pan enters the central tube,
" there to be conveyed to any desired spot for use or to a con-
" denser." Two or three or more of the above apparatuses may
be combined and employed for increasing the effect. " In some
" cases a double pipe is employed in the central part of the
" column, so as to leave an annular space, closed at its upper
" part, between the plates and the steam pipe." For raw sugar,
making use of two serpentine coils, one above the other, and
introducing, " slowly and gradually, a ' liquor ' or syrup already
" reduced and sufficiently dense not to dissolve the crystals which
" form in the apparatus."

[Printed, 8d. Drawing.]

A.D. 1860, November 16.—N° 2813.

WILLIAMS, CHARLES WYE.—" Improvements in steam boilers
" for increasing the evaporative effect thereof; applicable also to
" stills and other like vessels or apparatus." These are, first,
" the construction of steam boilers and other vessels for conveying
" heat to liquids, with flues or tubes enlarged and contracted at in-
" tervals, and having a series of face plates, or projecting surfaces,
" placed at right angles, or nearly so, to the direction of hot gases
" proceeding from the furnace." " This mode of construction is
" equally applicable to all surfaces designed for the conveyance
" of heat to liquids, such, for instance, as evaporating pans
" employed in the manufacture of sugar, and its application to
" such surfaces."

Second, " the construction of stills, refrigerators, and other like
" apparatus for cooling and condensing purposes, with surfaces
" placed at right angles, or nearly so, to the current of the fluid
" from which the heat is to be extracted."

[Printed, 8d. Drawing.]

A.D. 1860, December 11.—N° 3033.

TOWNSEND, JOSEPH.—(*Provisional protection only.*)—" Im-
" provements in obtaining animal charcoal and other products
" from bones and other animal matters." These are, " bones or
" other animal matters," such as " flesh, blood, hair, wool, woollen,

" and 'hard' rags, hoofs and horns, silk, feathers, skins, leather,
" old shoes, and gelatinous and albuminous substances," are
placed in retorts, such as are usually employed for the manufacture of animal charcoal, but a lower heat is used; in preference,
the heat is from 400° to 600° F., " distilling at or near these tem-
" peratures until nearly the whole of the condensable products
" are obtained. The heat is then raised to 800°, or higher, and
" is maintained until all the volatile products are driven off. The
" retorts are allowed to cool to 400°, or less, before the residue or
" charcoal is drawn out." The resulting charcoal has " superior
" decolorizing properties," and the liquid products are greater,
the incondensable gases less.

[Printed, 4d. No Drawings.]

A.D. 1860, December 11.—N° 3034. (* *)

CANU, ADOLPHE JOSEPH.—(Provisional protection only.)—" An
" improved pulverizing and bruizing machine."

The cylinders and plate which perform the grinding or bruising
in this machine are placed in a case hermetically closed, and
serving as a frame to the machine. On this case is a pipe having
a valve for the introduction of the matters to be operated upon.
The cylinders are suspended and turn on a plate, also revolving
and of which the exterior exceeds that of the cylinders so that
the latter act on the plate, with the whole of their lower surface
being placed in order round the centre of the plate. The speed
of the cylinders and plate varies according to the substance to be
pulverized. The substance passes through the pipe above-mentioned to the revolving plates, which by their rapid motion continually force the matters placed above to pass between the lower
face of the cylinders and the upper face of the plate (the latter
being adjustable by means of a screw) and falls when pulverized
into a chamber beneath the machine.

[Printed, 4d. No Drawings.]

A.D. 1860, December 11.—N° 3038.

TOWNSEND, JOSEPH, and WALKER, JAMES.—" Improve-
" ments in treating bye products arising in the manufacture of
" soda and potash for the obtainment of antichlores and other
" useful products." These are, in reference to this subject, in
making substances to be used " as purifiers in the manufacture of

" sugar," &c. :—Soda or potash waste is lixiviated with water,
and the solution oxidized by exposure to the air; in preference,
by percolating it down through a tower of coke, pumice stone,
pebbles, &c. while air is passed upwards through it. The solution
must be repeatedly percolated through the tower until it is fully
oxidized, "known to be so when it ceases to give a precipitate
" with sulphate of copper." In preference, the solution then is
concentrated to sp. gr. 1·15, or even higher; on repose the
precipitate is removed and the sulphates of soda, magnesia,
potash, or ammonia added, in which cases the simultaneous
products will be respectively hyposulphites of soda, of magnesia,
of potash, or of ammonia. "To obtain sulphite or hyposulphite
" of lime, the oxidized liquor (after being freed from some
" impurities in the manner already described) is evaporated to
" dryness " and used as an antichlore, or a concentrated solution
may also be used as an antichlore. Any of the foregoing products may be used for, among other purposes which are named,
" purifiers in the manufacture of sugar," &c. The remainder of
the Specification refers to the obtaining of " sulphur and certain
" useful sulphides " from the above lye products.

[Printed, 6d. No Drawings.]

A.D. 1860, December 22.—Nº 3151.

SAVAGE, ALFRED.—"Improved apparatus for separating, re-
" ducing in size, and mixing articles of grocery." These are, in
reference to this subject, in the apparatus described for cutting
" sugar in No. 1877, A.D. 1855, placing two springs, one on each
" side of the lever," sufficiently apart that the said lever may
not touch them, "also substituting for said lever and knives a
" knife or knives moved in right lines in or on suitable guides,
" and projecting to without said guides," this, in preference, is
" effected by fixing two cylindrical iron bars or guides on an
" iron base, so that the knife or knives, or the part of the appa-
" ratus to which they are attached, may slide freely on said
" bars or guides," and connecting "said knife or knives with a
" revolving or partially revolving crank by attaching the rod or
" other connecting material to such knives within such guides,"
in preference, this is effected " by forming the part which slides
" on the said bar or guides with an opening to receive a con-
" necting rod between the two sockets in the sliding piece or

" part which slides in the said guides," also, the supports for the
loaves to be cut up, by " casting in metal a bed " with " projecting
" supports on its upper surface, not more than two of which
" projecting supports shall be in the same plane, the third or
" third and fourth supports being the bed itself," so that the
" loaf of sugar may fall on each of the said supports in succession
" after the required portion or portions are cut off said loaf of
" sugar."

[Printed, 4d. No Drawings.]

1861.

A.D. 1861, February 5.—Nº 297.

WILLIAMS, GEORGE.—"Improvements in the construction of
" charcoal and other kilns." These are, first, in distilling bones
or spent charcoal, "hollow brick blocks or tile chambers are
" fitted the one above the other and socketted into each other
" by projecting ledges and sunk spaces at the inner surfaces, so
" that when built up an air-tight shaft is formed, which I bind
" together by dowals or other arrangement at the ends or sides
" and coat over with a non-absorbent of sulphur."

Secŏnd, "constructing at or near the bottom of these shafts,
" and at or near the top of the coolers cavities or chambers to
" collect the gases from the charcoal," &c. and providing said
cavities or chambers with plugs, cocks, or other moveable or
opening apparatus to allow the gases to escape at intervals."

Third, " constructing coolers with small permanent openings to
" allow the gases to escape at intervals into cooling tubes."

Fourth, the arrangement and construction of kilns as above.

[Printed, 10d. Drawing.]

A.D. 1861, February 11.—Nº 339.

MENNONS, MARC ANTOINE FRANÇOIS.— (A communication
from Julien Fradet.)—" Improvements in the construction of
" steam generators employed for heating, drying, evaporating
" and other purposes." These are, in reference to this subject,
first, " steam is supplied in the different processes of sugar manu-

S. T

" facture, a number (say two, three, or more) of iron tubes bent in
" a serpentine form, are passed into the generator, and are there
" secured by brackets, or other convenient support, in such a posi-
" tion as to expose the largest amount of surface to the action of
" the steam and hot water." In general, the tubes are distributed
longitudinally, but they may be vertically, or at any desired angle.
" The free extremities of each tube are carried through the sides
" of the generator at any desired point, and are connected with
" branch pipes leading to the different chambers or recipients of
" the factory to which steam is to be supplied."

Second, a superheating apparatus composed of serpentine tubes
is set behind the register in such a position as to be exposed
to the flame of the furnace. " The steam distributed, as above
" described, to the different points of the manufactory, is led
" on its return through this auxiliary apparatus, and being
" there superheated passes on to the serpentine mounted within
" the generator, so that loss from condensation is entirely
" avoided."

[Printed, 10d. Drawing.]

A.D. 1861, February 26.—N° 493.

BROOMAN, RICHARD ARCHIBALD.—(A communication from
Charles Emile André & Co.) — "Improved materials for and
" improvements in the manufacture of sugar moulds." It is
said that hitherto sugar moulds "having been made of sheet iron
" which have become oxidized in parts have produced yellow
" stains on that part of the sugar coming in contact with the
" oxydized metal," and to remedy this defect it is proposed to
construct "the moulds of materials which will not oxydize, such
" as of organic products, cardboard, pasteboard, leather cuttings,
" caoutchouc, vulcanized or not, and various rigid fabrics ren-
" dered impermeable by certain coatings." As an example,
when the moulds are made of cardboard, a strong sheet of this
material is divided into four parts, and the parts are united
together round a core in the hape of a sugar loaf by means of
glue, when dry, a ring of iron is fixed round the exterior " with
" a rib at or about the centre, on which the next mould rests
" when the moulds are piled over one another;" other rings are
attached. The moulds may be made in three pieces, and three
pieces of thin cardboard may be applied in the interior; they
may be made in one piece by moulding them in the state of pulp.

The glue preferred "is made of gelatine, rye flour, and fecula;" the varnish, of "essence of turpentine, well boiled oil, and " copal varnish, but any ordinary glue or varnish of commerce " may be employed." In some cases metal moulds are lined with cardboard and are painted, as are the moulds made entirely of cardboard, and "leather, paper, caoutchouc, or other suitable " material may be used instead of cardboard."

[Printed, 4d. No Drawings.]

A.D. 1861, March 30.—N° 781.

FIELD, JAMES JOHN.— (Provisional protection only.)—"Im-" provements in apparatus for evaporating in vacuo." These can, it is said, be carried out "with an ordinary vacuum pan " by making certain additions thereto," but it is preferred "to " employ a vacuum pan having a head or cover of different form " and construction to those ordinarily used." The cover, head, or " upper part of the vacuum pan is refrigerated sufficiently to condense the vapour arising from the fluid or within the upper part of the apparatus, some form of surface condenser may be used in substitution or in addition to the refrigerated head or cover itself. In either case the fluid resulting from the condensing action is prevented from returning to that operated in the lower part of the apparatus by one or more internal channels or gutters from which it is conducted "by means of a pipe or pipes opening " from the channels or gutters, passing from thence out of the " apparatus, and terminating in a receiving vessel kept in the " same state of atmospheric exhaustion as the vacuum pan itself. " This method of conducting evaporation in vacuo is particularly " applicable to the concentration of saccharine solutions and the " preparation of medicinal extracts."

[Printed, 4d. No Drawings.]

A.D. 1861, May 2.—N° 1095.

WILSON, JOHN CHARLES.—(Provisional protection only.)— " Improvements in machinery or apparatus for the manufacture " of sugar." These are, first, in a steam sugar cane mill, the mill and engine are both fastened together, either directly or by means of intervening girders or base plates, a boiler being employed requiring no brick setting and having an iron chimney, and burning wood or "megass" fuel.

Second, employing a boiler with a circular top and straight or curved sides which terminate in a flat base for it to rest upon. The fire-place is in a large internal flue suited to burn wood fuel. To increase the vaporizing power of the boiler without increasing its size, placing a water tube or tubes in the main flue.

Third, employing wholly or in part of "'bevil' gearing instead "of the 'spur' heretofore adopted," also constructing "gearing "of whatever kind used for sugar cane mills of wrought iron or "steel in lieu of cast iron as heretofore."

Fourth, making rollers of sugar cane mills entirely of wrought iron or steel or of both in place of cast and wrought iron as hitherto, also forging or casting solid "with the shaft the pinion "which actuates or drives the roller, instead of making it in a "separate piece and keying it on as heretofore."

Fifth, constructing side frames or standards which support the rollers of sugar cane mills of wrought iron or steel or a combination of both, with the lower rollers "adjusted in the direction "of a line passing through the centres of the top and lower rolls, "and are drawn out in a direction at right angles to the same;" also substituting in lieu of the adjusting screw a solid wedge "placed between the brass of lower bearing and the seat on the "side frame, the latter being at right angles to the line of thrust." "This wedge may be drawn out or pushed in by any convenient "mechanism, and the roller elevated or depressed accordingly." Constructing "the trash turner of sugar cane mills" with "a "steel spring which shall constantly press against the lower "roller in whatever position it may be adjusted." "Instead of "attaching a spring to the trash turner the latter may be so "constructed as to act a similar part itself, and the same arrange- "ment may be applied to the 'trash' as well as the 'front' "lower roll."

Sixth, in sugar cane mills driven by water power, "placing the "water wheel in an iron trough or breast so as to dispense with "the present brickwork breast or foundations and to attach the "cane mill to the same."

Seventh, "in regard to the cattle power or gear for transmitting "the power of animals to a sugar cane mill." In driving "a "bell or pan shaped casting for carrying the gearing" and when "desired to increase the speed of the horizontal driving shaft," employing "what is termed 'internal' gearing."

Eighth, in evaporating, &c, saccharine fluids "passing heated

" air over their surface" and "gradually removing the upper
" stratum of the liquid as it becomes concentrated" in place of
boiling.

[Printed, 4d. No Drawings.]

A.D. 1861, May 7.—N° 1155.

DAVIES, GEORGE.—(A communication from Hector Lavignac.)
—(Provisional protection only.)—" Improved apparatus for boil-
" ing sugar." These are, first, in place of the fire passing below
the 5 to 8 boilers employed in the colonies for evaporating sugar
liquids the fire passes through the boilers.

Second, " an apparatus called a ' simoon ' or hot blast, placed
" either in front of or behind the boilers, or partly in front and
" partly behind, which is for the purpose of drying the ' bagasse '
" or ' cane trash,' and feeding the furnace therewith."

" Each boiler comprises within it a second boiler, called a
" ' multi boiler.' " " The liquid to be evaporated is contained
" between the two boilers, and is heated by the interior wall of
" the multi boiler, which terminates at a lower elevation than
" that of the exterior boiler, in order that the greater part of the
" liquid may be placed above it. The top of the ' multi boilers '
" is corrugated." The cane trash is moved along an endless
chain over a perforated arch or vault above the furnace, by which
it becomes dry and falls into the furnace by a trap door at the
end ; the vapor is condensed and carried off by gutters or spouts.
The first two or three boilers have level tubes outside, and covers
with pipes which pass through two or more of the succeeding
boilers, and thence into smaller pipes which pass through the
cold cane juice and then into a vessel " where the vapor is con-
" densed, producing a sufficient vacuum in the covered vessels."
Below the two last boilers " are dampers, consisting of shallow
" vessels of water covered with trellis or perforated metal and
" moving on friction rollers or wheels. The flue leading from
" the boilers to the chimney is surrounded by an air chamber,
" the heated air from which is withdrawn by a fan or otherwise
" and fed into the furnace."

[Printed, 4d. No Drawings.]

A.D. 1861, May 14.—N° 1228.

BROOMAN, RICHARD ARCHIBALD.—(A communication from
Jean Baptist Joseph Quéruel.) —" Improvements in working

" sugar refineries and in sugar moulds, and apparatus for
" trimming the loaves therein." These are, " placing the pan
" or copper from which the sugar for filling the moulds is to be
" taken at the bottom or lower part of the building," and a shaft
with a " hoisting and lowering apparatus," "communicating
" with each of the floors on which the moulds to be filled are
" kept ;" fitting the pan with a valve commanding an outlet pipe
in the bottom, "from which the sugar is run into a jacketted
" filling pot formed, by preference, with a spout and provided
" with a cover. The filling pot, after being charged, is run upon
" a truck into the shaft and hoisted to one or other of the floors
" where the moulds to be filled are placed ; it is then put upon
" another truck, and is suspended from a tackle and blocks in such
" a manner that it may be tilted and the contents poured into the
" moulds." The moulds are formed at bottom with a threaded
aperture, closed by a pointed metal spile which rises a slight
distance inside the mould, forming a hole in the head of the loaf
of sugar ; the spike terminates inside the mould in a button, on
which a washer rests to make a tight joint. " The moulds with
" the spikes screwed in are held in frames constructed of wood,
" with apertures for the moulds to be supported in ; double lines
" of rails are laid on each floor, and the frames for the moulds
" are run about, for the purpose of filling, and otherwise, in
" carriages." For trimming the base of the sugar loaf placing
" a dome-shaped frame over the mould, which frame carries on
" the end of a spindle cutters or scrapers which, on being rotated
" make the base of the sugar loaf even, and at the same time
" give a bevel edge thereto."

[Printed, 10d. Drawing.]

A.D. 1861, June 26.—N° 1640.

COWAN, JOHN.—" Improvements in apparatus for re-burning
" animal charcoal." These are, " preserving it from the atmo-
" spheric air when it is to be removed from the cylinder or retort,"
by forming in the front or discharge end or cover a small opening
closed by a door, lid, or plate, also supporting the cylinder or
retort on wheels or rollers, on which it turns, preferring "two
" such wheels or rollers at each end of the cylinder, and forming
" the front end or cylinder cover with a flange broader than the
" rim of its supporting wheels to allow for expansion ;" also,

preferring, "to cause the flange at the back end of the cylinder to
" turn in a groove in its supporting wheels, to keep the end in
" gear with the driving arrangements." Likewise, driving or
turning "the cylinder or retort by forming or fixing teeth, or a
" cogged or toothed wheel, rim, or a flange on the end which I
" drive by a pinion or equivalent on a driving shaft, so that by
" having a number of pinions or wheels at intervals on a main
" driving shaft lying horizontally along the back ends of a row
" of cylinders or retorts placed side by side, and having toothed
" wheels or flanges on such ends, any desired number of cylin-
" ders may be driven by the same shaft. The cylinders or retorts
" are mostly formed with double plates or covers at the ends
" having a space between those of each end." Filling such space
with fire-brick, or other non-conducting media, to prevent the
radiation of heat. "Instead of employing a furnace to each
" moving or revolving cylinder or retort," sometimes constructing
an arch or chamber to contain more than one such cylinder or
retort; or a series or set of two or more arches or chambers so
connected one with the other that the fire of one furnace may
impart its heat to more than one such cylinder or retort.

[Printed, 10d. Drawing.]

A.D. 1861, July 24.—N⁰ 1856.

GEDGE, WILLIAM EDWARD.—(*A communication from Louis
Constant Bernard.*)—(*Provisional protection only.*)—" Improve-
" ments in the preparation and clarification of the saccharine
" matters obtained from beetroot, sugar cane, indian millet, and
" other sacchariferous vegetables or plants." These are, extracting sugar from solutions from the above substances "without
" the use of bone black or animal charcoal," using "alumina in jelly
" or viscious state," and subacetate of lead, as follows : the alumina
in a solution of about 2½ ozs. of alum is precipitated by ammonia,
washed, and pressed, the juice, say about 2 lbs. from beet root
" is then poured cold and by degrees into the jelly of alumina,
" and well stirred, then passed through a hair sieve," so as to
thoroughly mix it with the juice, the whole is boiled for one or
two minutes, thrown on a cotton filter, and the deposit pressed,
and the solution "precipitated without heating, by subacetate of
" lead, prepared as follows : "litharge 100 parts, acetate of lead
" 300 parts, boiling water 900 parts, leaving it at the tempera-
" ture of the bath until the litharge is completely dissolved;"

small leaden shot are added which precipitate any copper which may be in solution. By these means a clear solution of juice is obtained containing traces of lead which are precipitated by adding "forty or fifty grains of carbonate of ammonia, filter and " evaporate." " Either of the two preparations, alumina and " subacetate may act alone," but " their combined action leaves " nothing to be desired." " Certain very aluminous clays may " be used in some cases as an economical substitute for the " alumina."

[Printed, 4d. No Drawings.]

A.D. 1861, July 27.—N° 1886.

HARE, Sir John.—(*Provisional protection not allowed.*)—" Im-" provements in the manufacture of sugar." These are, " manu-" facturing and baking sugar in blocks, either square, round, " oval, hexagon, octagon, or other regular shape, and of various " sizes, and of uniform weight, according to the purpose for " which they are intended, and casting them in moulds of metal, " glass, china, or other suitable material, when they may be " colored to any desired tint as fancy may suggest, the object " being to produce sugar in suitably given sized blocks of uniform " weight, more particularly adapted for domestic purposes."

[Printed, 4d. No Drawings.]

A.D. 1861, August 6.—N° 1956.

CLARK, William.—(*A communication from Edouard Théophile de Gemini and Edmond Oswald de Gemini.*)—" Improvements in " bleaching & clarifying saccharine matters, & in apparatus for " the same." These are, first, adding to the juices and syrups " fuller's earth or other clay or earth mixed with fine black, and " subjecting the whole to a violent agitation either in a hot or " cold state."

Second, adding " fuller's earth or other clay or earth mixed " with the black, and employing one or several jets of steam for " producing agitation of the mass " as afterwards described.

Third, employing " fuller's earth or other clay or earth mixed " with fine black, together with one or several jets of steam " in combination with a mechanical agitator for stirring up the " mass."

Fourth, the following apparatus for operating the treatment :—
One apparatus consists of a vat, near the bottom of which, upon

the sides, are fixed a series of prongs or pallets, an agitator passing down through the centre of this vessel is made to rotate by any suitable gearing or by strap, and it has prongs or pallets fixed to it near the bottom, which pass between the fixed prongs or pallets on the sides of the vessel. Another apparatus is described which is a vat with a pipe of steam, to which is attached a series of pipes for conveying the steam nearly to the bottom of the vat. It is said that although only two forms of apparatus have been represented, " they may be greatly varied in order to produce the " desired effect without departing from the principle thereof."

[Printed, 8d. Drawings.]

A.D. 1861, August 15.—N° 2029.

CAREY, STEPHEN, and PIERCE, WILLIAM MORGAN.—" Im-
" provements in apparatus for re-burning animal charcoal."
These are, constructing a long cylinder or retort " of an octagon
" or other polygonal form with ribs or fillets bevilled on each
" side, and formed or cast upon the inner longitudinal angles of
" the said cylinder or retort; these ribs run nearly the whole
" length of the cylinder or retort," the " two lower ribs are made
" shorter in front near the door to facilitate the discharge of the
" contents. The cylinder is built in brickwork with arch over
" and heated by a furnace beneath. An iron plate is fitted on or
" into the back edge of the cylinder or retort and another plate is
" fitted a short distance within the same. A hollow trunnion or
" spindle is formed or cast on this inner plate through which a
" pipe or tube is passed and turned up inside the cylinder, with
" an aperture near the top for the purpose of carrying off the gas,
" steam, or vapour generated in the cylinder into a box below,
" and by flues into the main chimney. At each end of the
" cylinder or retort a ring or pulley wheel is fitted to receive an
" endless chain, the said chain passes round the same and over
" an upper and smaller pulley wheel, fixed over each; these
" pulley wheels carry the cylinder or retort and give it motion.
" The two upper wheels are connected together by a transverse
" shaft, running from front to back over the top, and are set in
" motion by an endless worm or screw keyed on to the said
" transverse shaft; a clutch with lever handle is fitted by the
" side of the worm to put the machinery in and out of gear.
" The machinery is set in motion by cog wheels, bands, or other
" ordinary means worked by steam or other power." Instead of

suspending the cylinder or retort as above sometimes casting
or fitting "a trunnion or spindle to one or both of the inner
" plates of the cylinder or retort," and carrying " the same thereby ;
" the said trunnion or spindle would be sometimes used at the
" back of the cylinder and sometimes at the front, and sometimes
" to both back and front. The said trunnion or spindle would
" run through the outer plates and be carried by a journal outside,
" and supported by a bearer or other ordinary means.". When
the trunnion or spindle is used at the back end of the cylinder it
is made hollow for a pipe to pass through to carry off gas, steam,
or vapour generated in the cylinder. A plate cap or covering is
fitted over the back end of the cylinder to keep the heat in.

[Printed, 1s. Drawings.]

A.D. 1861, August 20.—Nº 2074.

LAMBERT, RICHARD SYDNEY.— "An improved 'skipping
" 'dipper' or vessel for removing sugar and other liquids from
" boiling pans." This consists, first, in constructing a 'dipper'
with "external ribs or guards so applied as to prevent the contact
" of the exterior surfaces of the vessel with the interior surfaces
" of the pans from which it is intended to dip up the contents,
" whereby the heated surfaces of these latter, even whilst the
" improved dipper is actually within them, charged with a por-
" tion of their contents ready for removal remain covered with
" liquid matter, in quantity always sufficient to prevent injury
" from charring or scorching."

Second, in constructing a dipper with "a cover (by preference
" a dome or elevated cover) which enables the operator to 'charge
" 'over' or take into the pans in action, and before the improved
" dipper is in any degree lifted out or removed therefrom, suffi-
" cient liquid to replace that taken into the improved dipper for
" removal, so that when this vessel and its charge are actually
" lifted away the pan is found to be already replenished. By
" these means all possibility of injury from the destructive effects
" of large overheated surfaces acting on small quantities of highly
" inspissated substances is wholly avoided." The upper part of
the dipping vessel has a cone or dome shaped cover or inclosure
having an opening at the upper part of the passage of the valve
rod, and when the dipping vessel is lowered "into the teache or
" boiling pan, and the valve has been closed, a fresh supply is
" run into the teache or boiling pan."

[Printed, 4d. No Drawings.]

A.D. 1861, August 26.—N° 2129.

NEWTON, WILLIAM EDWARD.—(*A communication from Horatio Nelson Fryatt.*)—(*Provisional protection only.*)—"Improved ma-" chinery for filtering liquids, decolorizing saccharine and other " juices, and rectifying alcoholic liquors." This consists of a centrifugal machine "with a concentric annular compartment " outside of a central chamber, and closed at top and bottom," all the sides of which are formed of perforated metal so as to contain pulverized charcoal or other filtering or decolorizing or deodorizing substance, so that the liquid, introduced into the central chamber by centrifugal action, is forced through the pervious sides into the space between the filtering chamber and the outer case, from whence it escapes through an opening into a receptacle below. In combination with the annular filtering chamber a cap extends having a central opening for the intro- duction of the liquid, downward from this cap is an annular hoop, which leaves a " passage between its lower edge and the bottom " of the central chamber for the passage of the liquid to be " filtered, so that a body of such liquid may accumulate against " and inside of the inner periphery of the filtering chamber " without the danger of its escape through the central feed hole " in the cap plate."

[Printed, 4*d.* No Drawings.]

A.D. 1861, August 28.—N° 2146.

DUNCAN, JAMES.—(*Letters Patent void for want of Final Specification.*)—" Improvements in the manufacture of sugar, and " in the apparatus employed therein, also in the apparatus em- " ployed in re-burning animal charcoal." These are, first, in a vacuum pan arranging and constructing "a series of discs, either " solid or hollow tubes, or other generally similar surfaces," so as "to admit of rotary or other motion being communicated " thereto," the object, it is said, being to keep down "the " temperature during the boiling operation."

Second, "counteracting the alkalinity of the sugar solution " by means of sulphurous acid, which is added in sufficient " quantity to neutralize the amount of alkali contained in the " saccharine solution."

Third, in re-burning charcoal, "the retort or vessel in which " the charcoal is contained is fitted with a central longitudinal

" rod or bar having a series of scrapers or vertical discs fitted to
" it " the rod with its scrapers, by means of suitable gearing, " is
" caused to move both longitudinally and vertically, and so cause
" the charcoal to be pressed forward," rendering the process of
" burning continuous."

[Printed, 4d. No Drawings.]

A.D. 1861, September 3.—N° 2196.

ROBERTSON, PATRICK. — (*Provisional protection only.*)—
" Improvements in treating yeast, and in the manufacture of
" ammoniacal salts, and a substitute for animal charcoal."   These
are, in reference to this subject, in using yeast to manufacture " a
" substitute for animal charcoal.   The yeast in a moist state is
" mixed with clay and carbonate of lime or chalk; the mixture
" is to be dried and then calcined.   It is preferred that the
" mixture of the matters above mentioned should be such that
" the calcined product may contain at the rate of ten parts of
" carbon, eighty parts of clay, and ten parts of chalk, but these
" proportions may be varied.   When the substitute for animal
" charcoal is to be used for filtering acids, the chalk may be left
" out, and only yeast and clay employed."

[Printed, 4d. No Drawings.]

A.D. 1861, September 14.—N° 2288.

WALLER, RICHARD.—" Improvements in machinery and appa-
" ratus for manufacturing and refining cane juice and other
" saccharine substances."   These are, first, " the use of heated
" cylinders in combination with belts, webs, or straps " as afterwards described.

Second, " the hollow discs constructed with internal dia-
" phragms " as afterwards described, and " the several arrange-
" ments and combinations of machinery and apparatus."

Pans or troughs for evaporating saccharine solutions, of any
size or shape, may be cupped or corrugated, and arranged one
above the other or otherwise combined, and if required, inclined
so as to deliver their contents into a trough or receptacle of sufficient size to contain the same.   The liquid is to be pumped into
the pan or trough into which it was first introduced, and then
again conducted through the series of pans or troughs.   The pans
have partitioned jackets or the pans are constructed of tubes

placed side by side for hot water or other heating media to pass through. Flexible belts of metal or of textile or felted fabrics pass over hot cylinders and through the saccharine solution carries a film of it upon them into the air, they then pass between pressing rollers, and the saccharine matter is completely dried by passing over the heated rollers. In some cases "a perfo-"rated pan or trough sieve or filtering bag or other suitable "receptacle" is placed below the pressing rollers for the reception of the saccharine matter, those matters passing through the filtering medium are returned again into the trough. Hollow compound discs composed of two portions of a large sphere or spheroid united at their peripheries or edges, mounted on a hollow tube. Within such hollow disc and mounted concentrically therewith on the said hollow shaft is a thin diaphragm of metal, and steam is passed through this arrangement. A modification of this arrangement with hollow discs on a small scale is described.

[Printed, 6d. No Drawings.]

A.D. 1861, September 19.—N° 2333.

CONDROY, Louis Gabriel Auguste.—"An improved cen-"trifugal apparatus, intended for purifying, washing, drying, "moulding, or extracting from liquids, substances, or materials "of various kinds, which are deposited or poured for this pur-"pose in moveable baskets or boxes fitted in the said apparatus." These are, in reference to this subject, "in sugar manufactories "by pouring in the baskets equal quantities, either by weight or "bulk, of raw sugar, or sugar skums are poured into the baskets "lined with cloth," and rotating them in an apparatus having "any even number of horizontal arms on a vertical shaft, the said "arms being placed in a same plane, each one of them being "provided at its outer end with a hinged hanger suitably dis-"posed and shaped to receive either perforated boxes or baskets "which may be lined inside with either cotton, wollen, or metallic "cloth." "The filtering process can be rendered continuous by "pouring on the sugar scums in the central vessel of the appa-"ratus." For easily extracting the juice of triturated or mashed "beet root, washing the pulp by means of water through pipes "into the machines."

[Printed, 8d. Drawing.]

A.D. 1861, October 9.—N° 2526.

SCHWARTZ, John.—(*Letters Patent void for want of Final Specification.*)—"An improvement in the manufacture of sugar." This consists in a novel mode of obtaining sugar from molasses, as follows :—The molasses is placed in a tank fitted with a perforated coil of steam pipe, blood and water are added, and the whole brought to boil and maintained for a time at the boiling point, when it is run through fine bag filters into cylinders filled with animal charcoal, and passed slowly through them, it is now either colorless or of a light straw color, and boiled in vacuum till " a maximum amount of granulation is obtained," " discharged " into a heater (an open pan heated by a steam jacket), where an " even temperature is maintained while the operation of filling " into moulds is going on. The moulds after standing to cool " are raised into heated floors, where in due course the treacle " separates from the granulated sugar, and runs from the moulds " into suitable vessels." To produce a lighter sugar the surface of the sugar in "the moulds is loosened, and the first runnings " of the charcoal cylinders (the lightest in color) are poured on " the surface and run through the contents of the moulds, thus " washing or bleaching the crystals." To produce "lump sugar " (white sugar or nearly so) it is desirable to use refined raw or " bastard sugar in solution after passing the same through animal " charcoal. The process of liquoring must be repeated several " times."

[Printed, 4*d.* No Drawings.]

A.D. 1861, November 2.—N° 2754.

WILSON, John Charles.—"Improvements in machinery or " apparatus for the manufacture of sugar." These are, first, combining the horizontal class of steam boiler with steam engine, connecting gearing and sugar cane mill, all united together on the same iron foundation plate or girders, also the arrangement of bevil and spur gearing.

Second, in steam boilers of sugar cane mills employing "a " boiler formed with a circular top and straight or curved sides " which terminate in a flat base for it to rest upon, the fire-place " is placed in an internal flue of large capacity suited to burn " wood fuel." To increase the vaporizing power of the boiler placing a water tube or tubes in the main flue.

Third, "a combination of 'bevil' and 'spur' gearing, with " the fly wheel shaft placed against the foundation plate." "The " spur gear may be varied to 'single' or 'double' gear, and " to all 'external' gear if desired." By this combination " the fly " wheel is made to revolve very compactly alongside the founda- " tion plate."

Fourth, constructing rollers of sugar cane mills of wrought iron or steel discs or plates, "instead of with a wrought-iron shaft " and cast-iron cylinder keyed on to it as heretofore." With discs, the shaft has a solid end forged upon it, against which the discs are screwed up by a nut. At the other end keys are sunk into the shaft which drive the discs. With plates, thin plates are punched with holes to lighten them. Instead of the end nuts bolts may be used to keep them in place. The roll pinions may be forged solid with the shafts.

Fifth, constructing the side frames which support the rollers of sugar mills, of two external plates, and between them the centre of one or several plates rivetted together and wholly of wrought iron or steel, in lieu of cast iron, as heretofore, and the "mode " of elevating and depressing the lower rolls by means of a wedge " instead of the present adjusting screw and block," and a "self- " adjusting trash turner," consisting of a steel spring "fastened " to the ordinary trash turner, and constantly pressing against " the lower roll."

Sixth, in " sugar-cane mills driven by water power, and the " water wheel which actuates or drives them," "placing the " water wheel in an iron trough or 'breast,' so as to dispense with " the present brickwork, breast or foundations, and to attach the " cane mill to the same."

Seventh, "for transmitting the power of animals to a sugar- " cane mill," making the framing of the gear " of a spherical or " bell-shaped casting so as to cover in the wheels and protect " them from breakage, besides adding to the stability and com- " pactness of the machinery." " Further, when it is desired to " greatly increase the speed of the horizontal driving shaft, " internal spur gear is adopted."

Eighth, vaporizing and evaporating " the aqueous portions of " saccharine fluids, by forcing heated air in contact with the sur- " face of said fluids, at the same time gradually removing the " upper stratum from the action of the heated air, as it becomes

" concentrated.". An apparatus for effecting this consists of two cylinders with pistons moving up and down in the same, so as to cause the fluid to pass from the one cylinder into the other while heated air is driven into a compartment over the top of these cylinders, and the steam passes out at another part of the compartment.

[Printed, 1s. 4d. Drawings.]

A.D. 1861, November 6.—N° 2789.

SCHRÖDER, FREDERICK HILLS.—" Improvements in evapo-
" rating and in machinery employed therein." These are said
" chiefly to apply to the evaporation of the liquid parts from
" sugar when in a state of syrup," and consist " in placing the
" syrup or other matter to be evaporated in an open pan heated
" by steam or hot water let into a jacket or case in which the pan
" is placed, and causing a series of concentric cylinders " to
" revolve in the syrup, a portion of the cylinders being con-
" tinually in the syrup, and another portion revolving in the
" atmosphere. The cylinders are each formed with slots running
" in the direction of their length," and are used in combination
or not with a blast or current of hot or cold air " between and
" among those portions of the cylinders which revolve out of the
" syrup," directed by means " of a hood or channel."

[Printed, 10d. Drawing.]

A.D. 1861, November 25.—N° 2964.

COWAN, PHINEAS.—"A mode of utilizing the waste heat of
" furnaces used in reburning animal charcoal." This consists in
" the combination of a boiler or generator with a furnace for
" reburning animal charcoal, in revolving or moving vessels or
" retorts, or with the flues communicating with such furnace, in
" such manner that the said boiler or generator may be subjected
" to the waste heat of said furnace, so that the steam may be
" thereby generated in the boiler or generator and the waste heat
" of the furnace thereby utilized." The boiler is placed in the
flue which runs above the retorts. The pipe supplying water to
the boiler may be carried through a flue in the brickwork of the
furnace, so that the water will thus be partially heated before
entering the boiler. " Also the steam generated in the boiler may

" be superheated by carrying it away through a coil of pipes or
" ordinary superheating arrangements in a flue or flues in the
" brickwork."

[Printed, 6d. Drawing.]

A.D. 1861, December 12.—N° 3112.

MENNONS, MARC ANTOINE FRANÇOIS.—(*A communication
from Louis Marie Amand Achille de Courson de la Villeneuve.*)—
" An improved means of defecating and purifying cane and other
" saccharine juices." This consists in the application of albumen
" to the defecation and purification of cane, beetroot, and other
" saccharine juices," substantially as follows :—One ounce of
albumen (say the white of egg, is placed in a vat, and 750 gallons
of the raw juice is run into it, with stirring from time to time,
and " left to repose for about ten minutes, and is afterwards
" boiled in the ordinary way without the addition of lime or other
" reagent." " When in the evaporating batteries the liquid mass
" should be lightly sprinkled with cold water, in order to separate
" from the scums the extraneous matters which may still be
" present." The syrups from the first sugar should be reboiled.

[Printed, 4d. No Drawings.]

A.D. 1861, December 30.—N° 3257.

NEWTON, WILLIAM EDWARD. — (*A communication from
Gustavus Finken.*)—" Improvements in the manufacture of cube
" sugar." These are, first, " exposing the grains or crystals (pre-
" paratory to their introduction into, or while on their way to
" the moulds or cube-forming apparatus) to the action of steam,
" by which their surfaces are subjected to the necessary degrees of
" heat and moisture to give them the requisite degree of adhe-
" siveness to form the cubes." This it is said " may be per-
" formed in various ways, and by the aid of various apparatus ; "
the following " mode and apparatus " have " been found practi-
" cable and convenient." An upright trunk of quadrangular
or other form is fitted with a number of screens arranged one
above the other, and inclined in opposite directions, so that the
granular sugar delivered on to the top one rolls down to the
bottom of the next, and so on to the bottom of the lowest one,
from whence it may be delivered or carried to the moulding
apparatus ; below the last screen is a pipe delivering steam, which
ascends through the sugar and through the screens. An endless

band conveys the sugar to the uppermost screen. Second, " in
" the formation of the cubes by means of machinery composed
" principally of an endless or continuous series of moulds fitted
" with compressing and discharging pistons, and having applied
" in combination with them a cam or cams, or their equivalent,
" for operating the pistons one or more at a time in regular suc-
" cession throughout the whole series, so that if the moulds are
" regularly supplied with granular sugar, a continuous delivery
" of compactly compressed cubes will be effected."

[Printed, 8d. Drawing.]

A.D. 1861, December 31.—N° 3275.

BROOMAN, RICHARD ARCHIBALD. —(*A communication from
Absalon Hippolyte Leplay, and Jules François Joseph Cuisinier.*) —
" Improvements in revivifying animal black or charcoal, in col-
" lecting ammoniacal gases generated in the revivification, in
" the clarification of saccharine liquors, and in the apparatus
" employed in the revivification of the black, and in the filtering
" of saccharine liquors." These are, in reference to this subject,
first, " revivifying animal black or charcoal by wet processes, hot
" water or steam acting in the filtering vessels, whether for renew-
" ing its powers for absorbing lime, or renewing its properties for
" absorbing azoted ammoniacal and coloring matters " as follows :
—After the saccharine fluid is run off, at the lower part of the
filter a jet of steam is introduced, which penetrates through and
between the grains of black for from 15 to 30 minutes, sometimes
adding beforehand on the filter a few quarts of milk of lime.

Second, revivifying animal black by acids to eliminate lime
from defecated saccharine liquors, by using, by preference, hydrochloric acid with water for washing out the lime.

Third, "the clarification of saccharine liquors, juices, and
" syrups by means of phosphates," by bringing a solution of phosphate of lime with one proportion of lime in contact with the
black for some instants on the filter, when the solution is run out,
and the filter is ready for use.

Fourth, " the different methods of revivification of animal black
" which allow of the collection of the ammonia given off."

Fifth, " the apparatus for carrying parts of the invention into
" effect."

[Printed, 1s. 2d. Drawings.]

1862.

A.D. 1862, January 1.—Nº 13.

PATRICK, WILLIAM BARKER.—" Improvements in the manu-
" facture of sugar and in the apparatus employed therein."
These are, using a closed vessel or vacuum pan for heating
" saccharine syrups or solutions to a low temperature, consider-
" ably below the boiling point (or 212° F.), by means of hot
" water, air, or vapour, caused to circulate in pipes in or around
" such vessel, or in the jacket or outer case thereof, combined
" with the use of air heated to about the same temperature, forced
" through openings in a pipe or pipes applied so that the air may
" be distributed and pass through the syrup or solution, and
" thereby aid in driving off the aqueous particles contained
" therein in the form of vapour, which are then drawn off by the
" air pump or other suitable means."
[Printed, 6d. Drawing.]

A.D. 1862, January 11.—Nº 84.

MAC KIRDY, LAUCHLAN.— (*Provisional protection only.*) —
" Improvements in apparatus for re-burning animal charcoal."
These consist " of a series of trays arranged spirally and vertically
" over a fire-place or in a flue, the covers of the trays being per-
" forated to allow of the gases evolved in the burning passing
" readily off from the charcoal. The trays communicate at top
" with a hopper, in which the charcoal to be re-burnt is placed,
" and at bottom the trays open into a spout furnished with slides,
" through which the re-burned charcoal enters a closed receiver,
" in which the charcoal is allowed to cool without contact with the
" atmosphere."
[Printed, 4d. No Drawings.]

A.D. 1862, January 27.—Nº 209.

ORR, WILLIAM.—" Improvements in the machinery or apparatus
" for the manufacture of sugar." These relate to apparatus used
in melting sugar which " is technically known as ' blowing up.' "
A rectangular vessel is made about twelve feet by five and four
feet deep, with a curved or boat-shaped bottom; it has a steam
jacket with a heating pipe inside, and a steam pipe delivering the

steam to the lower part of the blow-up, and another steam pipe into the upper part. A horizontal shaft extends across the blow-up, carrying a fast and loose pulley, which is driven from contiguous motive power. This shaft has fitted to it floats like a paddle wheel, capable of adjustment to the depth of the liquor. Each float board has a series of adjustable knives. As the wheel rotates, the knives divide the lumps of sugar, and the boards put the fluid in motion. The steam pipe opening near the bottom of the blow-up, in preference, terminates in the form of a flattened tube.. If a circular vessel is used, the wheel and knives rotate towards one side of the vessel, giving a rapid horizontal motion. Another arrangement is to have a horizontal slot, or in a circular vessel a vertical slot, made in the blow-up through which the flattened end of the steam pipe is inserted, and covered by a broad guard. Attaching to this vessel a receiver arranged below the blow-up, a pipe from the lower part of the jacket of the blow-up conveys the water of condensation into the receiver. When required the water is carried up into the blow-up by a pipe from the bottom of the receiver. There is also a pipe conveying the water of condensation from all the pipes into the receiver. There is an escape pipe from the receiver. One receiver will serve for several melting vessels.

[Printed, 1s. Drawing.]

A.D. 1862, February 19.—N° 445.

PATERSON, JAMES.—(*A communication from George Alexander Drummond.*)—" Improvements in means or apparatus for reburn-" ing animal charcoal." These are, the retorts " of a cylindrical " form, may either revolve continuously in one direction, or partly " in one direction, and then in the other," and they are inclined, and "receive the charcoal to be re-burned at their higher ends " through suitable channels, and then by revolution of such " retorts the contained matter will progressively traverse to the " lower ends thereof, where there are vanes or cups, or other suit-" able means adapted to take up such matters, and discharge it " into a receiver projecting inwards from such lower ends, and " having a communication with the cooling apparatus," " a jet " or jets of water are employed to operate upon the external sur-" faces of the cooling chambers, or the passage or passages to " them to facilitate the cooling process." A " series of such

" retorts are also arranged to work together," " with the delivery
" end of the first, discharging by a suitable way into the feeding
" end of the next, and so on with the succeeding retorts of the
" series." The retorts when connected in a series, are arranged
" so that the last of the series is that to receive the most heat,
" and the next less and so on." The gas or vapour generated
flows back and escapes by the feed pipe, "so as to avoid
" re-admixture with the manufactured or re-burned charcoal."
[Printed, 10d. Drawing.]

A.D. 1862, February 20.—Nº 455.

PATERSON, JAMES.—(A communication from George Alexander
Drummond.)—"Improvements in the use of animal charcoal."
Animal charcoal it is said "is composed of particles of very
" different sizes, and according to the method generally adopted
" in applying the same as a filtering medium for sugar and
" various other matters, such particles become very unequally
" distributed, and imperfect filtration as a consequence ensues."
To remedy this evil "by which more effectually to ensure equal
" distribution of the different sized particles of the charcoal in
" the filtering vessels," the charcoal is fed on to revolving
" discs with projecting surfaces adapted in their rotation to
" throw off and distribute such matter equally in the filtering
" vessel. A fan or other suitable blowing or exhausting means is
" also employed to blow or draw away the finer particles or dust
" as it leaves the distributing means. Steam is also employed to
" act amongst the various particles of charcoal in the filters to
" purify, and also to reduce or equalize the temperature thereof.
" Cold air may also be employed in like manner after such
" steam."
[Printed, 4d. No Drawings.]

A.D. 1862, February 20.—Nº 456.

PATERSON, JAMES.—(A communication from George Alexander
Drummond.)—"Improvements in means or apparatus for facili-
" tating the evaporation of saccharine solutions." These are,
" in connection with vacuum pans to facilitate the evaporation of
" saccharine solutions," forming vacuum pans, with their sides
inclining outwards or widening towards the top, and applying in
the same " revolving discs, travelling belts, or other suitable sur-
" faces," also, " propellers or stirrers which may be of the form

" of screws with vanes," and revolving with a shaft, supported in
a step or bearing in the bottom of the pan. In the fluid under
operation, in whole or in part surrounding the propeller is " a
" stationary cylindrical or somewhat funnel-shaped tube or chan-
" nel, or other suitable conductor." The lower part of the pan is
of a curved form, and " in place of the coils of pipes being
" arranged close together near the outer surface of the pan," as
is usual, arranging them " in two or more series, one above and at
" a distance from the other, and with space between the several
" parts of the coil of each series sufficient to admit of the free
" motion of the solution for the evaporation."

[Printed, 4d. No Drawings.]

A.D. 1862, February 22.—No. 471.

ROSS, WILLIAM HENRY. — (*A communication from Edward
Beanes.*)—" Improvements in the manufacture of sugar." These
are, " the use or employment in the manufacturing or refining of
" sugar of phosphate of ammonia when used in conjunction with
" sulphurous acid, either gas or liquid, or with any of the sul-
" phites or bisulphites." The phosphate of ammonia is by preference made by adding to bones, calcined and ground, dilute
sulphuric acid, allowing the whole to stand with stirring for about
twenty-four hours, and filtering ; gas liquor is distilled with a
little lime into the filtered liquor until the solution is alkaline,
and the filtered solution " evaporated to any required density or
" crystallized." " The sulphurous acid gas is economically and
" conveniently obtained by burning sulphur in a close vessel, into
" which the necessary air is forced by a pump." The method of
operating is as follows :—The saccharine solution is made alkaline
by lime water or milk of lime, sulphurous acid is then passed into
it until it is slightly acid, and the whole is blown up by steam,
adding " about one or two pounds of crystallized phosphate to
" five hundred pounds of raw sugar, or more or less according
" to the quality of the sugar used." Should the acid not be
neutralized by the phosphate, more phosphate or a little lime
water is added ; whilst, if the solution is alkaline " more gas must
" be introduced, as it is important for the syrup to be neutral," it
is then boiled, filtered, and finished as usual. " Bisulphite of
" lime or other sulphite may be substituted for the sulphurous
" acid gas."

[Printed, 4d. No Drawings.]

BLAIR, Thomas. — (*Provisional protection only.*)—" Improve-
" ments in machinery or apparatus for cutting, chopping, and
" breaking refined lump sugar, and other substances." These
are, the employment for the above purpose " of an oscillating
" frame and sliding saw bench, each provided with an adjustable
" guage," the piece of sugar "is made to slide or oscillate against
" ribbon, band, circular, or vertical saws," whereby it " is sawn
" into slices of any required thickness. These slices are deposited
" upon an endless web, and brought thereby into contact with a
" pair of feeding rollers, which move the slice forward to a certain
" point where a knife descends and cuts the slice into long narrow
" strips. Motion is imparted to this knife by means of an eccen-
" tric crank or other suitable mechanical contrivance. The cut
" strips or pieces are pushed forward by the action of the feeding
" rollers over a grating," by which the dust is separated. " After
" this operation the strips are made to enter a series of tubes or
" channels placed at an angle of about 45°," so as to descend
until they come in contact " with a fence or guage placed in
" advance of a knife, which falls upon the strips and severs a
" piece from the end of each at every tube or channel between the
" fence and the knife. These pieces are subdivided by means of
" small knives fixed at right angles into the before-mentioned
" knife, the sugar being by this time broken into fragments of
" the desired size for use. The machine may be made of any
" size, and provided with any number of tubes or channels."

[Printed, 4d. No Drawings.]

MILLER, William.—" Improvements in the manufacture of
" sugar." These are, " evaporating saccharine solutions at tem-
" peratures below those at which they boil under the ordinary
" pressure of the atmosphere." A pan is described containing
the saccharine solution having long tubes passing through it.
The pan is surrounded with a jacket, which may be filled with
water, and this water may be heated by the direct action of the
fire or otherwise. In the saccharine solution a cylinder resembling
a water wheel revolves, which takes up these films and exposes
them to the atmosphere, or to a current of hot or cold air, which
can be transmitted by means of a fan through holes in the axis
of the cylinder. The cylinder is covered by a hood having an

outlet at the top leading to a chimney. In the hood are doors for the admission of anything to stir or skim the sugar, or clean the pan. "The pan should be made of copper, and the rest of the " apparatus of galvanized iron, except when the jacket pan is " placed over the fire, when it should be made of wrought iron."

[Printed 10d. Drawing.]

A.D. 1862, January 27.—N° 532.

TORR, George.—" Improvements in and an improved appa- " ratus for manufacturing and reburning animal charcoal." These are said to be improvements in the invention described in No. 621, A.D. 1858, and are said to be applicable to any revolving retort or cylinder, but it is preferred that the cylinder should be mounted and made to revolve in the way described in No. 13,954, Old Law ; and they consist, first, in attaching the thread of an archimedean screw to the inner surface of a revolving cylinder. In this cylinder, called the main cylinder, is a smaller cylinder, the axis of which coincides with the outer or main cylinder. The two cylinders are attached so as to revolve together ; in the inner cylinder is an archimedean screw, like the outer cylinder, " except that if the screw of the outer cylinder is a right-hand " screw that of the inner cylinder must be a left-hand one, and " vice versa." The pitch of the screws, &c., must vary with the diameters. A stationary hopper with a pipe supplies the inner cylinder with crushed bones or animal charcoal. The cylinder revolves so that the archimedian screws in the inner cylinder causes the materials to pass along to the back, and return by the main cylinder to a double revolving cooling box fixed to the front end of both cylinders.

Second, the cylinders, both of the same diameter, are placed one above the other, and arrangements are made by which the crushed bones or animal charcoal, after having been conveyed through the upper cylinder, are discharged into the lower cylinder ; a cooling box is attached to the lower cylinder. The cooling boxes are made of corrugated metal, in order to expose a greater surface for cooling, &c.

[Printed, 1s. 6d. Drawings.]

A.D. 1862, March 1.—N° 564.

ROBERTSON, Patrick. — (Provisional protection only.) — " Improvements in treating yeast, and in the manufacture of

" ammoniacal salts, and a substitute for animal charcoal."
These are, in reference to this subject, in preparing " a substitute
" for animal charcoal, the yeast in a moist state is mixed with
" clay and carbonate of lime or chalk. The mixture is to be
" dried and then calcined. It is preferred that the mixture of
" the matters above mentioned should be such that the calcined
" product may contain at the rate of ten parts of carbon, eighty
" parts of clay, and ten parts of chalk, but these proportions may
" be varied. When the substitute for animal charcoal is to be
" used for filtering acids the chalk may be left out and only yeast
" and clay employed."

[Printed, 4d. No Drawings.]

A.D. 1862, March 7.—N° 614.

WRIGHT, RICHARD.—" Improvements in heating and clarify-
" ing saccharine fluids." These are, " carrying on the heating
" and clarifying process in vessels or apparatus," such as are described in No. 2153, A.D. 1860, but not using the revolving discs or
surfaces described there. That is to say, " the pan or vessel con-
" taining the saccharine fluid is heated by the vapour of boiling
" water, which is in a vessel below, at the same time the water in
" this vessel is prevented from coming in contact with the
" bottom of the pan or vessel containing the saccharine fluid ;" by
means of an overflow pipe placed below the bottom of the pan or
vessel containing the saccharine fluid, to carry off superfluous
water, while at the same time the saccharine fluid is " also pre-
" vented from rising in temperature above 212° Fahrenheit, by
" reason of the vessel containing the water in the pan below
" being constantly open to the atmosphere."

[Printed, 6d. Drawing.]

A.D. 1862, March 22.—N° 807.

HENRY, MICHAEL. — (A communication from the " Société
" Coignet Père et Fils et Compagnie.")—" Improvements in kilns,
" ovens, and furnaces." These are, first, employing in kilns or
ovens " a sole or floor, directly or nearly directly under which
" furnaces or fireplaces are placed, and through which sole
" numerous openings are formed for the passage of flames and
" products of combustion from the fires into the kiln or oven."
Blocks or other contrivances may be used for opening and closing

the orifices.   For cooling down kilns or ovens, &c., passages,
&c. are formed in the side walls with doors, plugs, &c.; these
admit air which escapes by an outlet in the roof with a remove-
able cover.

Second, a receiver or matter to be subjected to heat is rested
on an arch extending over the fire, one or more flues are at each
corner of the apparatus, and lead up into and preferably to the
top of the oven, &c. which surrounds the receiver.   The heat and
gases escape through other flues, which are midway between the
ends or corners.

Third, "air may be forced in or introduced under pressure
" above and below the fire-bars.   The hot gases from the fire are
" brought into the kiln, oven, or chamber containing the materials
" to be treated, at the top or upper part thereof a valve register,
" damper, or similar contrivance is fitted up in the passage," so
as to maintain the hot gases "within the apparatus under a
" pressure slightly above the external pressure."   The hot gases
may escape through a box, &c. containing a charge to be fed into
the kiln, &c.   In preference, the main chamber is oval, and
opened and closed at top and bottom by slide plates.   These
furnaces, it is said, may be applied to calcining, among other
things, bones, and revivifying animal charcoal, &c.

[Printed, 1s. 6d. Drawings.]

A.D. 1862, March 25.—N° 822.

FRYER, ALFRED. — " Improvements in the manufacture of
" sugar, and in separating liquids from sugar and other sub-
" stances."   These are, first, " the method of crystallizing the
" sugar held in syrups," by placing them when boiled in a vessel
which "shall not hold less than fifty tons nor be of less depth
" than thirty feet, the larger and deeper the better.   The crystal-
" lizing vessel is kept nearly full, and the saccharine matter drawn
" off from the lower end " into a tank, and fresh boiled syrup is
added at the top.   " It is better not to allow the temperature of
" the sugar to fall below 105° Fahrenheit," and for this purpose the vessel is surrounded, " if needful, and especially if such
" be of small diameter, as the minimum aforesaid, with a casing,
" or employ other means to maintain the temperature."

Second, " separating liquids from sugar and other substances,
" with or without the employment of an atmosphere, artificially

" made moist and warm," as follows :—" in operating upon
" saccharine matter that has passed through the crystallizing
" vessel " into the tank above referred to, portions are folded in
cloths or other porous material forming parcels, which are packed
from about 20 to 50 feet high, or upwards, each parcel is thus
subjected to gradual pressure, the lower parcels are gradually
removed while fresh parcels are added at the top. " This method
" of extracting moisture is specially applicable to the removal of
" syrup from raw sugar. The sugar may either be pressed
" without previous preparation, or after admixture with a little
" water."

[Printed, 8d. Drawing.]

A.D. 1862, March 31.—N° 887.

MENNONS, MARC ANTOINE FRANÇOIS.—(A communication
from Martial Victor Jouannet.)—(Provisional protection only.)—
" Improvements in the manufacture from vegetable products of
" glucose or fermentable sugar." These are, first, applying " the
" 'carob beans' or fruit of the carob tree (' ceratonia siliqua ' L.)
" to the production of a species of glucose."

Second, the processes employed.

- The fruit is first reduced to a pulp by graters or otherwise ;
the pulp is digested for 24 hours at " 60° Centigrade (140° F.) "
in five times its weight of water, with one-thousandth part in
weight of the water of hydrochloric acid, after which the acid is
neutralized by " an alkaline carbonate (carbonate of soda, potash,
" or lime for instance) in such proportions as may be required,"
and the solid matters removed by filtration, the filtered liquid is
clarified and concentrated in the ordinary way."

[Printed, 4d. No Drawings.]

A.D. 1862, April 25.—N° 1217.

REED, CHARLES. — (Provisional protection only.) — " A new
" method of treating the sorghum saccharatum or holcus saccha-
" ratus, in order to obtain saccharine liquor and pulp therefrom."
This consists in " taking the plant after it has been cut," and
pressing it between rollers, or " in a hydraulic or other suitable
" press to express the juice, and partially prepare the fibre."
" After the juice has been expressed, or the dry plant may be
" taken unpressed, the plant is cut, subjected to a boiling caustic
" alkali, is bleached and converted into pulp in a pulping engine.

" The juice expressed may be evaporated and converted into
" sugar, or it may be distilled to produce a spirit, either process
" being performed in the ordinary manner of sugar making and
" distilling."

[Printed, 4d. No Drawings.]

A.D. 1862, April 28.—N° 1242.

FLETCHER, John. — (*Provisional protection only.*) — " Im-
" provements in the apparatuses for treating saccharine liquids."
These are, first, an evaporating apparatus, consisting of a pan or
vessel for containing the liquid with a curved bottom, from which
the sides rise diagonally to within a short distance of the top,
when they assume a vertical direction, the ends of the pan rising
perpendicularly from the bottom. The pan is jacketted, and
" near the bottom are several series or groups of hollow tubes set
" horizontally, the ends of which are secured to two circular flat
" drums, one at either end of the pan, and which with the tubes
" rotate in the liquid." Each drum is fitted with hollow spindles
for the induction of steam, and for drawing off the condensed
water, with other arrangements.

Second, evaporating apparatus, which consists " in the employment of a series of coils of metallic tubing, heated by steam,
and connected to a hollow shaft which passes through the
centre of the coils. This shaft has two compartments through its
whole length, one for steam and the other for condensed steam.
The top of the apparatus is mounted with a conical or curved
cover, and the vapour is carried off through one or more chimnies
on the top of the cover, employing " a small jet of steam in the
" chimney or chimnies," with other arrangements.

Third, a " cooling or granulating apparatus," the ordinary
wooden or metallic cooling or granulating trough has a triangular
bottom fitted with tubes running longitudinally. These tubes
have slits or apertures on their circumference at intervals, and
there are corresponding slits or apertures in the trough, by which
the molasses from the sugar runs off, with other arrangements.

Fourth, a cheap apparatus for " the clarifying, evaporating, and
" granulating operations" for making sugar from cane juice.
This consists of " a cistern having a corrugated bottom, and its
" lower part " jacketted for steam. The cistern is fitted with a
series of tubes side by side, the ends of which " are placed and

" fixed in orifices in the sides of the cistern, by which a communi-
" cation is obtained between the steam jacket and the tubes."
The top of the cistern is conical, and applied as described under
the second head, also employing as in it a jet of steam. " The
" water of condensation of the steam is carried off from the
" apparatus by means of an ordinary condensed water box. The
" bottom of the cistern is fitted with a sluice valve for discharging
" the contents of the cistern when required."

[Printed, 4d. No Drawings.]

A.D. 1862, May 20.—N° 1520.

MENNONS, MARC ANTOINE FRANÇOIS.—(*A communication
from Georges Serret, Paul Hamoir, and Achille Duquesne.*)—(*Provisional protection only.*)—" Improved processes for the conversion
" of amylaceous matters into saccharine and other useful pro-
" ducts." These are, modifying the processes now in use for the
above purposes, by dispensing with the disintegration of the raw
material, and replacing it " by exposure to the action of steam or
" boiling water," as follows :—The material, say maize, is placed
in water in a vessel with a false bottom, and is brought to boil by
a current of steam admitted through it, and kept at that temperature for about two hours with agitation, it is left to repose for
about four hours " at about 95° centigrade (203° Fahrenheit). The
" maize being thus thoroughly swollen, the acid is introduced in
" the same proportions as in the ordinary modes of treatment,
" and the liquid is from time to time tested with iodine until the
" non-coloration of the latter indicates the approach to com-
" pletion of the saccharification. This result is generally obtained
" within three or four hours. The purge cock at the bottom of
" the vat is then opened, and the quantity of liquid drawn off is
" replaced at the same time by boiling water admitted above,"
until " the complete separation of the soluble principles is effected,
" which point is ascertained by testing the density and acidity
" of the liquid ;" when, the water is stopped, the solid matters
drained and finally removed from the vat for use. The saccharification may be performed in a series of vats, the preliminary
scalding being effected in a separate vessel. These vats are ranged
side by side, and communicate with each other " by tubes passing
" from the bottom of the one to the top of the other, and so on
" throughout the set."

[Printed, 4d. No Drawings.]

A.D. 1862, June 24.—N° 1851.

CARR, Thomas.—"An improved machine for grinding, knead-
"ing, washing, and other like purposes." This machine it is
said is useful for a number of purposes which are named, among
which is "sugar making," and the arrangement is so, that—

First, "driving the edge runners independent of the revolving
" pan, either in opposite directions to it, or at a greater or less
" speed, and in such a manner as not to interfere with the rising
" of the edge runners when encountering a substance too hard
" for them to crush."

Second, "carrying of the edge runners independent of the pan,
" instead of their always resting upon it, and also the adjusting
" of their proximity thereto to any required gauge."

Third, "regulating the exact amount of pressure the edge
" runners should at any time be capable of imposing on the
" material passing beneath them."

Fourth, "the furrowing or grooving in some cases of the edges
" or peripheries of the edge runners, or the bottom of the revolv-
" ing pan, or both of them, instead of inserting therein knives
" or teeth."

Fifth, "the general arrangement and combination of the various
" parts of the apparatus."

[Printed, 3s. 6d. Drawings.]

A.D. 1862, July 19.—N° 2061.

BROOMAN, Richard Archibald.—(A communication from
Absalon Hippolyte Leplay and Jules François Joseph Cuisinier.)—
" Improvements in revivifying animal black, in apparatus
" employed therein, and in recovering a product employed in
" the revivification." These are said to be improvements upon
No. 3275, A.D. 1861, and consist as follows :—In treating
animal black, the animal black to be revivified is first washed
with hot water, and placed, by preference, in four cylinders with
perforated false bottoms "for permitting the inlet of steam from
" a steam supply pipe; each cylinder communicates with that
" next to it by a pipe extending from beneath the perforated
" false bottom of one to the upper part of the next. There is
" also a pipe leading directly from the upper part of the first
" cylinder to the lower part of the last in the series." The
washed animal black being placed in the cylinders, steam is

passed through "the perforated false bottom in the first cylinder,
" and a solution of caustic soda in a boiling state is admitted at
" the top of the cylinder." " The caustic soda solution being
" continued to be added passes into the second, third, and
" finally into the last cylinder in the series. When the action
" has been continued long enough in the first cylinder, the
" admission of caustic soda is prevented and communication
" closed to the next cylinder; hot water is now admitted, then
" hydrochloric acid, after which a solution of biphosphate
" of lime, then water to wash the black. The black now
" completely revivified, is withdrawn and dried. The like
" process is continued in all the other cylinders." The caustic
soda solution is finally dried and burned with nitrate of soda,
the coloring matter is driven off, leaving caustic soda.

[Printed, 1s. Drawing.]

A.D. 1862, August 11.—N° 2242.

CLARK, WILLIAM.—(*A communication from Theodore Augustus
Havemeyer.*)—" An improved carriage for conveying sugar moulds
" in sugar refineries." This consists as follows :—" A horizontal
" frame or plate " is mounted on three wheels, two wheels support
the back part of the frame, and they may turn loosely on a fixed
axle, or be rigidly attached to their axle, and the latter allowed to
turn. " One wheel supports the front part of the frame, and is
" what is commonly termed a castor wheel; this wheel has its
" axle fitted as usual in a fork at the lower end of a vertical arbor ;
" this arbor is fitted loosely in a vertical post, the lower end of
" which is permanently attached to the front of the frame, the
" upper end of said arbor bearing against the upper end of the
" opening in the post. The arbor is allowed to turn freely in the
" post, and the front end or part of the frame is bent or curved
" upward to form a recess so as to admit of a large sized castor
" wheel being used." " The draught pole is attached to the fork
" of the arbor or to the axle of the castor wheel. To the back
" part of the frame there is attached in a central line with the
" vertical post before-mentioned a second post. The two posts
" are of equal height, and they support two plates, which are
" placed one over the other, and are parallel with the frame.
" These plates are provided respectively at each side with curved
" arms of semicircular form, the arms of the upper plate having

" whole series. The first pan will, by preference, be made plain
" or with a flat bottom, but the eight following ones will be cor-
" rugated or wavy as above-mentioned." The corrugations may
vary, " so long as their efficiency is not materially interfered with.
" The five boilers will be separated by partitions somewhat lower
" than those of the others. The whole series are to be set in brick-
" work or masonry so as to present a raised rim at each side, and
" the seven boilers set nearest the fire-place are to be fixed in con-
" junction with two outside rims of greater elevation, which
" together constitute two troughs or channels at the sides of the
" boilers for the reception of any syrup which may overflow or
" boil over. The pans or boilers may be made to communicate
" by means of valves for the purpose of facilitating the operation
" which is carried on in this apparatus in the ordinary manner."

[Printed, 8d. Drawing.]

A.D. 1862, October 27.—N° 2894.

PEEK, ALFRED.—" Improvements in apparatus for evaporating
" saccharine and saline solutions." These are, applying heat
in such a way " that the temperature of the solution cannot be
" raised beyond the boiling point of water, but it may be kept to
" any required degree below that temperature." " Where steam
" is available " using " a hollow disc or agitator revolving slowly
" in the pan containing the solution to be evaporated; the
" agitator is half full of water, and it is supported on hollow
" trunnions, through one of which steam is admitted, by a pipe
" bent partly into the water, and through the other the overflow
" water makes its escape." To modify the above, under the
above arrangement, is a boiler, heated as required, " a pipe from
" the upper part of the boiler communicates with one of the
" trunnions of the agitators, and with a cistern above the level of
" the agitators, and another pipe passes from the other trunnion
" of the agitators to the lower part of the boiler." " The cir-
" cumferences of the agitators are connected by horizontal pipes."
" The form of the agitators can be greatly varied." The above
apparatus " may be used when the solutions to be evaporated are
" in open or covered pans having a chimney or fan to produce a
" draught or in vacuum pans." When the heating surface of the
revolving agitators is not sufficient, a series of steam pipes are
inserted in the vessel containing the solution, such pipes being

" surrounded by other pipes containing water, the water pipes
" being open at one point at least to the atmosphere."
[Printed, 10d. Drawing.]

A.D. 1862, November 28.—N° 3196.
ADAMS, James, and WHITE, William Cornwell.—"Im-
" provements in apparatus for boiling and evaporating." These
it is said, among other purposes which are named, are "appli-
" cable to the manufacture of sugar," and consist of a copper or
other metal pan with a false bottom, in the interior of the pan is
a coil of perforated tube supplied with steam or hot air which
enters at one of the trunnions of the pan formed hollow for the
purpose. "From the bottom and interior of the hollow pan a
" pipe ascends to the opposite trunnion in order that the water
" of condensation may be forced up the pipe by the pressure of
" the steam and driven out of the pan." When a great heat is
required applying a coil of pipe in the hollow space between the
pans or vessels, the coil being perforated and supplying superheated steam. "Below the coil, and intervening between it and
" the bottom of the outer pan, a curved or bent plate, according
" to the shape of the vessel, is to be fixed in order to concentrate
" the heat upon the upper or false bottom." "The pan is balanced
" or borne on a fulcrum" made "of a V or pointed form," resting
on a groove in the standard or framework; the short arm of a
lever is held down by a rod connected to a regulated spring
balance indicator, and other arrangements are made "so that
" when the pan is supplied with liquid to be boiled or evaporated
" the weight thereof may be known by the indicator, and the
" amount of evaporation or loss by weight continually shown."
[Printed, 8d. Drawing.]

A.D. 1862, December 15.—N° 3348.
BUCHANAN, George.—"Improvements in machinery used in
" crushing sugar cane." These are, the mode described of com-
" bining a number of mechanical parts, none of which are claimed
separately, "whereby the oscillating cylinder of the steam engine
" is placed intermediate of the fly wheel and the crushing rollers."
For these purposes "the oscillating steam engine is placed between
" the fly wheel and the large cogged wheel, which is, by preference,
" furnished with internal teeth, but this is not essential. The
" crank or driving shaft turns on a suitable framing between the

x 2

" engine and the fly wheel, and the crank or driving shaft by a
" pinion thereon communicates motion to a cog wheel on a shaft
" which has fixed on it a pinion which takes into and drives the
" the large cog wheel which gives motion to the crushing rollers
" as heretofore."

[Printed, 8d. Drawing.]

1863.

A.D. 1863, January 30.—N° 279.

GEDGE, WILLIAM EDWARD.—(*A communication from Ferdinand Wittmann.*)—(*Provisional protection only.*)—" Improvements in
" cones or forms for moulding refined sugar." These are, making
cones for the above purpose " of a single piece of sheet iron, of
" which the edges are tapered or thinned down so as to present
" on the outside as little ridge or wavyness as possible. The
" interior of the cone is perfectly smooth. The two edges of the
" iron are first adjusted by small rivets, the heads of which are
" filed or otherwise smoothed down ; then the joint is soldered
" and brazed with the usual brass and zinc soldering."

[Printed, 4d. No Drawings.]

A.D. 1863, March 2.—N° 584.

GARTON, CHARLES.—" An improved method of applying heat
" in the manufacture and refining of sugar, and in malting, hop
" drying, brewing, distilling, and vinegar making." This consists
in the means of applying hot water for the above purposes as
follows :—Connecting " a chamber or coil of pipes, or a series of
" such (fixed at the place or places where the heat is required to
" be applied), with a boiler by means of flow and return pipes,
" the former entering the chamber or coils and boiler at a higher
" level than the latter ; and, in the case of chambers being used,
" the same being fitted with partitions extending nearly through
" their entire width, and at the highest point thereof with pipes
" to allow of the escape of air." The whole is filled with water
or other fluid, and the heat is applied to the boiler. The circulation is regulated " at will by means of stop-cocks attached to the
" flow and return pipes."

[Printed, 10d. Drawing.]

GITS, JEAN FRANCOIS.—"An improved furnace for the revivification of animal charcoal." This consists of a " furnace of the
" reverberatory description ; it has at the back a series of drying
" chambers with covering plates, and in front of these chambers
" a series of upright cast-iron pipes jointed together, the upper
" ones having stoppers at the top and the lower being provided
" at the bottom with traps." When the animal charcoal has
been used for clarifying syrups, "it is withdrawn and placed in
" the chambers above-mentioned, around which pass the hot
" gases from the furnace; from thence it is removed to the upper
" cast-iron pipes, which are exposed to the direct reverberatory
" action of the heat, there revifying it, falls into the lower pipes,
" where it cools without contact with the air, and on the traps
" being opened the revivified and cooled animal charcoal falls
" into a waggon placed underneath to receive it."

[Printed, 10d. Drawing.]

A.D. 1863, April 4.—N° 857.

HANREZ, PROSPER.—"Improved machinery or apparatus for
" drying coal, grain, and other substances." This, it is said, is
" also applicable to sugar machinery in place of the centrifugal
" apparatus at present in use." A vertical shaft is supported by
bearings at each end, an annular casting is fixed on the shaft,
a perforated structure is fixed on the vertical shaft, below is a
reservoir to receive any liquid thrown off through the perforated
structure. Concentrically to the above shaft and turning in the
same direction is a screw shaft, the screw of which fits easily in
the perforated chamber. The material is fed in by a spout from
above. A perforated tube admits steam into the materials. In
order to produce the desired effect it is necessary that the vertical
shaft and the screw shaft should move at different velocities, the
latter moving the slowest. "This differential movement may be
" effected " by an arrangement of spur gearing which is shown,
and " by making the diameters of the wheels proportionate to the
" difference of speed required."

[Printed, 3s. 4d. Drawings.]

A.D. 1863, April 24.—N° 1027. (* *)

JOHNSON, JOHN HENRY.—(A communication from Edmond
Marquis de Ruffo Bonneval and Joseph Mouren.)—" A filtering

" apparatus for treating by pressure oils, syrups, and all sorts
" of liquids susceptible of filtering." This consists "in the
" employment of a peculiar arrangement of apparatus for filtering
" liquids, whereby such operation may be effected under pressure.
" The liquid to be filtered is contained in a vessel or tank, to
" which a suction and force pump is connected, which pump is
" also in communication (by means of a pipe) with a strong
" filtering vessel. The filtering vessel is provided with a man-
" hole at the top and a grating at the bottom, upon which grating
" is placed the filtering medium, consisting of felt or other suit-
" able material. Immediately beneath the grating, and secured
" to the filtering vessel, or otherwise, there is fitted a hemi-
" spherical or other receiver, in which the filtered liquid is col-
" lected and drawn off by a central pipe and stop-cock. The
" liquid to be filtered is pumped into the filtering vessel, and the
" pumping is continued after such vessel is full, in order that the
" liquid therein may be in a state of compression, which state
" facilitates and expedites the process of filtering by forcing the
" liquid through the filtering medium. In some cases this appa-
" ratus may be employed as a press, as for example, by introducing
" olives into the filtering vessel, and injecting and forcing therein,
" under considerable pressure, hot water, the oil will be expressed
" and thus the double effect of the press and the filter or purifier
" is obtained."

[Printed, 8d. Drawing.]

A.D. 1863, April 28.—N° 1060.

MARRIS, John, and MARRIS, William. — "An improved
" machine for breaking loaf sugar." This consists " of two metal
" frames about 4 feet high and about 2 feet wide, two circular
" saws, one 24 inches in diameter, and the other 13 inches in
" diameter, fixed one inside and the other outside these frames
" and working on a spindle, and two pair of metal rollers with
" steel knives fixed underneath the spindle." The 24-inch
circular saw cuts the loaf of sugar " down the centre from end to
" end into two halves. One of these halves is placed flat side
" downwards upon the table," and " then cross cut into semi-
" circular cakes by the 13-inch circular saw." The cakes as they
are cut off drop through a flat spout upon the upper rollers, which
rollers " are made of metal, and are furnished with horizontal
" knives 11 inches in length, by which the sugar is cut or

" chopped into long rectangular pieces or strips." These again drop into a lower pair of metal rollers, "furnished with circular " knives fifteen in each roller) which cut or chop the long strips " transversely into cubical pieces." These pieces fall into a double spout or screen, the top of which being wired allows the lumps to pass over into a receptacle below while the splinters or chips pass through the wires into a box underneath the table. This machine is worked by a handle.

[Printed, 10d. Drawing.]

A.D. 1863, May 13.—N° 1205.

HEUSNER, KARL LUDWIG.—(*Provisional protection only.*)— " Improvement in the manufacture of hydrate of barytes and in " the manufacture of sugar." These are, carbonate of barytes is mixed " with coal or charcoal, and a small quantity of a suitable " flux," in preference, an " alkali or alkaline earth," and heated from 800° F. gadually to a white heat, when cold water is added and the solution " is fit to be employed in the manufacture of " sugar as hereafter described." If required to be transported a distance the solution is evaporated to dryness, and " when about " to be used it is again to be dissolved." " In order to obtain " sugar from molasses, treacle, and other saccharine fluids, a " solution of hydrate of barytes is mixed therewith," and the precipitate "is mixed with water, to which carbonic acid is to be " applied," which precipitates the baryta as carbonate and leaves the sugar in solution "suitable to be treated in the ordinary " manner to obtain crystals of sugar therefrom."

[Printed, 4d. No Drawings.]

A.D. 1863, May 25.—N° 1307.

MUIR, WILLIAM.—" Improvements in machinery for cutting " sugar, and for assorting the lumps when cut." These are, " the arrangement and combination of self-acting machinery, " whereby a loaf of sugar may be cut up and the lumps and " dust assorted without intermission," as follows :—" The loaf of " sugar to be cut is attached to a moveable table or chuck, and " the saws are mounted on moveable slides; the distance between " the saws by which cuts are made in the face of the sugar-loaf " will correspond to the size of the lumps of sugar to be pro- " duced. These sets of saws are placed at right angles to each

" other, and act in succession on the sugar; after which another
" saw is brought into action to detach the lumps from the loaf;
" these lumps then drop in, or are conveyed to, a revolving or
" reciprocating sieve, through which the sugar dust and the
" smaller or irregular shaped pieces drop, while the cubes or
" lumps of a uniform size are collected into a suitable receptacle."
These details may be considerably varied; " the requisite motion
" may be given to the sugar for bringing it towards the saws, or
" to the saws for bringing them towards the sugar, both laterally,
" vertically, or horizontally; and instead of using two sets of
" saws for cutting across the end of the sugar loaf, one set may
" be used, and the loaf of sugar turned partly round before the
" second cut is given; the whole machine is made self-acting, so
" that when a loaf has been attached to the table or chuck it is
" cut up into cubes of the proper size, and the dust and irregular
" pieces separated from each other."

[Printed, 10d. Drawing.]

A.D. 1863, June 17.—N° 1516.

NEWNAM, James.—"Improved means of, and apparatus for,
" boiling in vacuo at a low temperature." This consists in " the
" employment of steam introduced into the heating coil or coils
" in an ordinary vacuum pan, and previously attemperated and
" maintained at the required temperature in a separate vessel,
" constructed, arranged, and acting " substantially as follows :—
Attached to a vacuum pan, by means of a pipe, is a vessel containing a worm, into which steam is admitted and maintained at
the temperature required by means of hot or cold water, supplied
to the vessel containing the worm through a pipe near the top;
there is an outlet pipe from near the bottom of this vessel for
drawing off water from it when necessary. A pipe from the
bottom of the worm conveys steam from it into the coil in the
vacuum pan, and any steam generated in the vessel containing
the coil is conveyed by means of the pipe mentioned above into a
separate coil in the vacuum pan. There are gauges for indicating
the water level and pressure of steam in the apparatus.

[Printed, 8d. Drawing.]

A.D. 1863, July 2.—N° 1649.

MILLER, William. — (Provisional protection only.) — " An
" improved mode of evaporating through the combined agencies

" of heat and centrifugal force, and the machinery employed
" therein, more particularly applicable to saccharine solutions."
These are said to be additions made to the apparatus described in
No. 525, A.D. 1862, and are said to consist, first, " in the intro-
" duction of a fan inside the rotating cylinder, which revolving
" on a separate axes, will, by a more rapid revolution, increase
" the current of air through the plates or ribs composing the
" cylindrical apparatus."

Second, "by the introduction of a fan connected to the hood
" covering the cylinder, the rapid rotation of the fan, acting by
" centrifugal torce, drawing the air through the open ends of the
" cylinder, and thence through and between the curved plates
" comprising the cylinder, so as to convey the vapors to any
" external outlet."

[Printed, 4d. No Drawings.]

A.D. 1863, August 18.—N° 2053.

BROOMAN, RICHARD ARCHIBALD. — (A communication from
Auguste Pierre Dubrunfaut.) — " An improved method of, and
" apparatus for, treating molasses, syrups, saccharine juices, and
" other products." This consists in the employment or adaptation " of the phenomenon of endosmose, that is to say, the
" double current which exists between two liquids of different
" density separated by a membranous partition." In some cases
employing " as the membrane paper, artificial parchment, paper
" covered with leaf or powdered metal, or with varnish, collodion,
" or other coating, close cloths of linen, hemp, or cotton trans-
" formed into membrane by sulphuric acid, collodion, coagulated
" albumen, or other coatings, plates or vessels of porous earths,
" skins, parchment, bladder, and all tissues or membranes of
" vegetable or animal origin." The " simplest arrangement " is
" a frame of a cubical form, closed at bottom, provided at its four
" sides with paper parchment, or other suitable membrane, and
" leave the top open; place this apparatus in a reservoir a little
" larger than it, in such manner that its top shall rise above the
" reservoir about 3 or 4 inches," and the bottom about the same
distance from the bottom of the reservoir; in the cubical reservoir
is an overflow orifice, in which is a pipe, and a syphon draws the
denser liquid from the lower part of the reservoir. By placing in
the inner vessel impure molasses, syrups, or saccharine juices, and

in the outer or reservoir water, salts pass through the membrane
into the water " to the almost complete exclusion of the sugar and
" colouring matter." " The analytical effect which eliminates the
" salts diminishes as the density of the molasses decreases and
" that of the water increases." The saccharine matter is supplied
continuously into the top of the inner vessel, and " that which has
" passed through the membranes " and " having accumulated at
" the bottom of the reservoir, has acquired the required density,"
is " drawn off by the syphon tube." The above " apparatus forms
" the basis of all the modifications," of which there are several,
described. This " apparatus may be usefully employed " in " the
" saccharification of amylaceous substances by acids ; " if " a syrup
" with all its acid " be put into the inner chamber " the acid will
" pass into the exosmose water to the exclusion of the glucose
" matter." " The acid may thus be employed indefinitely."
When molasses thus treated has furnished a crystallization of sugar,
the molasses which runs off may be again " osmosed," clarified,
and rendered crystallizable and so on. When molasses purified by
several osmatic treatments is placed in the "osmogene," sugar in
its turn may be eliminated and separated," and the watery solution
concentrated yields abundant crystallizable sugar, " and finally
" there remains in the osmogene a product non-crystallizable,
" and richer in extractive matter." " Osmosed molasses boiled
" with lime in excess, and treated with carbonic acid, are highly
" discolorized and purified ; this is a simple means of removing
" from osmosed molasses the coloring principle, but this process
" does not exclude the employment of animal charcoal."

[Printed, 1s. 8d. Drawings.]

A.D. 1863, August 26.—Nº 2106.

KESSLER, JACQUES LOUIS.—" Improvements in apparatus for
" evaporating and distilling." These are, in placing a series of
cast iron boilers or vessels one on another, the bottom boiler, containing the liquid or other matter to be evaporated, having a
second boiler also containing the same liquid, and similar in
construction, but with its bottom formed with a conical surface,
to allow the condensed vapours in contact therewith to run into
a circular trough fitted round the top of the bottom boiler, whilst
at the same time it serves to heat the second boiler. The circular
trough containing the condensed vapours also serves as a hydraulic

joint by dipping a rim pending from the bottom of the second boiler into it. When the circular trough becomes filled the liquid may be run off or passed into a worm placed on the bottom of the second boiler. In this manner the warmer liquid, in passing off, communicates its heat to the colder. A cock placed at the extremity of the worm allows a pressure of vapour to be maintained in the bottom vessel greater than that in the superposed vessel, without however forcing the non-condensed vapour into the upper vessel, and yet making all the condensed liquid of the worm pass into a circular trough fitted round the upper part of the second vessel. If another precisely similar vessel be placed on the second vessel the heat of the steam will heat and boil a third similar liquid, and so on to any extent, the liquids being fed to the vessels by pumps or other ordinary means. If the liquid in the bottom vessel boils under a pressure of six atmospheres, the other vessels superposed on it will, by the help of steam gauges, be maintained at pressures decreasing from six degrees to 0°.

This arrangement of evaporating apparatus is equally applicable in vacuo or under pressure, by merely inverting the steam-gauges to transform them into barometer gauges, and may be used in all cases where liquids or other matters are to be evaporated, such for instance as juice of beet-root or sugar-cane, wine-lees, salt water, extracts of plants, dextrine, glucose, syrups, saltpetre, sulphates of iron, zinc, and soda, fermented molasses, juices of vegetables, soda, mineral, and other waters, alum, chromates and prussiate of potash, gelatine, " tartaric, citric, and boric (boracic ?) acids," and for distilling and concentrating sulphuric acid and other matters.

[Printed, 2*s.* 6*d.* Drawings.]

A.D. 1863, September 17.—N° 2282.

COWAN, PHINEAS.—"An improvement in refining sugar." This consists "in promoting the 'settling,' 'covering,' or satu- " rating of the charcoal by exhausting, or partially exhausting, " the air from the charcoal and charcoal cistern or vessel," sub- stantially as follows :—It is said that, in both the modes of filtering saccharine liquids which are now employed, namely, one method in which "the syrup percolates the charcoal by natural gravita- " tion," and the other in which " steam or other artificial pressure " is applied at the top of the charcoal cistern or vessel, to drive " the liquor or syrup through the charcoal," the air " contained

" in the charcoal and charcoal cistern impedes the descent of the
" saccharine liquor; and to obviate this it is proposed to exhaust,
" or partially exhaust, the air from the cistern or vessel containing
" the charcoal, having first admitted " sufficient syrup or saccha-
" rine liquor into the cistern or vessel to cover the upper surface
" of the charcoal contained therein," and thereby preventing
" air from being drawn into or through the charcoal from above
" it, when the exhausting operation is afterwards applied," and
then pumping " the air from and around the charcoal, whereupon
" the saccharine liquor or syrup will descend and will saturate
" the charcoal with regularity," passing down through it with
rapidity, and presenting itself " at the bottom bright and clear."

[Printed, 4d. No Drawings.]

A.D. 1863, September 23.—N° 2342.

FONTAINEMOREAU, Peter Armand le comte de.—(*A
communication from Charles Hänsell.*)—" Certain improvements in
" centrifugal machines for treating saccharine substances, part of
" the same being applicable to other centrifugal machines."
These are, first, " in the purifying machine or apparatus, with
" which is combined a steam engine ; the self-discharging drums
" are turned over by means of moveable bearings, one of which
" turns on an axis keeping it in its place, and the other is so
" arranged as to be able to be withdrawn or disengaged from its
" shaft to set it free, so that it can turn on its axis, and thus
" overturn the drum to discharge it. To receive the liquid, the
" drums are provided with a casing, which receives and throws it
" into an external channel ; and for this purpose the ordinary
" recipient, which also serves as the framework, is replaced by
" framing, with which is combined a steam cylinder, and steam
" is introduced between the casing of the drum, which latter is
" balanced with a spring and counterweight for facilitating the
" discharge, and a proper opening is provided for charging the
" same."

Second, " in the preparing or extracting apparatus, which is
" composed of two parts or machines, the matters to be treated,
" are in the first, continuously introduced by means of a tube into
" a centrifugal drum, which is composed of three casings ; the
" saccharine matters pass from the tube between the first and
" second casings till they reach the bottom, where the drum is

" provided with spikes, by which the matters coming out of it
" are beaten for their final extraction. The juice extracted in its
" passage is guided by the third or outer casing into a circular
" channel placed round the recipient, and to facilitate the opera-
" tion water is pressed into the matters before they escape from
" the drums, from a circular channel properly placed inside the
" first casing, which is perforated for that purpose. The speed
" of the apparatus—that is to say of the operation—is regulated
" by means of a spiral screw, placed in the space between the first
" and second casings."

Third, " in the second part or machine the centrifugal force is
" assisted by other pressure, and for this purpose a piston is
" pressed on the matters under operation either by hydraulic
" pressure by means of a pump, or by a screw. The saccharine
" matters are conducted from the first to the second machine by
" means of a tube. The discharge is effected by hooking the
" perforated casing containing the pressed matter on to a crane,
" and drawing it out of its place."

[Printed, 1s. 8d. Drawings.]

A.D. 1863, October 1.—N° 2400.

SMITH, WILLIAM.— (*Provisional protection only.*)—"An im-
" proved process for re-crystallizing sugar." This consists "in
" the application of steam to the surface of sugar after it has
" been cut, sawn, or broken thereby re-crystallizing it," as follows :—The pieces of sugar fall "on to a wire revolving platform,
" which is kept constantly in motion by being set on endless chains.
" A continual supply of steam is kept up under the revolving
" chains, so that, as the sugar is carried along on the surface of
" the said revolving platform, it is brought into contact with the
" steam and re-crystallized, after which it falls into a trough con-
" veniently placed to receive it. Any suitable motive power may
" be employed to work the chains."

[Printed, 4d. No Drawings.]

A.D. 1863, October 1.—N° 2405.

REID, FRANCIS.—" Collecting and saving the spirit or alcohol
" generated by spontaneous fermentation in raw sugar, concrete,
" melado, and molasses, and thrown off during the process of
boiling or refining." This "may be carried into effect by con-

" necting the ordinary vacuum pans with a worm condenser,
" and collecting the products condensed and using them" for
" rectifying re-distilling or other purposes."

[Printed, 8d. Drawing.]

A.D. 1863, October 2.—Nº 2421.

SHEPHERD, George, and SHEPHERD, William Thomas.
—(*Provisional protection only.*)—" Improvements in restoring the
" crystals of lump or refined sugar which has been divided by
" saws and in apparatus employed for this purpose." These are,
" two rollers are mounted in a suitable frame and are caused to ro-
" tate with an equal surface speed, and are at such a distance apart
" that when a slice or piece of sugar which has been sawn from a
" lump is passed between them the slice will be subjected to
" pressure. Each of the rollers is coated with woollen or other
" suitable fabric or material which is constantly moistened and,
" by preference, with cold spring water. To each roller is applied
" a pressing roller or instrument by which any excess of water
" applied to the elastic and porous coating may be removed."
After passing through the rollers the sugar is dried.

[Printed, 4d. No Drawings.]

A.D. 1863, October 17.—Nº 2551.

DE WYLDÉ, Fedor.—(*A communication from Henry Schwarz.*)
—" Improvements in the separation of molasses and other im-
" purities from sugar crystals." These are, "the separation of
" the sugar crystals from the adherent molasses by the agency of
" alcohol, methylated or wood spirit acidulated," as follows :—
The matters to be operated upon are carbonized to ascertain the
quantity of ashes (salts of alkali and lime) they contain, and for
every part of ashes obtained, 10 parts of molasses are said to be
present in the raw sugar ; and for every part of ashes, a " half per
" cent. of pure muriatic acid, or one and a half per cent. of muriatic
" acid of commerce " is employed, so as to form with the inorganic
acids and salts of the alkalies, salts which are soluble in the
above alcohols. In operating upon a saccharine substance containing one part of ashes, " twenty parts of absolute alcohol or
" methylated spirit, and one and a half parts of muriatic acid
" of commerce, and four parts of water," are mixed with one
hundred parts of raw sugar, which dissolves the molasses and
the acid solution of the molasses is separated " from the sugar

" crystals by filtration or other process used for this purpose.".
A centrifugal machine may be used. The liquid still adhering
to the crystals may be removed by neutral alcohol, or methylated
or wood spirit, and as this operation proceeds the alcohol should be
increased in strength " until at last absolute alcohol or methylated
spirit should be used as the final menstruum." The crystals are
dried in a current of hot air. The spirituous solution, after
neutralizing, may be distilled or otherwise for the alcohol. By
adding an excess of lime a precipitate of sugar and lime is
obtained, which separated and decomposed by carbonic acid,
yields a solution of sugar. This process is applicable to crystallized
juice of the sugar cane, beet root, &c.

[Printed, 4d. No Drawings.]

A.D. 1863, October 20.—N° 2575.

GARTON, CHARLES, and HILL, THOMAS.—" Improvements in
" evaporating, cooling, and melting, and in apparatus employed
" therein." These are, in carrying out the above, "the construc-
" tion and employment of covered pans carrying at bottom cham-
" bers through which liquid is caused to circulate, and fitted
" with screw vanes or drums" to act as agitators upon the
matter evaporating. These consist " of archimedean screw vanes
" carried upon suitable spindles and wheels." These agitators
are mounted in sets of two or more, and in such a way " that in
" their revolution the vanes on one agitator shall clear those of
" the other," and they are driven in opposite directions. Also,
drawing off "through one or more apertures in the cover, the
" vapors and gases evaporated from the matter under treatment
" by fans, steam jets, pumps, or otherwise." In some cases
aiding the evaporation " by the introduction of hot air above the
" surface of the matter to be evaporated," or introducing pipes
" for containing a heating medium above such surface," or forming the cover or roof hollow and passing hot air or other heating
medium through it. For some purposes driving "a frigorific
" mixture through the pipes. In such case circulation through
" the pipes is continuous, and there is on outlet above the matter
" operated on.".

[Printed, 1s. 4d. Drawings.]

A.D. 1863, November 14.—N° 2855.

MACKIRDAY, LAUCHLAN. —." Improvements in saturating,
" washing, and cleansing charcoal and other matters, applicable

" also to the separation of syrups from sugar." These are, first,
" the saturating of charcoal in charcoal filters," as follows :—By
withdrawing the air from charcoal in air-tight cistern before the
liquor is run into the cistern, on opening the "brown liquor
" cock" the "liquor will enter into instant and complete con-
" tact with the charcoal." When the liquor is thoroughly
purified it is drawn off either in the ordinary way or into an air-
tight cistern. In either case air is allowed to enter in at the top
of the cistern. For "inferior or tough liquors," besides exhausting
the air as above, an air-pump is employed to force air into the
cistern on the top of the liquor, and " drive the liquor through
" the charcoal most thoroughly and effectively." When the
liquor cistern has done its work and remains full of charcoal
saturated with liquor the process of washing commences by
opening a hot water tap at top and drawing the liquor from the
bottom of the cistern ; this is called " sweet water."

Second, "washing and cleansing of charcoal," as follows :—
" After the sweet water has been drawn off, a vacuum pump
" which is in connection with the bottom of the cistern is brought
" into action, and a copious supply of hot or cold water is intro-
" duced to the top of the cylinder," and the water is continued
until it runs pure, when "the supply is cut off, and by opening
" an air valve air is drawn through the mass of charcoal and
" dries it."

[Printed, 4d. No Drawings.]

A.D. 1863, December 1.—N° 3010.

DODDRELL, GEORGE JOHN.—"Improvements relating to the
" manufacture or refining of sugar." These are, "the disposing
" of the concentrating vacuum pans of a sugar refinery at a
" higher level than that at which the goods are worked off or
" finished," so as save the labour and expense of elevating them
as is now the case. "The concentrated fluid or semifluid may
" be transferred from the concentrating pans or vessels by gravita-
" tion by means of pipes or ducts, or partly by pipes and partly
" by means of basins or filling vessels." If the filters deliver the
liquor below the level of the vacuum pans it "is pumped up
" through suitable piping to the elevated level at which the
" vacuum pans are placed according to this invention." "The
" passing of the concentrated fluid or semifluid into granulating
" vessels or receivers, and thence into centrifugal machines by
" gravitation," is not claimed "when the concentrating vacuum

"pans are situated at the ordinary low level," but is claimed when elevated "according to this invention."

[Printed, 4*d.* No Drawings.]

A.D. 1863, December 11.—N° 3123.

CORBY, JOHN.—"Improvements in centrifugal machines for " extracting the syrups from sugar." These are, first, dispensing with "the usual amount of multiplying gear, such as driving " wheels, drums, wall brackets, supports, axles, springs, and " leather belts," and replacing the whole "by a single horizontal " shaft imparting motion to the vertical spindle of the drying " cylinder through a pair of bevel friction wheels working in the " lower portion of the framework."

Second, " in rendering the vertical axle of the receiver, with its " friction pinion, capable of being moved up and down in a per-" pendicular direction, which movement is obtained by construct-" ing the vertical spindle to slide in suitable plummer blocks and " actuated by a lever carrying a footstep, into which the lower " end of the spindle is fitted, this arrangement being for the " purpose of rendering the starting and stopping of the whole " machine more simple and effective than usual."

Third, "the whole machine, including its driving gear, being " erected and attached to one sole plate or bed, is thereby " rendered of a steady, compact, and portable character, being " completely self-contained in construction, having all its parts " accessible to for the purposes of lubrication and lightening up " whilst the machine is in motion."

[Printed, 10*d.* Drawing.]

A.D. 1863, December 29.—N° 3294.

VANDERFEESTEN, JEAN MICHEL. — "Improvements in " apparatus for heating, boiling, evaporating, and distilling." These relate to apparatus for the above purposes " whether per-" formed in vacuo or otherwise and whether for the manufacture " of sugar or other purposes," and the apparatus consists of a close or open vessel in which are a number of coils; " these coils " instead of forming a single worm or coil rising in curved spirals " as is usual consist of a number of separate straight coils," that is " each coil consists of a certain number of straight or " rectilineal horizontal pipes (of circular or other transverse

S,   Y

" section) connected by curved junctions." "The coils are
" arranged side by side and can be removed from the receiver
" and replaced therein through openings in the front of the
" receiver closed by doors or plates secured by bolts and nuts.
" The coils of steam pipes are supported on rods or frames with
" hollow rollers. The ends of the coils are fastened and unfastened
" by exterior hollow nuts which work on pipes cast on the
" outside of the receiver, forming the joints or junctions between
" the coils and the steam inlet, and outlet for drawing off the
" waters of condensation therefrom, and to tighten the coils
" against the sides of the vessel to the required extent. The
" coils when required to be cleaned are removed into a vessel
" containing acid or other solution, other coils supplying their
" places." "The 'steam joints may be_fitted with nuts." A
modification of the above apparatus is shown in which the ends
of the coils instead of leading into horizontal passages communicate through junctions with vertical pipes.

[Printed, 8d. Drawing.]

1864.

A.D. 1864, January 19.—N° 144.

BROOMAN, RICHARD ARCHIBALD. — (A communication from
Leon Geauchez.)—"Improvements in machines for sawing and
" cutting sugar." These are, constructing such a machine that
performs "three operations which have hitherto not been effected
" without the employment of two or three separate machines or
" apparatuses." At one side of the machine is "a fly wheel
" worked by hand or otherwise, and with teeth cast on its inner
" circumference acting on a bevil wheel fixed on an axle. This
" axle carries one of two pulleys, round which an endless band
" saw passes kept in place by a guide. The revolution of the fly
" wheel thus imparts motion to the saw which divides the loaf of
" sugar into slices of the desired thickness. Instead of an
" endless band saw a saw with a to-and-fro movement may be
" employed. Other teeth on the axis of the fly wheel gear into
" a bevil wheel which transmits motion to an endless screw
" working in its turn a toothed wheel on the same shaft as six

" ratchets, more or less; these ratchets carry endless bands
" passing between circular saws, which act at right angles to the
" band saw and divide the slices of sugar into blocks as they are
" led up against the saws of the belts. To prevent the blocks
of sugar getting out of place, covering them "with a plate with
" slots therein for the passage of the saws, this plate is caused to
" press upon the blocks by springs. The tension of the belts
and of the endless band saw is regulated by two screws. On
leaving the circular saws the sugar comes below a guillotine,
chopper or cutter worked by a bar which receives motion from a
lever. On the driving axle of the machine is a socket which
carries a conical tooth or projection; the lever just mentioned also
carries a tappet, arranged so that when the projection carried
by the axle comes against it at each revolution it shall be
lowered. To give the pieces of sugar the proper size the parts
are so arranged so "that the chopper or cutter only descends once
" for every four revolutions of the driving axle " as follows :—
" Outside the projection before mentioned as being on the axle is
" a second projection, which on receiving motion takes into
" recesses in the periphery of a wheel; on this wheel is a metal
" band with parts cut away so as to form two curved bands, each
" tapered off at one end, into the spaces between the bands
" a hooked rod takes, working from a fixed point, and acting by
" pressure on the first-named projection. The wheel has eight
" divisions and it is only when the second projection takes into
" four of the recesses, and causes the wheel to make a semi-
" revolution, that the hooked rod can take into one of the spaces
" between the bands; in taking into the space it acts by pressure
" on the socket of the first-named projection, which is arranged
" only to come against the tappet on the lever, when it is caused
" to turn by the hooked rod, it is therefore only at the next
" meeting that the effect of the two projections, and consequently
" that of the chopper will be produced. When at the next
" revolution the projection which takes into the recesses in the
" wheel shall have caused the wheel to advance the hooked rod
" will have been forced out of the space between the bands by
" rising up the tapered end, it will thus allow the first-named
" projection to turn without coming in contact with the tappet
" on the lever until the driving axle shall have performed four
" revolutions."

[Printed, 10d. Drawing.]

A.D. 1864, February 3.—N° 283.

BEANES, EDWARD.—"Improvements in preparing or treating
" animal charcoal." These are, the charcoal dry and, in preference, heated, is impregnated with dry hydrochloric acid gas,
and the gas is allowed "to remain in the charcoal until the lime,
" the carbonate of lime, and other alkaline and earthy matters
" which may be contained in the charcoal have been converted
" into soluble chlorides." The excess of gas is expelled by heat
or a current of hot or cold air or by any other convenient means.
" After being thoroughly washed the charcoal (which may or may
" not be dried) is then fit for use." "Animal charcoal which
" has been used, or has become deteriorated in its properties
" may be subjected to the same process, and may be reburnt in
" the usual way, either before or after being submitted to the
" process."

[Printed, 4d. No Drawings.]

A.D. 1864, March 2.—N° 519.

MILLER, WILLIAM. — (*Provisional protection only.*) — "Im-
" provements in the manufacture of sugar and in the apparatus
" employed therein." These are, first, " evaporating saccharine
" solutions at temperatures below those at which they boil under
" the ordinary pressure of the atmosphere," and such solutions
never coming "in contact with metallic or other surfaces heated
" by fire or high pressure steam nor are subjected to any tem-
" perature which cannot be regulated to or below that of boiling
" water or 212° F." This invention it is said has reference to
No. 525, A.D. 1862, and " consists of a cylinder composed of curved
" blades arranged like the floats of a water wheel only overlapping
" each other considerably so as to increase the evaporating
" surface. This is made to revolve in the saccharine solution to
" be evaporated." The cylinder does not dip far into the pan;
the pan has large flat or slotted tubes passing through it. The
pan is surrounded by an outer casing or jacket the lower part of
which is covered by water heated either by direct fire heat or by
high pressure steam passing into it, an overflow pipe for carrying
off condensed water and a pipe open to the atmosphere carries off
any steam. In the colonies the cylinder may revolve open to the
air, but if desirable to get rid of steam, it is covered with a hood
attached to which is a fan.

Second, in dissolving sugar, making "use of cylindrical
" shaped pans," having "a coil of pipe with a condense cock
" passing round the lower part of the side of the pan, but suffi-
" ciently high up that the sugar when thrown into the bottom of
" the pan does not come in contact with the heated pipe except
" when it is in solution." Also using an agitator in the form of a
cross with knives attached to it and another cross with flat blades
to keep the solution in constant agitation.

[Printed, 4d. No Drawings.]

A.D. 1864, March 2.—N° 528.

LANGENARD, FREDERIC PIERRE.—"Improvements in cen-
" trifugal machines for extracting juice of plants, or drying up
" substances or materials of various kinds." These are, in reference
" to this subject, when beetroots or any other material of the kind
" have to be dried up," first, "a perforated or wire cloth cylinder
" or box, formed as in similar apparatuses, but without bottom
" and attached to the shaft by means of an armed ring, which is
" tied to the flanged upper end of the cylinder," is employed.

Second, "a funnel shaped upper cylinder fastened to the lower
" one with which it revolves."

Third, "a lower elastic or pliable bottom, made either of cloth,
" leather, india-rubber, or any other similar material" attached at
its outer and inner peripheries on rings and carried by arms hinged
on a central nave, the arms being carried near the other end by
rods, hinged on one side with tongues of the arms, and on the
other side on a moveable sleave, so that the bottom of the apparatus can be made quite free.

Fourth, the lower moveable sleave carrying a piece betwixt
its shoulders, guided in its up and down motion by rods, and
actuated by ropes winding up on pulleys; which contrivance is
intended to move the sleave up and down for unfolding and
folding up the pliable bottom, the upward motion of the sleave
being assisted by a helical spring carried by a shoulder of the
central shaft.

Fifth, the shaft carrying the centrifugal machine and its ap-
pendages is partly hollow, and is provided inwardly with a rod,
capable of moving up and down into the shaft, which shaft
carries it along in its rotary motion by means of two keys driven
fast into the rod which moves up and down within the hollow
shaft.

Sixth, the combination of an armed ring connected by a keyed sleeve to the rod which carries along in its up and down motion in the hollow shaft to rub off the dried material adhering to the periphery of the cylinder.

Seventh, the following contrivance, to slide up and down in the hollow shaft. The rod at its upper end is fitted with a bullet, held betwixt two jaws, hollow shaped to fit the bullet without coming close together; the jaws are tied close together by screws. At the top part of the jaws, in the middle, there is an eye piece formed with a hole in which runs a rope, which is notted where passed through the eye, so that when pulled upward the rope will draw the rod. The two jaws above are pierced through on each side with a small hole, through which runs a rope; these ropes are wound round two pulleys fixed one on each side of the rod; the ropes are continued further on to allow of their being wound around a drum. When the said ropes are pulled downwards they cause the rod to move down; the jaws are guided in their motion by a rod on each side fastened at their lowest end on the frame.

[Printed, 1s. Drawing.]

A.D. 1864, March 4.—N° 552.

MANBRÉ, ALEXANDRE.—"Improvements in the manufacture " of glucose sugar." These are, first, effecting "the trans- " formation of starch or fecula into glucose sugar" by mixing them with 5 per cent. sulphuric acid and submitting them "to " the action of a high degree of heat not under 275° F., but pre- " ferring the temperature of 320° F. for quickening the process," and by this means producing "a pure glucose free from gum, " essential oil, and empyreumatic fatty matters."

Second, employing in the process a strong wrought iron oblong boiler with steam pipe inside along the bottom. The boiler is lined inside with lead and outside with a jacket between which and the converter is a space of 4 inches which is filled with sand or any other non-conducting matter. On the top is a pipe with a cock through which diluted starch may be gradually intro- duced, also " safety valves, steam gauge, water gauge, thermometer, " pipe for escape of steam, exit cock at bottom, and a worm or " distilling pipe through which the high pressure steam is allowed " to blow off."

Third, allowing "the high pressure steam to blow off out of
" the vessel while the heat is maintained " at 320° F. " to distil off
" and separate from the product the vegetable empyreumatic and
" acrid oils."

Fourth, the " use of superheated steam as an agent for heating
" the mixture." This mode of heating is more convenient than
other means but " is not essential to the success of the process
" provided a sufficient degree of heat is obtained." Another
method to obtain from starch glucose free from dextrine or gum
is to use a much larger proportion of sulphuric acid ; up to
the present time from 2 to 5 per cent. has been used and it has
never reached 10 per cent., it is proposed to use at least 20 and
as much as 60 or 70 per cent., heating in the ordinary manner,
but this process is inferior to the first " although preferable to the
" ordinary process now employed."

[Printed, 8d. Drawing.]

A.D. 1864, March 7.—N° 568.

NEWTON, WILLIAM EDWARD.—(*A communication from Louis
Pierre Robert de Massy and Louis Robert de Massy.*) — " Im-
" provements in refining sugar and molasses." These are, in
reference to this subject, first, treating " syrup or saccharine
" liquid with baryta or strontia whereby the syrup may be refined
" with greater speed and economy and an increased yield obtained
" than by the processes now in use." The juice " is first
" defecated by simple or double saturation," with lime " with the
" assistance of carbonic acid," treated with about sixty per cent.
of caustic baryta, heated to form a saccharate of baryta which
precipitates and is separated " by decantation, washing, or any
" other suitable means." "This saccharate is put on one side
" and the liquid is treated either with a current of carbonic acid
" or by sulphuric acid in order to precipitate the baryta which
" is in excess in the process of saccharification." "The pre-
" cipitate thus separated may be used over again in the same
" manner." The water may " run away or be sold or it may
" be used as a manure for which it is peculiarly applicable."

Second, " after saturating the syrup with lime it is to be
" evaporated to any desired density and the caustic baryta then
" added in the same proportions as before. Then proceed as in
" the other process, but instead of throwing away the water or

" liquid after separating the insoluble salts, the liquid is to be
" evaporated again in order to obtain the salts of potash or soda."
" The saccharate which has been put on one side as before-
" mentioned," when required " to be operated upon for sugar is
" mixed with four times its weight of pure water and submitted
" to the action of carbonic acid under pressure until the baryta is
" completely saturated. The sugar obtained in this manner is
" chemically pure and the insoluble carbonate of baryta may be
" separated therefrom in the manner already mentioned. By
" decomposing the carbonate of baryta by means of sulphuric
" acid carbonic acid may be obtained for the purpose above set
" forth. The decomposition of the carbonate of baryta may also
" be effected by means of sulphate of soda or sulphate of potash,
" when an insoluble sulphate of baryta will be obtained and
" soluble salts of the other bases." " Strontia may be used in
" place of baryta thoughout the whole process, if preferred, and
" will produce precisely the same result."

Besides the foregoing there are several modes given for obtaining
carbonic acid, also for obtaining some salts of potash, soda, and
baryta, and it is stated "that in case any traces of baryta should
" remain in the syrup, they must be precipitated by an addition
" of sulphate of lime. There will then be a precipitate of sul-
" phate of baryta which may readily be removed."

[Printed, 4d. No Drawings.]

A.D. 1864, March 8.—N° 580.

NEWTON, WILLIAM EDWARD.—(A communication from Louis
Pierre Robert de Massy and Louis Robert de Massy.)—" Improve-
" ments in the manufacture or production of baryta of strontia."
These are, in reference to the subject, as follows :—The sulphates
of baryta and strontia either natural or artificial, or " produced
" from the sulphurous acid obtained by this process, or by other
" means, or by double decomposition," are mixed with charcoal
and roasted, this process, it is said, " is so well known that there
" is no necessity for describing it; when, however, a sugar manu-
" factory or refinery is within reach, it is better to operate in
" closed vessels, the reaction being more certain, and the carbonic
" acid produced by the decomposition is utilized in order to
" carbonize the saccharates of lime, baryta, and strontia." A
solution of the sulphuret of barium or strontium thus made being

boiled with oxide of copper, insoluble sulphuret of copper is formed, and the hydrate of baryta or strontia is in solution, "the " baryta may be used direct in this state for the manufacture and " refining of sugar," &c. " The precipitated sulphuret of copper " is dried and then roasted," the gas produced (sulphurous acid) being conducted into lead chambers for the production of sulphuric acid. "The deutoxide of copper is put on one side to " serve for the succeeding operation, so that with the same " quantity of copper, an indefinite quantity of sulphurets can be " decomposed."

[Printed, 4d. No Drawings.]

A.D. 1864, March 19.—N° 700.

JONES, DAVID.—(*A communicaion from Simon Labayen.*)— (*Provisional protection only.*)—" Improvements in sugar funnels " or moulds." These are, strengthening sugar funnels or moulds which consist of hollow cones of sheet iron open at top and bottom, by rivetting a hoop around the wide end " immediately above " the ordinary turned-over edge, in which a ring is inserted." It is preferred, to make the hoop " of a strip of iron, the edges of " which are bent into a trough-like figure, the said strip being " bent into a hoop, having the concave sides of the trough-like " edges in the inside of the ring." " The figure of the hoop may " be varied without departing from the nature of the said in- " vention. The narrow end of the mould may also be protected " and strengthened by a hoop similar to the hoop herein-before " described."

[Printed, 4d. No Drawings.]

A.D. 1864, March 26.—N° 754.

BROOMAN, RICHARD ARCHIBALD.—(*A communication from Jean Baptiste Felix Trolliet.*)—"An improved method of, and im- " proved apparatus for, revivifying animal black." This consists " in progressively drying and heating animal black until revivified, " and then in cooling the same." The apparatus "consists of " an oven with retorts placed horizontally and superposed; the " fire is inside the lower retort, and the upper retort is heated by " flues from the lower retort. The top of the apparatus serves " for drying the black, and it is progressively heated by passing " through passages provided for that purpose around the upper

" retort, and between it and the flues surrounding it, and then
" around the lower retort and its flues. On leaving this passage
" the black is revivified, and continues on into a cooling chamber
" furnished with cold air flues, and traversed by air pipes below
" the lower retort. At the bottom of this chamber doors are
" provided, through which the revivified black is withdrawn."

[Printed, 8d. Drawings.]

A.D. 1864, April 23.—N° 1024.

WORSSAM, George Jarvis.—" Improvements in machinery
" and apparatus for expressing liquids or moisture from sub-
" stances." These consist, of a screw or worm, which is covered
with a lining or casing " of copper, brass, iron, or any suitable
" material, or wirework, being perforated, grooved, or ribbed to
" admit of drainage ;" around this again is a case " in halves or
" parts made of copper, brass, iron, wirework, or any suitable
" material for the purpose," perforated recommended ; between the
outer and inner casings are placed " a cloth or cloths, also wirework,
" if required, through which the substance to be operated upon
" is pressed by the means of a force pump worked by hand or any
" other suitable power " feeding the machine by four pipes in the
lid of the machine. At the other end is a cover with a discharge
door, " through which the material or matter after being pressed
" is forced by means of the screw or worm." " In some cases
" it may be necessary to use blades, scrapers, or knives in place
" of the screw or worm," or together with it ; above the cover is
a worm wheel fixed on end of a screw or worm, and a worm for
working the worm wheel. There is a trough below the machine,
" into which the liquid runs after being pressed." The machine
is fixed between upright columns, and among a number of
purposes named to which, it is said, it is applicable is for " filtering
" sugar."

[Printed, 1s. Drawings.]

A.D. 1864, April 23.—N° 1038.

BRINJES, John Frederick.—" Improvements in apparatus
" for the reburning of animal charcoal." These are, applying
and using for the above purpose cylinders or retorts as follows,
and a rotating cooling drum or chamber provided with an inner
or outer skin, and supplied with water for the purpose of more

readily cooling its contents," as follows :—The retorts of " a
" cylindrical form are intended to revolve or reciprocate in a
" circular direction backwards and forwards, a complete revolu-
" tion, or nearly so, on their longitudinal axis, and to be in a
" horizontal position. The motion of the retorts may be pro-
" duced by means of a mangle wheel and pinion." In the
interior of the retorts a series of flanges or rings are cast, or they
may be bolted therein " about 4 to 6 inches deep, and 4 to 8
" inches apart " in the direction of the cylinders' length, each ring
or flange having openings left through them, the several openings
being in a line with or opposite each other. Between each of the
flanges or rings, and opposite each of the openings above referred
to, is provided a kind of door or vane mounted in the centre
upon a spindle, and connected to a rod passing from one end of
the cylinder or retort to the other, " and through its ends or
" covers, which rod is intended to move backwards or forwards
" in the direction of the cylinder's length, so as to actuate and
" move the doors or vanes to their required positions simul-
" taneously. The doors or vanes are moved from the outside of
" the retort by any suitable self-acting contrivance." Another
modification of fitting the vanes or doors is described. After
the charcoal has traversed from one end of the chamber to the
other ; " it is discharged into a cooling chamber fixed on the end
" of the cylinder or retort, and rotating backwards and forwards
" therewith ;" it is provided with a jacket for cold water passing
through it. A pipe at one end of the cylinder conveys the steam
or gases into the chimney.

[Printed, 1s. Drawing.]

A.D. 1864, May 4.—N° 1119.

BEANES, EDWARD.—(*Provisional protection only.*)—" Improve-
" ments in the treating animal charcoal." These are, " treating
" animal charcoal which has lost its decolorizing and absorbtive
" properties " with certain chemical compounds which have the
" power of removing or rendering soluble the lime and other
" metallic, earthy, and alkaline substances with which it is
" contaminated, using for this purpose chloride of ammonium
" (sal-ammoniac) in the liquid or solid state," also using " nitrate
" of ammonia, nitrous acid, hyponitric acid, and carbonic acid.
" The effect of these substances is to render the lime and other

" earthy, metallic, and alkaline impurities soluble, and which are
" afterwards removed by washing."

[Printed, 4d. No Drawings.]

A.D. 1864, May 5.—N° 1136.

BEANES, EDWARD, and FINZEL, CONRAD WILLIAM.—" Im-
" provements in sugar boiling." These are, in place of using
steam "at temperatures of 225° F. and upwards, equal to a
pressure of 4½ lbs. and upwards, whereby there has always been
more or less carbonization and consequently coloring of the sugar,
" using hot water continuously kept at or as near as may be the
" boiling point, or steam at a temperature not exceeding 215° F.,
" or at a pressure not exceeding 1½ lbs. to the square inch," or
as near as possible, "so as to boil the syrup or liquor without
" carbonization." To do this "instead of using the long tubular
" vacuum pan and tubes, or the ordinary vacuum pan with worms,
" as at present in use," using "a tubular vacuum pan, but
" reducing the length of the tubes therein, increasing at the same
" time the number of tubes according to the evaporating surface
" required, so that it will only be necessary for the heating thereof
" to use hot water or steam as aforesaid at a temperature below
" the carbonizing point of saccharine syrups, while from the
" shortness of the tubes in the pan, the water or steam will
" continue sufficiently hot during its passage through such tubes,
" so as to be perfectly effective for the purpose of boiling the
" saccharine liquids and causing the proper evaporation through-
" out the pan without causing any carbonization and colouring
" of the sugar."

[Printed, 4d. No Drawings.]

A.D. 1864, May 14.—N° 1228.

FRYER, ALFRED.—" Improvements in treating animal charcoal
" in the process of revivification, and in apparatus employed
" therein." These are, first, bringing the charcoal from the
reburning apparatus while hot in contact with steam. The
charcoal when reburned is placed within a conical vessel having
an aperture in the apex or lower end; the stream of hot charcoal
issuing from the apex comes in contact with a jet of steam provided for that purpose, reducing its temperature, separating the
grains, exposing them to air, and spreading them.

Second, hot charcoal after reburning is brought into contact with a rapidly rotating horizontal plate or disc, whereby the hot grains of charcoal are dispersed, separated from each other, and brought into immediate contact with cold air ; or the same object is effected by discharging the red hot charcoal from the kiln or revivifier into a rotating horizontal cylinder open at both ends, in which are vanes. "The current of air acts upon and cools the " charcoal, and carries it along the cylinder, and eventuually out " of it."

Third, "the employment of an apparatus composed of conical " pipes enclosed in a heating chamber placed vertically, and of " larger diameter at bottom than at top for the purpose of drying " animal charcoal previous to revivification." The pipes are most economically heated by the waste heat from the reburning apparatus; but they may be otherwise heated.

[Printed, 4d. No Drawings.]

A.D. 1864, May 30.—N° 1336.

PATERSON, JAMES.—(*Provisional protection only.*)—" Certain " improvements in the cooling and preparation of charcoal to be " used for refining sugar, and in the machinery, apparatus, or " means employed therefor." These are, first, "the application " of a cold blast or current made to impinge against or on the " coolers by means of a fan or other convenient apparatus, or the " cooling may be effected by drawing or forcing air through or " amongst the charcoal."

Second, applying " machinery driven by steam or other power " for the purpose of drawing the slides of the coolers." "The " slides consist of two parallel flat bars or plates perforated with " holes of sufficient size to enable the charcoal to fall through when " any one of them is placed in connection with any of the coolers " containing it." "The vibrating motion of the rod or quadrant " is obtained by connecting it by link work to one eccentric or " crank on an adjoining shaft, or it may be produced in any other " convenient manner."

Third, "substituting a travelling or moving table or tables in " lieu of the hand barrows at present employed for conveying the " charcoal from the coolers as they are emptied." In preference, the tables are below the coolers.

Fourth, "imparting a vibratory or oscillating motion to the " travelling tables, which is obtained by causing the rollers,

" barrels, or supports over which the tables pass to move in their
" bearings horizontally as they revolve, thus causing the tables
" to act as a riddle, thereby causing the dust to be separated
" from the larger pieces of charcoal. This motion may be im-
" parted to the rollers by means of a rod attached to their ends,
" and connected to a revolving crank, cam, eccentric, or any
" convenient mechanism."

[Printed, 4d. No Drawings.]

A.D. 1864, May 30.—Nº 1342.

NEWTON, WILLIAM EDWARD.—(*A communication from Eugène
Bertholomey.*)—" Improvements in the treatment of the low or
" poor products obtained in the manufacture or refining of
" sugar." These are, " a part of the low or poor matter is baked
" by the ordinary process in a sugar pan either in a vacuum or
" not. This baked product if left to itself will, at the expiration
" of some time, become crystallized. When this matter has
" arrived at this commencement of crystallization the baking is
" continued in a vacuum pan, and is conducted in such a manner
" that during the process of evaporation the crystals already
" formed are fed with successive supplies of syrup or clarified
" sugar of the same nature. The commencement of crystallization
" above mentioned may be effected in the apparatus which has
" served for the first baking, or by preference, in a separate vessel.
" In any case, the second stage of the operation, which consists
" in feeding the crystals, must be conducted either in a vacuum
" or in apparatus open to the air, and at a low temperature." In
place of the vacuum apparatus being charged as above, it may
even be charged " with matter prepared with powdered sugar, or
" matters completely crystallized and diluted with syrups, clarified
" or not, or water, and the apparatus having been charged, the
" working is to be continued by feeding the crystals successively
" with clarified or other syrup."

[Printed, 4d. No Drawings.]

A.D. 1864, June 30.—Nº 1631.

CORBY, JOHN.—" Improvements in centrifugal machines, such
" as are used in separating syrups from sugar, and in apparatus
" for making the same." These are, first, " the making of the
" outer shells or strainer plates of centrifugal machines with

" corrugations running round horizontally." The internal metallic
cloth bears against the internally projecting ridges of the corrugations. Perforations, in preference, " elongated slits are formed in
" and along the externally projecting ridges of the corrugations."
" It is preferred to construct the centrifugal basket with a hollow
" conical centre, by which it is fixed on its carrying and actuating
" spindle, and this centre is formed with a checked flange, upon
" which an annular iron bottom plate is shrunk and rivetted, the
" cylindrical shell being fixed to the rim of this bottom plate."
" The form and proportions of the corrugations may obviously
" be varied, and three varieties are shown."

Second, "the making of such corrugated shells by means of
" apparatus substantially," as follows : — This apparatus comprises four rollers, carried between two massive standards, or side
frames. Two of the rollers " form the corrugations, whilst the
" other two assist in curving the sheet to the cylindrical form of
" the basket. The two corrugating rollers are placed one above
" the other in the middle of the machine, and the two bending
" rollers one on each side, and a little above the bottom corru-
" gating roller, to give the required curve. All the rollers are
" grooved to correspond to the required corrugations, and the
" bottom one and side ones are formed with flanges, to guide the
" sheet when passing through." The driving power is applied
by suitable gearing to the bottom corrugating roller, and the top
corrugating roller is forced into closer contact by wedges acted on
by a lever, the roller being elevated by springs when not so
depressed.

[Printed, 10d. Drawing.]

A.D. 1864, July 12.—N° 1727.

CAREY, STEPHEN.—" Improvements in apparatus for calcining
" bones, and for reburning and revivifying animal charcoal."
These are "making a long retort, of a **D** shape in preference,"
and setting the same " at such angle as will allow the contents
" to run down the retort by their own gravity or by the force of
" supply at the upper end, the angle I prefer is about 32½
" degrees." The flue is carried " from the furnace under nearly
" the whole length of the retort and return the same over the
" top and one side of the retort and back again in the other
" side, and there lead it off near the top to the main flue or

" chimney, and sometimes when extra heat is required, I would
" construct another furnace about half-way between the present
" furnace and the upper end of the retort and connect the same
" with the flues before described." The retort is fed from a
hopper in which is a valve to regulate the supply, and the
charcoal after running down the whole length of the retort is
discharged into a cannister or receiver, through a mouth-piece
with a valve to regulate the discharge. In some cases when it is
required to burn the contents of the retort more slowly, valves or
flaps are placed accross the inside of the retorts and these are
raised or lowered by means of a shaft or rod connected therewith
and passed through a stuffing box at the end of the upper mouth-
piece or otherwise. In order to feed the retort when required for
reburning charcoal, a hopper is fixed into the mouth-piece and
this is supplied by means of an endless chain ladder on which are
fixed a number of small buckets.

[Printed, 10d. Drawing.]

A.D. 1864, July 26.—N° 1855.

DIXON, Thomas.—(A communication from Manuel Bea.)—
"Improvements in sugar funnels or sugar moulds." These are,
strengthening the wide open end of sugar funnels or sugar
moulds consisting of hollow cones of sheet iron open at both
ends as follows :—"The edge of that portion of the sheet iron
" which in the furnished sugar funnel or mould constitutes the
" wide end of the sugar funnel is folded or doubled, so as to give
" two thicknesses of metal at that part. The said wide end of
" the sugar funnel is further strengthened by a metallic hoop
" rivetted thereon. The said hoop is conical both within and
" without, excepting at the extremity of its widest part, at which
" part is a strong beading of a nearly cylindrical figure in cross
" section." The beading has in its inner side a shoulder upon
which the doubled or folded edge of the funnel or mould rests,
and by which it is protected from injury. It is stated that " the
" narrow end or top of the funnel or mould may be protected
" and strengthened " in a similar manner.

[Printed, 6d. Drawing.]

A.D. 1864, August 25.—N° 2101.

DAVIES, George.—(A communication from Jacques Jacquier
and Vincenz Danek.)—"An improved system of filter presses."

This consists as follows :—" In these filter presses the pressure is
" obtained either by the liquid to be filtered, that is to say, by
" hydrostatic pressure, or by compressed water, or by steam with
" the intervention of a strainer, or by steam direct, or by air, or
" by any gas, cold or hot, or lastly by employing a vacuum, by
" atmospheric pressure, or by a pump." The apparatus " is com-
" posed of a certain number of filtering surfaces placed in juxta
" position, covered with cloth and of moveable frames or moulds
" interposed between the filtering surfaces." The entry of the
material to be filtered " into each frame is effected by means of an
" internal duct, formed by the juxta position of the plates of
" filtering material and the frames, and this duct has an opening
" over each frame." In some cases it may be necessary to exhaust
" and wash the cake or matters in the frames or moulds, for this
purpose, another duct is arranged in the interior of the apparatus
" like the preceeding by the juxta position of the filtering surfaces
" and the frames. This duct, instead of opening like the former
" into each frame, opens into the space between the filtering
" surfaces, that is to say, between the plates of filtering material
" and the frames or moulds containing the matter. This duct
" introduces either hot or cold water, steam or gas, or any suitable
" liquid, and these agents flowing through the duct, pass through
" the cakes of matter and deprive them of the last traces of
" liquid which they may contain, and dry them also if required."
" In the application of these ' filter presses ' to the clarification
" and refining of sugar (which application is more particularly
" claimed) the hot water and steam injected effect a perfect
" cleansing of the cakes of animal black and albuminous coa-
" gulum formed in the frames," labour will also be saved, as
also the production of weak syrup diminished, " which is so easily
" spoiled in the ordinary method of cleansing the animal black
" employed in the clarification."

[Printed, 1s. Drawing.]

A.D. 1864, September 21.—N° 2317.

BROOMAN, RICHARD ARCHIBALD.—(*A communication from
Adolphe Lequime.*)—(*Provisional protection only.*)—" Improve-
" ments in the manufacture of sugar, and in apparatus employed
" therein." These are, obtaining at one operation from dried beet
root a large per centage of sugar, with a small quantity of

s.   z

molasses, and the residual matters of greater value than hitherto. This is effected by alcohol, heated, by preference, in a copper vessel fitted with brass tubes, and the mother liquors are heated in a similar manner; any convenient number of extracting vessels are arranged and enclosed " in a wooden or metal case or jacket. " These vessels have bottoms and covers provided with stuffing " boxes; a vertical shaft furnished with a fly wheel passes through " each, and carries inside the vessel a mixer or agitator and a " brush. The extractors are also provided with man-holes and metal " cloth and are each fitted with a four-way cock for the entrance " into and exit from the vessel as required of alcohol, saccharine " extract, and mother liquors." Also fitting " to the exit side of " the vessels indicating apparatuses or gauges, for showing the " levels of liquid in the vessels. Pipes lead from the alcohol " heating vessel and the mother liquor heating vessel to the four- " way cocks, and other pipes from the four-way cocks to the " bottom of the extracting vessels, and others again from the " extracting vessels to draw off the liquids when required; there " is also a pipe leading from the top of each extracting vessel to " the four-way cocks, and another pipe from the top of each " extracting vessel for carrying off the evaporated liquid, if " required, to any convenient receptacle."

[Printed, 4d. No Drawings.]

A.D. 1864, October 14.—N° 2538.

WRIGHT, RICHARD.—" Improvements in preparing saccharine " matters." These are, farina or flour usually of indian corn, although, "flour of barley, or other grain, or other farinaceous " substances, may be used," are " converted in the ordinary well- " known manner into saccharine matter or sugar," and with this is mixed or combined " golden syrups, molasses, treacle, and such " like syrups of cane sugar, or beet-root sugar, in large quantities," and evaporated, in preference, " by means of the process and appa- " ratus " described in No. 2153, A.D. 1860, but the evaporation may be otherwise performed. When the compound syrup has been evaporated to a thick syrup or " proof," it is placed in sugar moulds or other suitable vessels in a moderately warm atmosphere, say, 70° to 80° of Fahrenheit, as when refining cane sugar. In a few days " a film will appear on the upper surface," when the con- tents should be well stirred or " hauled," and in a few days the contents will be solidified and fit for market. The sugar from the farina or flour may be dissolved in water before combining as above, and " then evaporated and finished as before described " but it is better that the sugar manufactured from farina or flour " should not be first crystallized." The syrups of cane or beet-root sugar which are used are those which will not crystallize, or will not freely crystallize. With the better class of such syrups mixing from five to ten per cent. of the sugar from farinaceous matters or flour, with an in inferior quality of syrup, employing " a greater proportion of the sugar obtained from the farinaceous " substances."

[Printed, 4*d*. No Drawings.]

A.D. 1864, October 29.—No 2678.

SMITH, ALEXANDER, and SMITH, WILLIAM.—" Improve-" ments in and relating to centrifugal apparatus, such as is used " in the manufacture of sugar." These are, first, in driving the horizontal shaft of such a machine " directly from one or more " steam cylinders which may be either oscillating or fixed, and " with common piston rods or with trunks. The cylinder or " cylinders is or are curved or fixed upon the casing or framing " of the machine in either a vertical, inclined, or horizontal posi-" tion." In stopping the machine the steam is cut off, and a duplex frictional brake actuated by a lever is applied. This brake comprises two frictional blocks, which are applied to opposite sides of the boss of the cone, by means of crossed levers centred on a fixed stud and connected to the lever by links and a rod.

Second, " fitting on the driving shaft placed nearly horizontal, " two conical driving pulleys facing each other, one of which is " in contact with the cone on the turbine spindle, whilst the other " is in contact with a similar cone on a short intermediate spindle, " the intermediate cone being in contact with the cone on the " turbine spindle on its other side. By this means the turbine " spindle will be driven from both sides, and the friction, in its " bearings, will be greatly reduced, whilst it will run more " smoothly and continue longer in good order than when driven " on one side only."

[Printed, 10*d*. Drawing.]

A.D. 1864, November 1.—No 2695.

BRINJES, JOHN FREDERICK.—" Improvements in apparatus " for the reburning of animal charcoal." These are said to relate to improvements in the apparatus for the above purpose, described in No. 1038, A.D. 1864, and " it is proposed to replace the " moveable or self-adjusting doors or vanes inside the cylinders " or retorts by double inclined pieces fixed immoveably to the " retort between each of the internal rings or flanges, and " opposite to the openings therein, the circular, reciprocating, or " to-and-fro motion of the retort causing the charcoal to slide " along the inclines, and so pass from one space to another " throughout the length of the cylinder or retort. Or, if pre- " ferred, the internal rings or flanges may be either wholly or " partly dispensed with, the double inclined pieces fixed to the " interior of the cylinder or retort, serving by the aid of the to- " and-fro circular motion thereof to cause the charcoal to travel " from end to end of such retorts."

In the Provisional Specification it is said, " in lieu of making the " cooling drum with an inner and outer skin " as in No. 1038, A.D. 1864, it is proposed that it should be made " with a single " skin only, and caused to revolve in a trough or tank of cold " water, or a stream of cold water may be caused to flow over " its surface thereby more rapidly cooling its contents."

[Printed, 10d. Drawing.]

A.D. 1864, November 22.—N° 2906.

NEWTON, ALFRED VINCENT.—(*A communication from Charles Rostand.*)—" An improvement in the manufacture of sugar, and " in the machinery to be used therein." This consists, " in solidi- " fying granular or raw sugar, and the apparatus " for facilitating the same, as follows :—The mixture consists of about equal parts by weight of the granular sugar and raw syrup. The apparatus is intended to work in a boiler which is substituted for the vat now employed, called the settling vat or reheater. It consists, of a vertical shaft which carries two horizontal beaters formed of radial arms which are fitted with vertical teeth. Between these is a fixed bar similarly fitted with teeth. The mass to be broken up being obliged to pass between the fixed and moveable teeth, is thus reduced to a perfectly homogeneous mass without its crystals being destroyed." " The boiler in which the heaters are mounted " is made, by preference, of sheet iron of a cylindrical form." It is heated by a steam pipe inside, or a steam jacket, so as " to " maintain the mass at a high temperature, which should be about

" 70° centigrade." By these means two difficulties attending the operation are overcome ; a perfect mixture is obtained by purely mechanical means of the two kinds of sugar which were separated by their difference in condition and density, and the required degree of heat is ensured for the operation without the attendants being exposed to inconvenience or injury."

[Printed, 10d. Drawing.]

A.D. 1864, November 28.—N° 2965.

MONTAIGUE, LODI. — (*Provisional protection only.*) — " An " improved turbine for drying sugar and other watery matters." This consists, in "causing to turn within the drum acting as sieve, " a screw, the edge of which nearly touches the sieve, and which, " receiving a motion slightly more rapid than that of the sieve, " forces the sweet juices or other matters to descend with a speed " which may be regulated at will by giving more or less speed to " the screw. The sieve is open at the lower part to permit the " solid matters to fall freely while the liquid matters are received " in a gutter made near the casing which surrounds the sieve, so " that it is only requisite to pour the matters to be treated into " the top to gather separately at the bottom the liquid and solid " parts." The matters are fed into the machine by means of a hopper, and a tube in the centre "serves for injections on to the " matters to be treated."

[Printed, 8d. Drawing.]

1865.

A.D. 1865, January 6.—N° 42.

LEBAUDY, JULES.—"A new system of boiling grain sugar in " vacuo." This consists as follows :—" After having run into " the apparatus (a vacuum pan) a certain quantity of syrup, " dissolved and clarified by the usual process, and when suffi- " ciently condensed," adding " a quantity of grain sugar, such as " is produced from the cane and beet root manufactures, or from " the refineries, varying the proportion according to the quality " of the syrup, that of the grain sugar introduced therein, and " the quality of the sugar desired to be produced." "The intro-

" duction may be accomplished at once or by successive charges."
" The sugar before being introduced may be bleached by the
" centrifugal or other process, but it is not absolutely necessary."
The crystals "can be enlarged almost without limit either by con-
" ducting the operation slowly, or causing the same crystals to
" undergo several operations by being alternately pressed in the
" centrifugal apparatus and re-introduced into the boiling appa-
" ratus over and over again." "The grains being thoroughly
" separated" are washed and treated "as in the ordinary process."

[Printed, 4d. No Drawings.]

A.D. 1865, January 7.—N° 57.

BEANES, EDWARD, and FINZEL, CONRAD WILLIAM.—"Im-
" provements in the construction of vacuum pans." These are,
to prevent the syrups, &c. from being acted upon by an excess
of heat, and to secure a sufficient extent of the tubular heating
surface, so as to secure "moderate and comparatively equable
" temperatures with rapidity and success," the "heating medium
" is caused to pass through one half or portion of the tubes in one
" direction, and through the other half or portion of the tubes in
" the opposite direction." In carrying out the above, the evaporating pan, in preference, is "comparatively narrow in proportion
" to its length," and "the evaporating tubes are placed across
" the pan." The heating agent, such as hot water or steam, is
introduced "into chambers placed on two sides of the pan," and
it is transmitted "through one half of the evaporating tubes from
" one side, and through the other half of the tubes from the other
" side," so as "to obtain great uniformity of action in the heating
" operation."

[Printed, 1s. 2d. Drawings.]

A.D. 1865, January 26.—N° 221.

HASELTINE, GEORGE. — (*A communication from Frederick
William Goessling.*)—"A new process of manufacturing syrup
" and sugar from maize and other cerial grains." This consists
as follows :—First, "three thousand pounds of maize or other
" cereals" are soaked in "water or a caustic alkali liquid of about
" one and one half degree specific gravity for about one week,"
when it is crushed, passed through sieves, washed by agitation in
weak alkaline water, and finally in pure soft water, leaving the
milky mass at the bottom of the tanks.

Second, about two hundred pounds of pure soft water, and about twenty four pounds of sulphuric acid are boiled in a tank by a steam pipe passing through the boiler, and the milky mass, above referred to, is gradually added so as not to stop the boiling, and the whole afterwards boiled for about six hours, when the steam is shut off, and about ten pounds of fine pulverized charcoal (preferring animal charcoal) are added, and then add very gradually about thirty pounds of finely pulverized carbonate of lime, and in addition thereto "about three per cent. of prepared carbo-" nate of quicklime," thoroughly neutralizing the whole; after which about twenty pounds more of the above charcoal are added, when the mass is boiled for about five minutes, and filtered through bags or cloth filters, and evaporated in a vacuum pan to thirty-three degrees of Beaumé while hot; if not clear it is run through flannel or cloth filters and placed in crystallizing vessels in a room from 75° to 100° F., and when crystallized it is pressed in cloths. The pressed sugar is re-dissolved in the " heater vessel," crystallized in moulds, and the loaves dried.

[Printed, 4d. No Drawings.]

A.D. 1865, January 28.—N° 246.

HASELTINE, GEORGE. — (*A communication from Frederick William Goessling.*)—"An improved process of manufacturing " syrup and sugar from maize, starch, and other cereal grain " starch. This consists as follows :—Put into a tank in which is a coil of lead pipe through which steam is passing, about two hundred pounds of pure water and about two pounds of oil of vitriol, when this acid water boils, about one hundred pounds of starch in about one hundred pounds of pure water and half a pound of oil of vitriol mixed together in another tank are then drawn off slowly into the boiling acid water in the tank first named, and the tank covered closely and the whole boiled for about two hours, when about one pound of pulverized wood charcoal is added, and the whole again boiled for about one hour, when the solution is passed into another tank, and neutralized with carbonate of lime, filtered through suitable bag filters, transferred into the first tank with coil of lead pipe as before described, where " it is treated with half a pound of carbonate of quicklime and " two pounds of bullock's blood," and the whole heated to the " boiling point, whereupon all the impurities and foreign matter

" in the liquid will then rise to the surface, when they must be
" removed." The liquid is filtered through charcoal "boiled
" down in vacuo to forty-five degrees specific gravity, and then
" transferred to suitable moulds in a crystallizing room at a tem-
" perature of about 90° or 100° F." When crystallized the loaves
are dried " on plaster of Paris plates in a room at about 100° F."

[Printed, 4d. No Drawings.]

A.D. 1865, January 28.—N° 249.

BURQ, VICTOR. — "Improvements in filtering apparatuses."
These are, in reference to this subject, as follows :—For the filtration of "saccharine or other solutions" employing thin blades.
plates, slabs, diaphragms, or other similar mediums, constructed
of either natural or artificial suitable porous stone or suitable
" porous filtering compounds." The artificial porous compound
is made by mixing with " porcelain or other suitable clay or
" ceramic mass," "wood dust, powdered wood, or animal char-
" coal, peat, or other suitably minutely divided combustible
" matters," and burning these combustible matters during the
baking or firing; or if it is desirable that these "porous slabs,
" plates, or diaphragms should act not only as a filtering, but
" also as a disinfecting medium, and consequently require to
" contain a suitable quantity of carbonaceous matters ;" they may
be formed, as examples, "of finely pulverized bituminous coal,
" coke, breeze, wood, or animal charcoal to be thoroughly mixed
" with tar, clay, or other suitable aglomerating material, and after
" being moulded they are baked in air-tight vessels." These said
thin blades, plates, slabs, or diaphragms, rest on, and are cemented
by means of suitable prominent parts, or to a suitable resting plate
or plates, or resting on and cemented to each other by the said
prominent parts, or kept together by means of suitable brackets,
bolts, rods, or tubes, to give them strength to resist the pressure
of the filtering liquid, and also for allowing the filtering to take
place for some time in an opposite direction to that in which the
same had taken place before, so as to cleanse the diaphragm, which
cleansing may be aided by brushes or by a continuous scrubbing
or cleansing of their surfaces by the current of the water in which
the diaphragm is immersed.

[Printed, 1s. 6d. Drawings.]

A.D. 1865, February 14.—N° 418.

FRYER, ALFRED.—" Improvements in the mode of treating, for
" evaporating and concentrating purposes, cane juice and saccha-
" rine and other solutions and liquids, and also in machinery or
" apparatus for the concentration of cane juice and saccharine
" and other solutions, and for the evaporation of liquids." These
are, first, "treating cane juice and saccharine solutions and
" liquids " in order to obtain what is termed "' Fryer's concrete,'
" or sugar, in a non-crystalline homogeneous state, by exposing
" it or them to heat in shallow trays, and then to heat in a
" revolving cylinder," with or without the subsequent treatment,
which is, when "the saccharine mass is required more highly con-
" centrated "than it is as it issues from the revolving cylinder, it
is caused, " while in a sufficiently fluid state, to flow into a trough
" in which a heated drum is made to revolve slowly."

Second, "treating solutions and liquids for evaporating and
" concentrating purposes, by exposing them to heat in shallow
" trays, and then to heat in a revolving cylinder."

Third, "treating cane juice and saccharine and other solutions
" and liquids ·for evaporating and concentrating purposes, by
" passing it or them through or into a revolving cylinder heated
" from the inside, and through the inside of which heated air is
" forced or driven."

Fourth, " the machinery or apparatus for the concentration of
" cane juice and saccharine and other solutions, and for the evapo-
" ration of liquids consisting of shallow trays," corrugated and
inclined, in " combination with revolving cylinders."

[Printed, 1s. 6d. Drawings.]

A.D. 1865, March 3.—N° 599.

BROOMAN, RICHARD ARCHIBALD.—(A communication from
Alfred Guillon.)—" Improvements in refining sugar, and in appa-
" ratus employed therein." These are, first, " the aglomeration
" of the crystals of sugar by means of a sweet liquid or syrup of
" a density below 40° Beaumé," substantially as follows :—The
sugar to be made into loaves is mixed in a cold state in a mill or
mixer with a syrup " of a density of 36° Beaumé in the proportion
" of about forty-three quarts of syrup to two hundred pounds of
" sugar," until the mass is perfectly homogeneous, when it is
placed in a double-bottom pan and heated," not " exceeding

" 131° F., and run into moulds. Eight hours (more or less)
after filling the loaves are set to drain without receiving any dis-
" colouring, or being in any way worked ; after draining about
" forty-eight hours the loaves are removed and stoved."

Second, constructing a mixing mill composed of an oblong case
closed, except to an inlet and outlet for the syrup and sugar,
extending from one end of the case to the other are screw blades
" over the inlet aperture is a hopper.

Third, a double-bottomed pan heated by steam, into which the
mass is directed by a pipe or shoot from the mixing mill for
heating the saccharine matters up to the desired temperature.

[Printed, 8d. Drawing.]

A.D. 1865, March 15.—N° 730.

BRINJES, JOHN FREDERICK.—"Improvements in apparatus
" for cooling animal and other charcoal." These are, first, the
constructing of coolers for the above purpose "of a spiral or
" volute form with free air spaces between the several convolu-
" tions," these are adapted to the discharging end of a revolving
or reciprocating retort employed in the manufacture or reburning
of animal and other charcoal. The spiral or volute conduit is of
any convenient sectional form and number of convolutions, and
" communicates with the interior of such retort by an opening or
" mouth made in the inner or first coil of the volute." As the
retort revolves, the charcoal travels "the entire length of such
" volute, and is finally discharged at the outer end of the last
" coil thereof (which may or may not be provided with a door
" as required), in a sufficiently cool state, so as not to be
" injuriously affected by contact with the atmosphere."

Second, another modification of the above "particularly
" adapted for use with partial rotatory or reciprocating retorts,
" such as are referred to " in No. 2695, A.D. 1864. "In place
" of employing a continuous volute or spiral, I employ a series
" of annular passages or conduits, each communicating with the
" other at any convenient point in its circumference." This is
attached as the above, and acts in a similar manner. In both
these arrangements the outer face of the cooler has a plate or
cover in which is a bull's eye "for facilitating the inspection of
the interior of the retort.

[Printed, 10d. Drawing.]

A.D. 1865, April 4.—N° 955.

NEWTON, WILLIAM EDWARD.—(*A communication from Louis Pierre Robert de Massy.*)—" Improved apparatus for expressing " liquids from pulpy and semi-fluid substances." This consists in " the industrial employment of water or gases for the purpose " of subjecting various matters to pressure by means of a fabric, " or sheet, or plate, consisting of any flexible and impervious " body," such as india rubber, which " is interposed between " the pressed product and the pressing agent." The armiture of the press consists of a perforated metal cone lined with metallic cloth covered with other fabric; inside this cone is the flexible cone of caoutchouc for containing the liquid or fluid which is to exert pressure; into the annular space between these cones is introduced the substance that is to be pressed. Water or air is forced into the flexible cone, and by this means the liquid contained in the substance will be forced out, through the perforated sides of the outer cone, and will run down into a gutter which surrounds the bottom, and from whence it is collected. There is a steam cylinder in which there is a moveable diapragm or piston with pipes in the same to supply the pulpy substances into the press. It is said that " this apparatus is more particu-" larly intended for expressing the liquid parts from the pulpy " matters, floating scum, and barytic salt used in the manufacture " of sugar."

[Printed, 10d. Drawing.]

A.D. 1865, April 29.—1198.

WHITE, THOMAS.—" Improvements in apparatus employed in " the reburning of animal charcoal." These are, first, " forming " on the inside of cylinders placed horizontally or nearly so, a " series of scoops or buckets which, on the cylinders being made " to revolve, either continuously or in the same direction, or partly " in one and partly in the reverse direction, pick up, disperse, and " transfer the black from one end of the cylinder to the other. " The cylinders are heated and motion is communicated to them " in the usual manner. I prefer to employ more than one cylin-" der, the cylinders being placed over one another, and the outlet " from one being made to open into the outlet of the cylinder " next below it, and so on."

Second, the employment "of a cooling cylinder made of thin
" metal, into which the black is received after being sufficiently
" burnt, and in travelling through which it is cooled."

[Printed, 1s. 4d. Drawings.]

A.D. 1865, May 13.—N° 1336.

OGSTON, George Henry.—(*Provisional protection only.*)—
" Improvements in the manufacture and reburning or revivifica-
" tion of animal charcoal." These are, " animal charcoal is at the
" present time manufactured and revivified in retorts, and it is
" discharged from these into coolers or vessels which are covered
" or closed to exclude the atmosphere;" and it is now proposed
to introduce steam "into the coolers and into the passages leading
" thereto from the retorts, so that the heated animal charcoal,
" after leaving the retorts, is caused to pass into and through an
" atmosphere of steam by which the charcoal is improved."

[Printed, 4d. No Drawings.]

A.D. 1865, May 22.—N° 1409.

MÜLLER, Richard, WELD, Arthur Thomas, and POWELL,
John Folliott.—" Improvements in the preparation of materials
" to be used as substitutes for animal charcoal." These are, a
porous artificial matrix is prepared, in preference as follows :—
Using "pipe clay without admixture with other matters, though
" other clay may be used, and wood charcoal or other matters
" may be mixed therewith, in order to obtain porosity with the
" requisite hardness in the artificial matrix." When rendered
plastic by mixing with water, it may be made into thin small tubes,
but in preference it is formed into slabs about three-sixteenths of
an inch thick, which are dried and burned and broken into pieces
about a quarter of an inch square, and "impregnated with car-
" boniferous matters in a state of solution in caustic alkali," preferring that, obtained by charring woollen rags, and known as
ulmate of ammonia. This is best done by boiling the pieces in a
saturated solution of the ulmate till air bubbles cease to come off,
the saturated substance is removed from the solution, allowed to
cool and drain by spreading and stirring it on perforated plates,
when it is " dried as quickly as possible on a hot plate, or in an
" oven without charring." Afterwards, in preference, it is charred
in upright retorts heated gradually. When the charred product

ceases to give off gas or vapour, it is removed and "broken up
" into grains of the size of those of ordinary charcoal, which grains
" are again saturated with the animal solution, and again treated
" as above explained." If on breaking some of the grains they
are thoroughly saturated, this process need not be repeated. The
grains are then washed with water to remove all alkali, then in a
dilute hydrochloric acid (7 per cent. of acid), finally washed with
water and dried, are fit for use as a decoloring agent as in sugar
refining, or as a purifying agent for the filtration of water. After
use it may be revivified in like manner to animal charcoal. In
place of the ulmate, solutions of glue, size, resin, coal tar, or
blood may be used, but not so advantageously.

[Printed, 4d. No Drawings.]

A.D. 1865, August 2.—Nº 1996.

McEWAN, JAMES, and NEILSON, WILLIAM.— " Improve-
" ments in the raising, lifting, or drawing and forcing of water
" and other liquids, and in the apparatus and means employed
" therefor." These are, " the general construction and arrange-
" ment of apparatus or mechanism," and " the application and
" use of high-pressure steam in the form of conical jets caused
" to pass through narrow or fine slits of the same shape formed
" round and through the body of conducting pipes, so as by the
" direct action and force of steam to draw, raise, and force
" water or other liquid in a constant stream through the full
" bore of the said pipes," as afterwards described; and among
other applications which are named, "for sugar houses, and many
" other purposes where large quantities of liquid have to be
" raised." The apparatus may be made by "forming a short
" length or piece of pipe with flanges at each end similar to
" those of the main column of pipes through which the water is
" to be raised, so that it can be jointed between any two of the
" main pipes and form part of the whole column; this short pipe
" is made somewhat larger in diameter than the main pipe, and
" has formed at or near the centre of it an enlarged annular
" space or cavity, with a small lateral branch pipe attached or
" leading into it. A considerable portion of this short pipe
" at each end is bored out parallel, and fitted with two rings
" or tubes, by preference, formed of brass, composition, bush, or
" gun metal, the two making up nearly the whole length of the
" short pipe, and having their internal diameter the same or nearly

" so as that of the main pipes. The two adjacent ends of these
" tubes are turned to the same or nearly the same annular
" bevilled or truncated conical surface, and as one or both of the
" tubes are made to move a short distance to or from each other,
" the width between these bevelled ends or surfaces is contracted
" or widened, and as these form the slit or annular orifice through
" which the steam passes to act on the column of liquid being
" raised within the pipes, the adjustment of the width of this
" annular orifice or distance between these surfaces regulates to a
" considerable extent the volume and force of the jet or stream
" of steam passing through it and acting on the liquid inside the
" pipe. This motion and regulation of the distance between the
" ends of the tubes, and the fixing of them when so regulated,
" may be effected by various simple mechanical contrivances,
" one only of which need here be described, and consists in
" having one or more short spindles with excentric points, each
" passed in through a small stuffing box formed in the outer
" short pipe, the excentric point working into an annular groove
" or slot, which may be formed by rings or ' snugs ' on the outer
" circumference of one or both of the tubes inside the outer short
" pipe, so that when this spindle is turned by its square head
" outside, the excentric point working in the groove of one or
" both of the tubes regulates the width of the annular steam
" orifice between them, and when properly regulated and set, the
" spindle or spindles is or are fixed by screwing up the stuffing
" box gland upon a steam-tight collar formed on each of the
" spindles." Modifications of the above are described.

[Printed, 1s. Drawing.]

A.D. 1865, August 4.—N° 2023.

LEON, JEAN ADOLPHE, TESSIMOND, GEORGE, and KIS-
SACK, JOHN.—" Improvements in apparatus for filtering sugar
" and other liquid solutions." These are, "the combination in
" filtering apparatus of textile filter bags " with " a sloping or
" declivous lower part or bottom." " Constructing the delivery
" cock " with " a moveable nozzle." The combination in filtering
apparatus " of " the sloping or declivous bottom " and " the
" delivery cock."

It is said that when pulverized bone black has been used in the
clarification of cane juice or solutions of sugar, the filters used

to remove the black colour from the liquids have hitherto been ineffective, and the use of " the said bone black has been aban- " doned, and costly granulated charcoal used as a substitute." In constructing a filter to allow of the employment of the said pulver- ized bone black, the internal lower portions of the filter box or case should be so arranged that no insoluble matter can rest upon them, and the outlet from such box or case should be large, in order not to be obstructed by the black. This is effected as follows :—A rectangular or other shaped box has a perforated plate supported in it, to the under side of which is attached " cotton or other textile fabric filter bags. The internal lower " portions of the said filter are formed or fitted with sides or " plates which slope downwards to, or have a declivous direction " towards the outlet cock," which is double the size of ordinary outlet cocks. The liquor is run upon the filter, the first discharge of liquor is impure, and is returned by a gutter or way into the foul cistern ; but when it runs clear it is directed into another gutter or way, and conveyed to suitable cisterns or pans. The cock is made with a hollow plug and with a swivel or other jointed pipe on the nozzle of the same.

[Printed, 10*d*. Drawing.]

A.D. 1865, September 7.—N° 2296.

DAWSON, James.—" Improvements in supplying charcoal to " sugar, decolorising vessels, and in apparatus therefor." These are, " distributing the charcoal uniformly" and "without the " workman having to go inside the vessels, as has hitherto been " necessary." " The decolorising vessel being of the usual " cylindrical form with a central opening in the cover," there is fitted to such " opening, and so as to turn therein, a distributor, " consisting of a hollow piece formed according to one modifi- " cation with two diametrically opposite hollow arms or ducts, " which are inclined slightly downwards, and reach to within a " few inches of the sides of the vessel. The distributor is fitted " with bevil or conical pulleys which run on the flange of the " opening in the cover, and is turned by means of a vertical " spindle in any convenient way whilst the charcoal is being " thrown into the hopper shaped mouth. The charcoal falls into " the vessel through openings in the bottoms and ends of the " arms, and is distributed by them in a uniform manner. The

" distributing arms are fixed together when in action, but are
" made to separate so as to be easily removed one after the other
" before fixing on the cover and admitting the syrup." If preferred, the distributor "may be made with three or more hollow
" arms instead of two."

[Printed, 8*d*. Drawing.]

A.D. 1865, September 19.—N° 2385.

FLETCHER, JOHN.—" Improvements in the machinery or
" apparatus and in the process for the treatment and manu-
" facture of sugar." These are, first, dissolving apparatus,
consisting of a circular metallic tank, on the bottom of which is
an arrangement of copper pipes or tubes, perforated on the under
side, for the egress of steam and water for dissolving the sugar;
over these tubes is a perforated plate of iron. In the centre of
the tank is a step carrying two spindles one inside of the other,
the upper parts being carried on a block on a cross plate. These
spindles carry two arms revolving with them in opposite directions. The steam and water are admitted into the pan or tank
through a chamber cast on one side of the pan or tank.

Second, " facilitating the passage of the syrup through the
" filter," by exhausting the air from the filters preparatory to
the syrup entering the charcoal filters; also, washing the charcoal
after the process of filtration by steam or steam and water
exhausting the filter of air. The syrup tank or tanks are also
exhausted.

Third, employing " a closed vessel or vessels of any suitable
" shape," and attaching " to the same an air pump to exhaust
" it of atmospheric air, whereby the liability of the syrup to
" acidify and fermentation is very much lessened," for " keeping
" or storing the syrup when in the stage of being converted into
" sugar."

Fourth, a vacuum pan of " a square or rectangular shape with
" a curved top and a sloping solid bottom. The steam chambers
" or heating surfaces in the inside of the pan are of a flat form,
" and appear in section of a link shape; they run from end to
" end or side to side of the vacuum pan, and are placed side by
" side of each other with a space left between them to afford
" room for cleaning and for sugar." These steam chambers are
placed one above another in tiers, " each tier has its separate

" steam connection, by which they can be worked separately or
" collectively as required; a steam box is fitted to each end of
" each tier of chambers, into which is admitted the steam." A
large pan requires admission of steam at both ends of the boxes,
while a small pan requires steam at one end only. The boxes at
one end of the pan are secured to it by rings which serve for
admitting steam into the boxes. The boxes at opposite ends of
the pan have tubes connected and pass through stuffing boxes on
the end of the pan to allow for the expansion of the steam
chambers; the above tubes or boxes are arranged to take the
condensed steam from chambers and boxes. To this pipe also is
connected a condensed steam box, to pass off water but not the
steam. On the top of the vacuum is an arm pipe with a
receiver attached to it, to this receiver is also connected a measure
which receives the syrup prior to its being passed into the vacuum
pan. The measure is fitted with a series of metallic tubes,
through which the vapour and air from the syrup under treatment in the vacuum pan pass, to this measure is attached a
cylindrical condenser placed vertically; this condenser is fitted
in its centre with a metallic tube perforated with small holes.
" The water for condensing is admitted into the tube either at
" the top or bottom as may be most convenient."

Fifth, " self-acting apparatus applied to the condensed steam
" box for allowing the escape of air but not of steam." On
the cover of the condensed steam box is a valve to which is
attached a lever, the other end of which is held by a stud pin
fixed in the cover, to this lever, near its fulcrum, is connected
the end of another lever by a short link, the opposite end of
which lever is also held by a stud pin fixed in the cover; this
lever has also near its fulcrum another link attached, the opposite
end of which is secured to the cover by a screw or bolt; the hole
in this end of the link is made of a slotted or elongated form to
allow of the adjustment of the levers. The second link " is made
" of zinc, brass, or other metal easily affected by heat or cold,"
preferring, " however, to employ zinc, the expansion and con-
" traction being greater in that metal." The air will pass out
of the open valve, but when steam arrives in the box the heat
causes the second link to expand which, acting on the two levers,
closes the valve.

[Printed 2s. Drawings.]
s.   A A

A.D. 1865, September 21.—N° 2409.

CLARK, WILLIAM. — (*A communication from Chrétien Jean Gaade.*)—(*Provisional protection only.*)—"Improvements in the " manufacture of materials for decoloring sugar and other " saccharine and liquid matters." These are, partially or entirely utilizing the refuse resulting from the revivification or otherwise of the grains of animal black. The refuse is the powder re- sulting from animal black, and the object is to transform it into grains. For this purpose the powder is mixed with argillaceous earth, or that used for the manufacture of glazed ware, made with water into a pasty mass, and moulded into the form of small bones or otherwise, and dried in the sun, and calcined in a similar manner to bones, with a piece of bone, and of green wood in the centre, and the whole is heated " until the wood and bones are " completely carbonized," when it is reduced to "grains of a " similar size to those now used." This composition can be revivified after use. It is of greater density than ordinary animal black, but " a material exactly similar to that ordinarily " in use " is obtained " by mixing 50 per cent. of animal black in " fine powder, 25 per cent. of vegetable charcoal in fine powder, " and 25 per cent. of pottery clay, the whole being mixed and " prepared as before described. This compound may, if desired, " be mixed in any suitable proportions with soot, gelatine, " potash, or other matters."

[Printed, 4d. No Drawings.]

A.D. 1865, October 7.—N° 2590.

CAMPBELL, TOMLIN.—" Improvements in evaporating and " distilling liquids, and in the apparatus employed therein." These are, the application and use as above " of one or more air- " tight vessels in combination with a steam heated, hollow, or " tubular revolver or revolvers," and the system or mode of effecting the above operations, substantially as follows :—" One " or more closed vessels are employed capable of sustaining " pressure, each vessel containing a coil of piping made to " revolve on a horizontal axis passing through stuffing boxes in " the ends of the closed vessel. The liquid to be evaporated is " placed in the vessel so as only partially to fill the same, and " waste or low-pressure steam (if the evaporation is to be con-

" ducted at a low temperature), is admitted into and through
" the first of the coils, and its heat is imparted to the liquid to
" be evaporated, which liquid as the coil revolves is taken up in
" a thin film, thereby greatly facilitating and expediting the
" process of evaporation. The non-condensed steam, if any, in
" the first coil or 'revolver' of the series passes off to the
" atmosphere, but that from the subsequent coils, if any, together
" with that condensed passes to an air pump. The steam
" generated in the first vessel (when more than one are used) is
" drawn out of that vessel through a pipe to the second coil or
" 'revolver' of the series by the air pump attached to that coil,
" and that in the second vessel to the third coil of the series, and
" so on, each coil except the first having an air pump attached
" to it. The vessels should be provided with regulating valves,
" in order that the vacuum in the several vessels of the series
" may be successively greater from the first to the last of the
" vessels. When a low temperature is not required in the process
" of evaporation, high pressure steam may be introduced into
" the first of the coils, and the several coils should each be in
" communication with a steam trap to allow of the condensed
" steam passing off." The principle of this invention is, "that
" when the pressure in the vessel is less than in the coils of pipe
" or 'revolver' the steam or vapour in the coils or 'revolvers'
" will be condensed, and the liquid in contact with the coils will
" be evaporated, and the greater the difference in pressure between
" the coils and the vessels, the greater will be the rapidity
" of evaporation." There is a plan view given of " apparatus
" for the manufacture of sugar, which manufacture consists
" mainly of the operation of evaporation " conducted on the
principle described above.

[Printed, 2s. Drawings.]

A.D. 1865, October 10.—N° 2606.

LÉON, JEAN ADOLPHE.—(*Partly a communication from Absalon Hippolyte Leplay.*) — (*Provisional protection only.*)—" Improve-
" ments in means and apparatus to be employed in the manu-
" facture of sugar." These are, treating " cane juice and other
" solutions of sugar so as to produce therefrom in a simple, &c.
" manner, crystallized sugar fit to be sent into the market with-
" out further refining," first, by employing " a purifying agent
" consisting of finely pulverized bone black and phosphate of

" soda," these are mixed "with the cane juice or solution of sugar
" conjointly or one after the other."

Second, running this semi-liquid mixture "into a closed vessel
" of any suitable external form, in which there is a perforated
" bottom covered over with a porous fabric," and when a sufficient quantity has been admitted shutting off the supply and
admitting "steam or other vapour at pressure over it to force
" the liquid through the straining cloth and perforated bottom,
" and when desired into a cistern or vessel at greater elevation
" than the strainer," "to remove most of the bone black,
" phosphate of soda, and other insoluble matters from the cane
" juice or solution of sugar," and to prepare such "for final clarifi-
" cation," in the filter described in No. 2023, A.D. 1865. "The
" fine bone black and other insoluble matters removed from the
" liquid by this straining, when ammonia is added thereto, makes
" an excellent manure for the growth of sugar." The same
apparatus and pressure may be used for "pressing scum and
" residue collected and obtained in the manufacture of sugar by
" the above or the ordinary process."

[Printed, 4d. No Drawings.]

A.D. 1865, October 17.—N° 2679.

BEANES, EDWARD.—(*Provisional protection only.*)—"An im-
" provement in treating animal charcoal." This consists in
" allowing animal charcoal that has been heated or reburned (and
whilst in the filter) to absorb carbonic acid gas by simply admit-
" ting the said gas into the filter," which treatment, it is said,
has been found not only to neutralize "any caustic alkali that
" may be present in the charcoal," but also to remove "the air
" which exists between the grains of the same, thus allowing the
" liquor when run on (by expelling the carbonic acid gas) to come
" directly into perfect contact with the whole of the charcoal in
" the filter."

[Printed, 4d. No Drawings.]

A.D. 1865, November 7.—N° 2872.

JASPER, GUSTAVUS ADOLPHUS.—" Improved means and appa-
" ratus for the cleansing and bleaching of sugar, which said means
" and apparatus are also applicable for other useful purposes of a
" like character." These are, " liquoring sugar " in a centrifugal

machine, applying the liquor while the machine is in revolution the liquor being supplied under pressure, or its equivalent, substantially, as follows :—The centrifugal machine is of the simplest form, " an open cylinder without a cone in the centre, and driven " from below by suitable mechanism." Two centrifugal machines are shown, and at a suitable altitude above them there is placed a tight vessel capable of bearing a pressure of about 125 lbs. to the square inch, with a filling pipe, having a stop cock, a safety valve, and a gauge tube. A pipe from an air forcing pump enters the upper part of the vessel. Another pipe having a stop cock " near its upper end leads off the bottom of the vessel," and communicates with a flexible pipe arranged over each centrifugal machine, each pipe terminates with a rose nozel, which may be provided with a stop-cock. There may also be a stop-cock at the lower extremity of the pipe. Strong or thick cleansing liquor or syrup is charged into the vessel above, and air is forced into it " under a pressure of 25 lbs. and upwards to the square inch." The centrifugal machines are charged with a mass of sugar and revolved rapidly, and the stop-cocks opened, the syrup is delivered with force in minute streams against the inner surface of the mass of sugar. Open tanks at a high elevation may afford pressure in place of air as above.

[Printed, 6d. Drawing.]

A.D. 1865, November 30.—N° 3078.

CLARK, WILLIAM. — (*A communication from Chrétien Jean Gaade.*)—(*Provisional protection only.*)—"Improvements in the " manufacture of materials for decoloring sugar and other saccha- " rine matters." These are, " combining soot with other matters, " and reducing it to a granular form," and calcining and apply- ing it in " decoloring sugar." Further carbonizing blood and animal flesh, and mixing the result " with soot and with other " ingredients herein-after mentioned, and reduce it to the granular " form desired." The proportions which are suitable " are about " sixty per cent. of soot to forty per cent. of clay, but the pro- " portions may be much varied, while at the same time three or " four per cent. of potash may be added." " The decoloring " agent instead of being nearly all soot, may consist, of soot mixed " with animal and vegetable charcoal, and with a proportion of " clay or earthy or other matter," and formed into a plastic mass.

This mass is expressed "and cut off in small nodules or grains
" by rotary knives or wires, in the manner of making vermicelli,
" or in any other manner." These "nodules or grains" are then
calcined, "when they are fit for use in the decoloration of saccha-
" rine matter in the process of filtering or otherwise," and "which
" may be revivified and used again for the decoloration of sugar
" indefinitely."

[Printed, 4d. No Drawings.]

A.D. 1865, December 29.—No 3372.

CORMACK, WILLIAM.—(*Provisional protection only.*)—"Im-
" provements in revivifying, deoderizing, and calcining animal
" and vegetable charcoal and other matters." These are, first,
in revivifying and deodorizing the above substances, introducing
a jet, stream, or current of superheated atmospheric air or steam,
singly or combined, into retorts, ovens, or other vessels containing
the matter to be operated upon," thus employing "a black heat
" instead of a red one as at present employed."

Second, in calcining the above matters, in addition to the external heat now used, introducing "superheated steam at a tempera-
" ture of not less than 650° Fahrenheit into ovens or other vessels
" containing the materials to be operated upon, by which I am able
" to calcine a larger quantity of charcoal or other material and in
" a shorter space of time than has hitherto been accomplished."

[Printed, 4d. No Drawings.]

1866.

A.D. 1866, January 17.—No 162.

ANDERSON, MARK FRENCH.—(*Provisional protection only.*)—
" Improvements in refining sugar." These are, employing "sul-
" phate of tin in treating syrups of sugar, and in neutralizing the
" sulphate of tin after heating the syrups by an alkaline earth."
The quantity of sulphate of tin added varies "with the quality of
" the syrup of sugar to be treated," it is generally employed "at
" the rate of about one pint of a saturated solution of sulphate
" of tin to each cwt. of sugar in solution," and the whole

intimately mixed, and heat applied "in the blow up pan; after
" which the syrup is allowed to stand in order to settle. The
" clear liquor is separated, and the requisite quantity of the car-
" bonate of an alkaline earth is stirred in; the syrup is after-
" wards allowed to stand, or is filtered in order to separate the
" syrup from the other matters, and the process of refining is then
" completed in the ordinary manner."

[Printed, 4d. No Drawings.]

A.D. 1866, January 23.—N° 225.

BENSEN, GERD JACOB.—" Improvements in the manufacture
" of beet root sugar." These are, " combining fine liquor or syrup
" with raw beet root sugar," as afterwards described, and " then
" separating the liquor therefrom," the object being to remove
from the crystals of raw beet root sugar impurities, and " vege-
" table matter, which is offensive to the smell," " by which a very
" superior manufacture of beet root sugar is obtained." Into
" any quantity of fine liquor or syrup of cane sugar of about
" 35° of Beaumé is stirred a quantity of raw beet root sugar, so as
" to bring the mixture to the same density as usual in the manu-
" facture of stove goods, the object being that the crystals of the
" raw beet root sugar shall not be dissolved, or as little as may
" be. The mixture when heated, by preference, to about 170°
" of Fahrenheit, is then run into moulds, where the same is
" allowed to stand for about 36 hours. Fresh quantities of fine
" liquor are afterwards to be applied, as is well understood, and,
" finally, the sugar is to be dried in stoves." " The liquor
" drained from the crystals of raw beet root sugar, after being
" filtered through animal charcoal may again be used as above
" described."

[Printed, 4d. No Drawings.]

A.D. 1865, January 26.—N° 258.

MONTCLAR, JEAN MARIE ARMAND.—" Improvements in the
" manufacture of materials or compositions for decolorising or
" purifying saccharine or other liquids, and for making paint,
" blacking, and foundry blackening and apparatus therefor."
The materials used for the above purposes " are composed of car-
" bonaceous elements, earthy elements, and in some cases phos-
" phates," and a number of these are named. And it is stated

" by way of example," "that a very good composition is formed
" by combining 60 parts by weight of bone flour, with 25 parts
" of bone tar, coal tar, pitch, or soot, and with 30 to 35 parts of
" the best fire-clay; a second composition contains 20 parts
" vegetable charcoal, 40 parts animal charcoal, 50 parts soot, and
" 80 parts clay; a third composition contains 40 parts vegetable
" charcoal, 80 parts animal charcoal, and 75 parts clay; a fourth
" composition contains 50 parts soot, 80 parts animal charcoal,
" and 80 parts clay; and a fifth composition contains 75 parts
" carbonized animal body (bones?) and 40 parts clay. It, is how-
" ever, to be understood that the proportions of the various
" ingredients may be altered as practice may indicate." In
making a composition, each ingredient in a dry state is ground
separately to a fine flour, and being mixed together they are
ground with liquid (water or urine, or water containing ammonia
or glutinous matter) to form a paste, and " dried sufficiently to
" enable it to retain the form imparted to it by the granulating
" apparatus." "This apparatus consists, of two hollow cylinders
" placed in a vertical or inclined position, which are made to revolve
" in contact, whilst the plastic material is pressed against a portion
" of their surfaces by a revolving spindle with screw blades work-
" ing in a casing adapted to the cylinders" perforated with holes
through which the material is forced and formed by them, being
cut off in suitable lengths as the cylinders revolve by a stationary
knife. The grains fall through the bottom open ends of the cylinders, and they are dried, by preference, in a revolving cylinder.
" A very powerful decolorising material is obtained by applying
" the manufacturing processes herein-before indicated to bones
" alone or to bones combined with some soot. The bones are
" reduced to flour, which is made into a paste with water rendered
" milky with a little fire-clay, and after granulation the material
" is carbonised in the ordinary way."

[Printed, 8d. Drawing.]

A.D. 1866, February 27.—N° 594.

GEDGE, WILLIAM EDWARD.—(A communication from Frederick
James Vivian Minchin.)—"An improved process for extracting the
" juice from sugar cane, beet-root, and other plants." It is stated,
that, macerating the above materials with heat (a "temperature of
" 80° Réaumur") "was employed in certain sugar houses" to

extract from them their saccharine matter, but this heat injured the products, and that cold maceration and centrifugal force has also been employed; and it is said, " by these methods more juice " can be extracted than by the press, but they have the incon- " venience of carrying with them many particles of the beet and " cane into the extracted juice," and the following method, termed " diffusion," is a more practical method, and is the subject of this invention, which is, that the plants should be cut into slices from $\frac{1}{12}$ to $\frac{1}{6}$ of an inch thick and from a $\frac{1}{4}$ to $\frac{1}{2}$ an inch in breadth, and the equal heating of these slices is obtained by filling the extractors simultaneously with the slices and the warming liquid in corresponding proportions. "The temperature of the warm- " ing liquid (which may be water or juice) depends upon the " temperature and thickness of the slices." The materials in the extractor, when combined, should not be higher "than 35° to 40° " Réaumur." After about half an hour the liquid is drawn off, after which the diffusion is several times repeated with juice of a lower temperature, which is continued till all the juice is extracted. The final diffusions may be made with cold water instead of juice. The "heating of the juices which is intended to effect " the beginning of the first diffusion" is always effected in a " separate vessel called the heating apparatus."

[Printed, 4d. Woodcut.]

A.D. 1866, March 6.—N° 683.

NORMAN, JOHN.—" Improvements in apparatus for reburning " and in apparatus for washing animal charcoal or charcoal sub- " stitutes." These are, first, forming the reburning vessel, " or " the principal portion thereof, in the form of a hollow basin, " having from three to ten sides, and, by preference, six, seven, or " eight sides." It may be arranged singly, but, if preferred, it may be arranged in a set, one over the other, with appliances for leading the charcoal from one into the other. The vessel is slightly inclined from the horizontal, and is made to turn slowly round by suitable gearing, which motion causes the charcoal " to " gradually move towards the lower end. The vessel is mounted " so that it can be adjusted to what is in practice found to be the " most suitable angle, which may vary with different charcoals." A mouth piece is " supported by rods from the roof, and formed " with a horizontal cylindrical mouth, which is fitted into a con- " centric circular opening in the end of the reburning vessel, so

" that the vessel can turn round without the mouth piece turning."
The mouth piece is fitted with a valve, and is formed with a branch
pipe for the escape of gas or vapor, which passes into a main pipe.
At the lower ends there are also fitted inclined blades. Prismatic
vessels as above may be formed with angular ribs.

Second, a cylindrical or prismatic cage or vessel, formed with a
permeable material, as wire cloth or perforated sheet metal, is
formed internally with helical or screw-shaped blades or flanges,
and it is set at a slight inclination from the horizontal. The
materials are introduced at the lower end, and the cage being
made to rotate carries them gradually through water in a trough,
and delivers them at the upper end just above the level of the
water. The water is admitted into the upper portion of the
trough, and flows out at the other end. The shell of this cage
may be made in some cases of impermeable materials.

[Printed, 1s. 2d. Drawing.]

A.D. 1866, March 21.—Nº 842.

ELLIOTT, EDWIN DEMAS.—" Improvements in treating animal
" charcoal used in refining sugar." These are as follows :—It is
said that, after a time the animal charcoal used in filtering the
liquor or syrups, requires to be washed with hot water and to be
reburnt, and that after washing " much water and offensive matter
" is retained in the charcoal, which has to be dried before reburn-
" ing," and this invention consists " in exhausting the bottom or
" lower parts of the cisterns or vessels containing the animal
" charcoal by means of air pumps or other exhausting apparatus ;
" by this means streams of atmospheric air are caused to pass
" through and amongst the grains of animal charcoal, and thus
" the removal of the moisture and offensive matter is facilitated
" and the animal charcoal partially dried." In addition, several
advantages which are detailed are said to be " obtained by this
" process."

[Printed, 4d. No Drawings.]

A.D. 1866, March 29.—Nº 909.

MYERS, MORRIS. — (Provisional protection only.) —" An im-
" proved apparatus for sifting sugar, or flour, or other meals, or
" separating the smaller from the grosser particles in sauces and
" condiments." This consists, " of an upright spindle made to

" revolve in a supporting socket," which socket "is fitted with
" arms to press against the sides of the vessel in which the opera-
" tion of sifting or separating is to be carried on in order to
" strengthen and support the apparatus. On the bottom of the
" spindle are three or more arms, each of which carries a metal
" scraper which presses against the sieve or sifter below and urges
" the material" through it. "Pressing against each scraper is
" a small spring, which allows of the scraper passing over any
" lump or accumulation of material should the sieve become
" choked." "The sieve may be of wire cloth, or hair cloth, can-
" vas, or other suitable or usual material." In some cases, instead
of the metal scrapers, using "brushes inserted in the arms, which
" answer the same purpose as the metal scrapers where the
" material composing the sieve is soft or more easily worn, as
" hair cloth or canvass."

[Printed, 4d. No Drawings.]

A.D. 1866, April 3.—N° 952.

ROBEY, JAMES.—"Improvements in reburning animal char-
" coal." These are, mixing the charcoal to be reburned with
sal ammoniac, applying heat to this mixture, and subsequently
washing of the reburned matter with water. In preference, "the
" crude sal ammoniac of the manufacturers in the state of powder,
" or of small crystals, or the equally crude liquid sal ammoniac
" obtained from muriatic acid, and gas water" is employed.
These are mixed, and the mixture heated, in preference, in upright
pipes or retorts heated externally. The lime, as well as other
objectionable matters, form soluble chlorides with the acid of the
sal ammoniac, and are afterwards washed away, "whilst the
" ammonia is driven off and may be collected." Instead of
applying external heat to the retorts, superheated steam may be
passed freely through the retorts, but "upright retorts heated
" externally by fire" are preferred.

[Printed, 4d. No Drawings.]

A.D. 1866, April 4.—N° 960.

JOHNSON, JOHN HENRY.—(A communication from François
Joseph Chauvin and Frederic Mathurin Légal, the younger.)—"Im-
" provements in the treatment of sugar." These are, "the
" moulds containing the sugar are each connected at their lower

" ends with a suction pipe, similar to those already in use in
" sugar refineries, formed with a cap or mouth-piece which is
" packed with a washer of india-rubber fitting tightly upon the
" mould. These pipes are in communication by means of several
" cocks with different air pumps, for the separation of the various
" qualities of syrup." Between the suction pipes and air pumps
are large vessels or reservoirs to receive the syrup drawn from
the loaves. "The refining and decoloring or washing of the
" crystals are thus effected by the sole action of the air pumps,
" and may both be accomplished in one day." In place of subjecting the loaves "to the subsequent operation of desiccation in
" stoves for from 7 to 8 days," the "complete desiccation is
" effected at one operation by simply continuing, but with in-
" creased vacuum, the action of the air pumps, and thereby caus-
" ing dry and hot air to pass through the crystals in sufficient
" quantity to effect their complete desiccation." The "refining,
" decoloring, and desiccation are effected by this system in the
" space of from 4 to 5 days with a cheap and simple plant in lieu
" of requiring, as heretofore, from 15 to 18 days, and the use of
" costly plant and extensive premises."

[Printed, 4d. No Drawings.]

A.D. 1866, April 7.—Nº 1005.

GORDON, GEORGE—(*Provisional protection only.*)—" Improve-
" ments in treating animal charcoal." This consists, " in placing
" the charcoal to be treated in a close vessel, and creating a vacuum
" by means of an air pump, or otherwise, and then admitting the
" dilute acid (hydrochloric acid) to the charcoal in the vessel
" whilst the air is exhausted. It will be evident that owing to
" the extraction or exhaustion of the air from between the particles
" of charcoal the saturation will be much quicker, and it will also
" be found that the exhaustion of the air from the interior of each
" particle will cause the coating of mucilagneous substance "
upon charcoal which has been employed in filtering saccharine
liquids, to " become porous, and admit the dilute acid to act upon
" the interior of the particles, and hence the saturation will be
" more perfect, at the same time that a more attenuated dilution
" of the acid than that hitherto employed will produce the desired
" effect," which is the neutralization and removal " of the lime
" or other alkaline salts " from the charcoal. " It will also be

" obvious that this Invention is equally applicable for the neu-
" tralization of acidulated charcoal by the admission thereto of
" and saturation with an alkaline solution."

[Printed, 4d. No Drawings.]

A.D. 1866, April 16.—N° 1073.

JOHNSON, JOHN HENRY. — (*A communication from Joseph Wells.*)—" Improvements in apparatus for sifting flour and other
" substances." These are, the employment of the hinged wings
and radial arms, also the combination of the radial arms, hinged
wings, springs, and wire gauze, substantially as follows :— For
sifting "flour, sugar, and other similar substances, or for sepa-
" rating the smaller from the grosser particles in sauces and con-
" diments," and consists "in the employment for that purpose
" of one or more curved wings or blades hinged to one or more
" horizontal revolving radial arms, which are rotated upon or over
" a finely-perforated or wire gauze surface, forming the bottom of
" the receptacle in which the substance or substances to be sifted
" is or are placed. The wings are curved in transverse section,
" and are kept in uniform contact with the wire gauze or perforated
" sifting surface by means of springs carried by the radial arms
" and bearing against the wings or blades. In order to prevent the
" wings or blades from passing too far under the radial arms stops
" are placed or formed upon the under side of the radial arms
" next to the hub or central boss." By hinging the wings or
" blades, and by the use of the springs, "the wings or blades are
" enabled to rise or yield slightly and pass over any hard sub-
" stance that would otherwise injure the wire gauze or other sifting
" surface. A rotatory motion is imparted to the wings or blades
" by means of a vertical spindle, the lower square end of which
" fits into a square socket in the central boss of the horizontal
" radial arms. A cross piece provided with a collar bearing or
" socket serves to maintain or support the vertical spindle in its
" proper central position ; this cross piece is fixed in its place by
" being sprung into the interior of the receptacle, and is provided
" with pins or studs at its opposite extremities, which enter corre-
" sponding holes in the sides of the receptacle. A winch handle
" is fitted on the upper end of the spindle for the purpose of ·
" rotating the same, and thereby keeping the substances well
" stirred or rubbed over the sifting surface."

[Printed, 8d. Drawing.]

NEWTON, WILLIAM EDWARD.—(*A communication from Louis Pierre Robert de Massy.*)—" Improvements in the mode of and
" apparatus for expressing liquids from pulp and semi-fluid sub-
" stances." These are, first, " introducing a small quantity of
" defecating or purifying agents into the matters to be operated
" upon before pressing or washing, and also of effecting the com-
" plete defecation or purification of the liquid before pressing,
" filtering, or washing." No mention is made as to what these
agents are, but it is said that, they " may be introduced either hot
" or cold for the purpose of opening the pores, and thus facili-
" tating the work of separation, so that in the case of the pulp of
" beet-root or in the extraction of oil or other matters," the liquid
is more abundant, well defecated or filtered, much clearer, purer,
and less injuriously affected, and " the separate and distinct
" operation of defecation is avoided, and also the expenses and
" trouble consequent on it." " In all cases where the apparatus
" admits of it a jet of steam can be made to pass through the pulp
" after the pressing takes place," so as to force out any juice that
may remain.

Second, the arrangement of apparatus " by which the filtering
" surface of all kinds of filtering presses is considerably augmented
" by increasing the number of points at which the liquor flows
" out." " Particularly for this purpose the employment as filter-
" ing partitions of an assemblage of bars, rods, tubes, or projec-
" tions of any kind inside the press, and set either in front of or
" behind the metallic cloth or perforated metal which forms the
" filtering surface in whatever way the pressure may be applied."

[Printed, 8*d*. Drawing.]

A.D. 1866, May 2.—N° 1242.

CORMACK, WILLIAM.—(*Provisional protection only.*)—" Im-
" provements in deoderizing, revivifying, and calcining animal
" and vegetable charcoal, and the apparatus employed therein."
These are, a cylinder is fixed in an upright position, a pipe passes
through the lid of this cylinder at the bottom is a moveable
perforated sieve or filter, on which the charcoal rests. Superheated steam, air, or gas is driven through the above pipe on to
the top of the charge, and drives out the impurities downwards
through the " sieve or filter at the bottom of the cylinder to be
there received and conveyed either to condensers, or furnaces to

be consumed. By these means, "I am enabled to revivify and
" purify charcoal at a black heat instead of a red heat, as now
" practised, and also to calcine at a much lower temperature than
" has hitherto been accomplished."

[Printed, 4d. No Drawings.]

A.D. 1866, May 9.—N° 1332.

ROWLAND, ROBERT. — (*Provisional protection only.*) — " An
" improvement in the manufacture of metallic acetates and
" carbonates simultaneously with the production of vinegar and
" glucose." These are, in reference to this subject, manufacturing
" glucose by the use of about two per cent. of sulphuric acid
" boiled in water contained in one vessel, and to which is added
" starch, sawdust, or other vegetable matter; the liquid is drawn
" off into a second vessel, and neutralized by carbonate of lime.
" From the first vessel an acidulated vapour arises containing
" sulphurous acid and oxygen, while from the second vessel car-
" bonic acid will arise from the neutralization of sulphuric acid
" by carbonate of lime. The gases arising as aforesaid are con-
" veyed by air-tight covers and pipes to a chamber in which the
" metal to be acted upon is contained," thus utilizing "the
" vapours which have heretofore been wasted in the manufacture
" of glucose; try cocks are to be introduced in the boiling
" vessels to determine when the operations are completed, and
" the glucose or grape sugar are produced." The "refuse of the
" manufacture of glucose," and other vegetable matters, are
fermented, the carbonic acid used for the carbonization of lead,
the solution filtered through a layer of "marble dust, which
" retains the gummy and slimy portions of the liquid," after
which the liquid is converted into vinegar, by passing through
wood shavings, and the gases from this process again conveyed as
above into chambers containing metals.

[Printed, 4d. No Drawings.]

A.D. 1866, May 12.—N° 1366.

JASPER, GUSTAVUS ADOLPHUS.—"A new or improved process
" of cleansing animal black or bone charcoal, after or before its
" use, for the purpose of filtering a saccharine syrup." This
consists in " boiling the acidulated solution holding the charcoal,

" and thereby so sets in motion the particles of the carbon as to
" thoroughly or very effectively neutralize the alkali and separate
" from them the foreign matters, and cause such matters to be
" floated off or expelled from the mass," and this treatment, it is
said, is also advantageous for "bone charcoal before being used
" in a filter." A wooden or metal tank, sufficient to hold one
filter charge of charcoal has a partition, parallel with and raised a
few inches above its bottom, perforated with numerous holes. A
pipe for discharging water below the perforated partition and a
pipe to discharge water into the upper part of the tank. The
perforated partition is covered with a blanket to prevent the
escape of charcoal which is placed in the tank. Muriatic or
acetic acid is used for neutralizing the lime in the charcoal, "but
" little acid is required in this process, one part of acid to four
" parts of lime being usually sufficient. Care must be taken not
" to use an excessive amount of the acid, as this will injure the
" coal." A pipe into the bottom of the tank supplies steam so
as to boil the liquid, and a pipe at the top and at the bottom
draws the liquid off as may be necessary when washing the
material or otherwise after treatment.

[Printed, 6d. Drawing.]

A.D. 1866, June 14.—N° 1623.

KNAGGS, WALTER.—"Improvements in the manufacture of
" sugar and in the apparatus employed therein." These are,
first, "the injection of sulphurous acid gas into the juice as it
" comes from the mill before tempering;" this is effected by
means of an air pump, and the acid is generated by burning
sulphur.

Second, boiling of the "cane juice prior to the addition of
" lime or tempering." After treating the juice with the acid it
is brought as rapidly as possible to the boiling point, the scum
removed "as soon at it cracks," and the heat continued until
it boils, "removing any scum that may be thrown up thereby."

Third, "the application of a combination of manganese and
" oxygen (combined or uncombined with a base.)" Any excess of
acid in the juice is neutralized by adding "a small quantity of car-
" bonate of lime or calcareous earth, clay, or marl," and "then
" add sufficient manganic acid to throw down a distinct floculus
" precipitate," by manganic acid is meant "any of the combina-

" tions of manganese and oxygen with or without a base,"
preferring "permanganate or manganate of soda," the boiling
is then stopped, the juice is drawn into a subsiding vessel and
milk of lime added until on subsidence of the precipitate the
liquor is bright and transparent.

Fourth, " revivifying animal charcoal " by allowing it to ferment
under water until it becomes acid and offensive to the smell,
assisting this process, if necessary, by carrying on the process in
an artificially heated room, and by the addition "of a small
" quantity of acetic acid, if the fermented liquor is not suffi-
" ciently acid," when the fermentation has ceased the liquor is
drawn off from the charcoal or the charcoal separated from the
liquor, and washed with water acidulated with acetic acid, allowed
to continue the fermenting process uncovered by liquor, and
when the offensive smell is gone, it is washed by a stream of
water.

Fifth, " heating an evaporating tray, covered or uncovered,
" divided into serpentine gutters or channels or compartments
" communicating with one another, by steam."

Sixth, the combination of apparatus "for concentrating or
" granulating sugar, consisting mainly of a revolver, supported
" by means of pulleys and guides acting upon its outer and
" inner circumference, so that the interior of the revolver being
" free from all fittings allows of tubes being passed through it
" by which sufficient surface may be attained to admit of suffi-
" cient heat being communicated to the syrup to cause its
" concentration at a low temperature, waste steam may thus be
" utilized."

[Printed, 2s. 2d. Drawings.]

A.D. 1866, June 18.—N° 1640.

PATRICK, WILLIAM BARKER.—(*Provisional protection only.*)
—" Improvements in the treatment of animal charcoal used by
" sugar refiners or others in order to its re-use." These are, the
spent charcoal whilst still in the filter is washed by passing hot
water through it, and " then dried or partially so, by causing hot
" air to pass through or amongst it, aided by exhaustion." It is
then allowed to remain for fermentation with the taps or valves
of the filtering vessel " open to the atmosphere for, say 24 hours

s.   B B

" or more." When fermentation has taken place, carbonic acid gas is admitted " to act on the charcoal for the purpose of ren- " dering soluble or neutralizing the lime or alkaline matters " which have combined or mixed with the charcoal in its use," and " then pass hot water or water and steam" through it to " cleanse it of matters rendered soluble by the action of the gas." The filtering vessel, if of iron, should be " lined with copper, " tinned, or otherwise coated, to avoid the otherwise injurious " action of the metal." "In some cases the charcoal thus acted " upon may now be allowed to flow from the filtering vessel by " suitable channels over inclined shelves or other supports in fine " streams or films, and subjected to the action of jets or sheets of " gas or other flame; it is then acted upon by fans or winnows " to remove loose particles therefrom." In place of the charcoal " being treated in the filtering vessels, it may be treated in other " vessels, or in vessels with agitators, or in vessels revolving " rapidly and having reticulate outer surfaces, so that the hot " water may be caused to flow through the charcoal by centrifugal " action."

[Printed, 4d. No Drawings.]

A.D. 1866, June 25.—N° 1693.

AUXŸ, GASTON CHARLES ANGE, MARQUIS OF.—" Improved " apparatus for treating corn and other materials." This con- sists of a cylinder which, it is said, is applicable to a number of purposes, among which is named sugar works. The frame of the cylinder " may be made of iron, of wood, or of wood and metal," and staves or boards are fixed upon it, and may be of wood, or of metal, or of metal and wood, of perforated boards, or of wire gauze, or otherwise. The fastening may be combined in different ways. It is important that it should be capable of being fastened by a key. Openings may be made in the heads. The head may be made of plates of metal or of wood, with beams or ribs cor- responding with the beams or ribs of the cylinder properly so called. The axle may go from one end to the other or not. This cylinder is made to oscillate and at the same time to receive shocks, and for this purpose placing at the extremity of the cylin- " der hooks or catches capable of receiving the levers intended to " work it. These levers, if the cylinder is of inconsiderable size " or but little loaded, may suffice to produce the desired oscil-

" lation, but if the mass to be turned or shaken is considerable,
" a rope must be added at one of the ends of each lever, in order
" that by working in a similar manner to bell ringers, the matters
" to be treated may be more easily moved and shaken." "Finally,
" they may receive movements, varied as before stated, but also
" cams, eccentrics, and any similar and well known means em-
" ployed with the same object." "Inside the cylinder there
" should be placed one or more framings or skeletons in proportion
" to the lengthening of the cylinder."

[Printed, 1s. 2d. Drawings.]

A.D. 1866, July 23.—N° 1910.

SOVEREIGN, LÉVI LEMON.—(*Provisional protection only.*)—
" An improved implement for beating and cutting meat and
" other substances." This consists, of a beater made of some
tough metal and "with teeth or projections on its face" which
may be " of any shape, but are preferably, of a square oblong
" form slightly rounded at their extremities. The shank of the
" beater is provided with a hole for the insertion of a handle,"
secured in any way. "The shank is extended beyond the said
" handle, and is formed with a lug or projection for securing the
" cutting or chopping blade, which may be of any convenient
" size and form, according to the substances intended to be cut
" or chopped thereby. The said blade is formed of steel, and is
" secured to the shank of the beater by screws or rivets in such
" manner as to be easily removable from the said shank when
" required ; this allows blades of different size and form to be
" used instead of a blade. When two hammers are thus combined
" the teeth of each are differently formed and act in a different
" manner upon the fibres of the meat, thereby producing the
" required effect thereon. It is said that by the use of the blades,
" which may be varied according to the substance to be operated
" upon, meat, suet, ice, sugar, or any other description of food
" may be cut or chopped as quickly and finely as desired."

[Printed, 4d. No Drawings.]

A.D. 1866, August 11.—N° 2067.

ENSLEY, JOHN ISRAEL.—(*Provisional protection only*).—" An
" improved apparatus for manufacturing illuminating gas and

" producing bone black and other valuable residuum." A retort
has a fire chamber and smoke pipe as usual; within this retort is
a chamber closed except at its outer end. " A set of cylinders is
" constructed, any one of which fits the chamber in the retort,
" and is made open at one end to receive the charge to be
" subjected to the heat." The cylinders are perforated for the
escape of gaseous matters. When one of these cylinders is placed
in the chamber its mouth is closed by means of a screw like a gas
retort, and a pipe conveys the gas to purifying arrangements, and
finally to a gasometer. It is said, "the apparatus is very effective
" for producing bone black, also for charring wood and retaining
" its essential principles."

[Printed, 4d. No Drawings.]

A.D. 1866, September 3.—N° 2255.

VICKESS, SAMUEL.—(*Provisional protection only.*)—" An im-
" proved method of and apparatus for facilitating the moving of
" moulds and the draining of syrup from the sugar solutions and
" sugar in the said moulds in the manufacture of sugar." This
consists as follows :—On the floors of the rooms or places where
the moulds are used, fitting or placing "lines of rails or ways,
" turntables, and carriage carriers," and providing " suitable
" carriages, each constructed to carry a number of sugar moulds
" to run thereon ;" " these carriages have guides or supports to
" maintain the moulds in proper vertical position, the lower ends
" of the moulds project through the bottom of the carriages."
When filled and plugged, the carriage load of moulds is run into
the drying or working-off rooms over gutters or runs consisting
of " grooved pieces of metal or other material placed at a slight
" incline and in rows, so that one gutter or run serves for a line
" of several moulds resting in their carriage just over them."
These gutters discharge into main ones leading to receptacles for
the syrup. " When it is desired to collect the different qualities
" of syrup in separate receptacles a ' swivel end piece ' or a system
" of taps, both of which are well known, could be used. It will
" be obvious that drip pots could be used in combination with
" the moulds, but this arrangement would not be worked so
" economically as the gutters or runs above described."

[Printed, 4d. No Drawings.]

A.D. 1866, September 24.—N° 2453.

KUNTSMANN, Robert. — "Improvements in drying solid
" substances, and in the apparatus or means employed therein."
In carrying out the above, a chamber is described made " of sheet
" iron or other suitable material," in which are a number of
shelves. At the top is a fan driven by a belt "from a steam
" engine or other motive power." The fan chamber is divided
from the drying chamber by a perforated plate; another perforated plate or sheet of wire gauze is placed at the lower end of
the drying chamber. The perforations or apertures are finer in
the lower than in the upper plate, which " allows the desired
" partial vacuum to be maintained " in the chamber. An
arrangement of pipes, tubes, cocks, dampers, &c. admit hot and
cold air, properly mixed, to enter the drying chamber. It is said
that the above arrangement " is more especially adapted for
" drying articles and substances which will stand a high
" temperature without being injured thereby, but for drying
" others which require a more gradual treatment the drying
" chamber is constructed of brickwork and is preferably made
" larger," and the cold air is conducted through an arched
passage which extends along the top of the drying chamber, the
top of which passage is perforated with holes for the admission of
cold air, and whose bottom is also perforated with holes to admit
the same to the drying chamber. The hot air passes through
pipes which extend through the said passage, and these pipes are
perforated to allow the heated air to escape and mingle with the
cold air in the same before it passes into the drying chamber,"
the bottom of which is perforated, and communicates with a
passage in which is a fan or exhausting apparatus. " Super-
" heated or dry steam may be admitted into the drying chamber."
Among a number of substances which are to be dried by the
above means is sugar.

[Printed, 8d. Drawing.]

A.D. 1866, September 25.—N° 2460.

CORMACK, William. — (Provisional protection only.) — " Im-
" provements in means and apparatus for effecting the revivi-
" fication of animal charcoal." These are, to effect the revivification of charcoal in less time and at less cost than by calcining
or burning, "passing superheated steam through amongst a mass

" of inert charcoal the said charcoal being contained in apparatus
" constructed substantially as herein-after described." The charcoal is placed in a cylindrical or other shaped metallic or other
vessel, "provided with doors at the filling and emptying ends,
" and is placed preferably in a vertical position, and is set in
" brickwork, so that heat can when desired be applied externally.
" At the top of the charcoal vessel superheated steam is admitted
" through pipes into a perforated distributor or its equivalent, so
" as to cause the steam to permeate the whole mass, and at the
" bottom there is a pipe or pipes for conveying away the steam,
" gases, colouring, and other matters, as well as a pipe for any
" water of condensation."

[Printed, 4d. No Drawings.]

A.D. 1866, October 6.—Nº 2571.

GORDON, GEORGE.—"An improved system or process of treat-
" ing animal charcoal used in the purification of sugar, and also
" improved apparatus to be employed in such process." This
consists, first, in "the automatic handling or conveying of the
" 'char' in small and constant quantities upon carrier belts from
" the char cisterns, or from near the entrance end of the washing
" cylinder, through the various processes of purification, which
" allows the effect of these processes to be exercised continually
" on the smallest regular quantities at a time sufficient to treat
" the required daily amount to be purified."
   Second, "the washing and boiling of the char in a rotating
" cylinder in such manner as to cause it to pass through and be
" washed, boiled, and thrown about in successive quantities of
" water, commencing with dirty or previously used water, and
" finishing with clean water, while the impurities disengaged
" from the char which float to the surface can flow off con-
" tinuously at one end, the washed char being discharged con-
" tinuously at the opposite end and without emptying away, the
" main body of the water being used."
   Third, drying "the char in open rotating drying cylinders
" having ribs or projecting plates on their inside surface at an
" angle of from 30 to 60 degrees to a line passing across their
" axes sloping upwards, so as to carry the char well round with
" the cylinders and shower it down as they revolve, while currents
" of air, previously heated, or otherwise, pass freely through the

" cylinders from the end where the dry char issues to the end
" where the wet char enters, and thence by a flue into the
" chimney to carry off the vapours formed."

Fourth, "the feed motion in the washing and boiling and
" drying cylinders, consisting of vanes or plates of iron hung on
" spindles between two stationery side plates running through the
" cylinders, which vanes or hung plates are connected with a rod or
" rods or chains either at the top or bottom edge, which enables
" the operator outside either to keep them vertical or to slope
" them to any angle, backwards or forwards, as he may desire to
" feed the char."

Fifth, "the employment of a chamber to contain char, whence
" the air can be exhausted by means of an air pump from its
" pores, and the use of perforated pipes distributed through the
" body of the char in such chamber, by some of which pipes
" purifying substances or hot air can be rapidly disseminated,
" and by others of which the same substances can be rapidly
" drawn off through the air pump."

Sixth, "drying the char either by open rotating cylinders," or
" by drawing or forcing hot air through the char in a chamber"
" previously to re-burning, or so as to dispense with subsequent
" re-burning."

Seventh, re-burning char in a kiln "in which the char is re-
" ceived in hoppers on the top of the kiln, round which the air
" circulates so as to prevent red heat ascending to such portions
" of the char as are exposed to the open air, in which the char
" descends through retorts or pipes in the heated chamber of the
" kiln, of such form and dimensions that no part of the char shall
" be more than three-fourths of an inch from the red-hot exterior
" surfaces of the said pipes; and lastly, in which the char made
" thus red hot shall pass down into chambers under the kiln in the
" midst of, or surrounding which, are water chambers, in which a
" current of cold water enters, which while it cools the char is
" itself heated for use in the washing or boiling cylinders, or for
" other purposes, thus expeditiously cooling the char and at the
" same time utilizing the heat."

Eighth, "the improved system or process of treating animal
" charcoal used in the purification of sugar," with the " con-
" struction and arrangement of the apparatus employed," or
" any mere modification thereof."

[Printed, 2s. Drawings.]

A.D. 1866, October 12.—N° 2645.

BEANES, EDWARD.—" Improvements in refining or decolorizing
" sugar and syrup." These are, " the application of ozone in
" treating saccharine matters " for the above purposes substantially as follows:—Ozone may be pure or otherwise ; it may be
obtained " by passing dry oxygen gas through an ozone tube or
" generator in connexion with an induction coil and galvanic
" battery, or by various other means." For this purpose air, by
preference, previously dried may be used instead of the oxygen.
When acting on dry or moist sugar, the sugar is placed in a deep
vessel having a perforated plate near the bottom, under which the
ozone is forced, and passing upwards through the sugar decolo-
rizes it. " For syrup, molasses, or other saccharine solution, a
" a similar vessel to the above may be used, but in some cases it
" would be preferable, to keep the solution warm, also to par-
" tially fill the vessel above the perforated plate with animal char-
" coal or other granular substance, so that the ozone may be
" brought more effectually into contact with the saccharine
" matters or liquid to be acted upon and decolorized." For
evaporating solutions the temperature does not exceed 212° F.,
and " the ozone is introduced in the lower part of the liquid by
" means of a perforated pipe or pipes, or a perforated double
" bottom." In the melting or blowing up of crude sugar ozone
is employed much in the same way as above, " the temperature
" required being lower than that hitherto usually employed."

[Printed, 4d. No Drawings.]

A.D. 1866, October 26.—N° 2773.

WAGENER, JOHN, and FIRMIN, GEORGE JORDAN.—" Im-
" provements in sugar refining." These are, employing " a sand
" filter in combination with pressure by atmospheric air or steam,
" in order to force the syrup through the filter." In preference,
fine silver sand forms " the upper part or surface and coarser and
" coarser material below." In using the word sand it is intended
" to convey the meaning of a fine granular matter not acting
" chemically on sugar, such as river sand, broken flints, or stone or
" glass, vitrified or like material, which materials are sifted into
" different degrees of fineness, and the coarsest is placed on the
" false perforated bottom of the vessel constituting the filter, and

" finer and finer layers are placed in strata above the coarse layer.
" The syrup of sugar to be filtered is placed on the upper surface
" of the filter, and pressure of the atmosphere is obtained thereon
" by exhausting the space below the bed of the filter, or by forcing
" atmospheric air above the syrup on the filter, or by means of
" steam admitted above the syrup on the filter."

[Printed, 6d. Drawing.]

A.D. 1866, November 30.—N° 3146.

HUGHES, EDWARD THOMAS.—(*A communication from Frederick Jünemann Pierre du Rieu and Edouard Roettger.*)—(*Provisional protection only.*) — " A new or improved mode of saccharatifica-
" tion of sugary substances." This consists, " in producing a
" granular or gelatinous precipitate of saccharate of lime," by
adding say 20 or 30 per cent. in weight of quicklime in powder
through a sieve into a cool solution, " such as evaporated juices,
" cane juices, syrups, or molasses from sugar manufactories or
" refineries." " As soon as a precipitate is formed the coloured
" liquid is decanted, and the residuum washed with boiling
" water." The liquids by decanting and washing are boiled,
and the precipitate of saccharate of lime "is then placed in the
" first boiler with the precipitate of washed saccharate of lime ;
" a small quantity of water is now added, and the whole is satu-
" rated with carbonic acid, steam being admitted in order to
" precipitate the carbonate of lime more speedily. The dense
" syrup thus produced may be at once boiled, or it may be pre-
" viously filtered." " Barytes, strontia, and magnesia may be
" used instead of hydrated quicklime, and instead of carbonic
" acid, it may in some cases, be preferable, to use a solution of
" alum to precipitate the lime in a solution of saccharate of lime
" and to decompose the saccharate."

[Printed, 4d. No Drawings.]

A.D. 1866, December 4.—N° 3187.

KOHN, FERDINAND.—(*A communication from Frederick James Vivian Minchin.*)—"An improved machine for cutting sugar cane."
This machine, " has for its object, the production of small pris-
" matical slices from sugar cane for the purpose of extraction by
" the diffusion process," described in No. 594, A.D 1866, and

consists of a revolving disc carrying a number of knives, which in their rotation pass a fixed knife or cutter, and cut the cane introduced between the fixed and the revolving cutters into slices of a prismatical form. The revolving cutters are shaped so as to form a series of rectangular stops of equal width and depth. Each stop is sharpened so as to present two cutting edges placed at right angles to each other. The whole series of cutting edges will thereby have the appearance shewn in the following sketch |____|____|____|____. A knife of this description will produce a clean sharp cut, both in slicing off thin discs of cane, and also in simultaneously cross-cutting the discs into slices of a prismatic shape. The cane is fed into the machine by an inclined hopper formed with clean flat sides, with its bottom placed at an angle of 65 or 75 degrees to the horizontal level. The cane will by its own weight slide down the inclined plane, or be pushed down and present itself to the action of the knives at a suitable angle."

[Printed, 8d. Drawing.]

A.D. 1866, December 29.—N° 3422.

SLATTER, Joseph.—(*Provisional protection only.*)—" An im-
" proved screen or sifter for screening cinders, corn, tea, sugar,
" gravel, malt, and other granular substances." This consists,
of " a long cylindrical or rectangular or polygonal casing of metal
" or other suitable material," and fixing " it in an inclined posi-
" tion a suitable distance from the floor," for placing receptacles
beneath. " The upper end of the casing is closed, and a diaphragm
" or screw of perforated metal or wire gauze or grating, arranged
" so as to have a space between it, and the lower surface of the
" casing, extends from the closed upper end of the casing to its
" lower end, which is also partially closed, but which has an
" aperture on a level with the upper surface of the screen, so that
" the granular substances passing down upon the latter will pass
" out through such aperture." On the top of the upper end of
the casing is a hopper for introducing the substances, " the dust
" or smaller particles of such substances being caused to pass
" through the holes or meshes of the screen into the lower space
" in the casing as such articles travel downwards by gravity. An
" aperture is formed in the under surface of the lower end of the
" casing, and is provided with a spout through which the dust

" or smaller particles passing through the screen fall into a recep-
" tacle, while the larger particles remaining on the top of the
" screen pass through the end aperture, as before described, into
" another receptacle." The apparatus, in preference, is fixed
" simply by forming a projecting flange with one or more holes
" on the upper end surface, whereby it is hung upon one or more
" nails or hooks in the wall." The upper end surface of the
casing is made " at such an angle to the casing as to cause the
" latter to assume the requisite inclined position when hung
" up."

[Printed, 4d. No Drawings.]

APPENDIX.

A.D. 1788, November 6.—N° 1673.

RUMSEY, JAMES. — "Certain new methods of constructing
" boilers for distillation and other objects, and for steam engines
" for various purposes." These are, in reference to this subject,
in connecting "the receivers of my stills or boilers with common
" stills, and causing the vapor rising from them to act upon
" vessels moveable within the receivers, with such power as to
" give or assist in giving motion to sugar mills, or to raise water,
" grind grain, or any other necessary purposes, after which the
" vapor passes into the still worm, and is condensed as usual with-
" out receiving damage or waste. As the vessels moving within
" the receivers of my stills or boilers perform the office of pistons,
" I shall therefore hereafter call them hollow pistons."
  [Printed, 1s. Drawing. Rolls Chapel Reports, 6th Report, p. 179.]

A.D. 1798, February 1.—N° 2212.

SHANNON, RICHARD.—"A new method of improving the
" process of brewing, distilling, boiling, evaporating, raising,
" applying, and condensing steam or vapour from aqueous,
" spirituous, saccharine, and saline fluids, which expedites the
" process, improves the quality, and causes a great saving of time
" and fuel in each, with suitable utensils, on improved principles,
" correspondent to these intentions, part of which improvements
" are applicable to the utensils now in use." These are, in
reference to this subject, sugar pans and vessels for evaporating
saccharine solutions, &c., are "long cylinders, double at the
" ends, and proper arms, doors, and tubes with the usual
" means of charging and discharging them at present used, and
" of working them off, cleansing," &c. They are to be "one,
" two, or three or more times longer than broad," and in place
" of one immense fire (as now practised) two or more are to be
" placed, in all of which together not a quarter part of the fuel
" now used is applied. These cylinders laying lengthways over

" the fires employed to work them the flame and smoke will be
" made to pass round them in a spiral direction " in their passage
to the chimney. Over each of these coppers can be placed
another which serves "as a covering in of the flues of the under
" one, receives a full heat from them at its bottom, after which
" the remaining heat, instead of passing into the chimney, makes
" a circuit round the sides of the upper copper, still, &c., and at
" length arrives at the chimney." " By the addition of one or
" two partitions these long coppers may be divided into two or
" three short brewing coppers, and the flame of the fires placed
" near each end of the copper will cause the middle one to boil
" as soon as either of those under which the fires are placed.
" By means of proper registers the boil in any of them is regu-
" lated. By this contrivance these coppers may be alternately
." used as occasion requires." The vessels may be constructed
single or double at the ends.

[Printed, 4d. No Drawings. Repertory of Arts, vol. 10, p. 17.]

A.D. 1801, May 2.—Nº 2495.

TICKELL, HENRY.—"An apparatus or refrigerator for more
" speedily and effectually cooling the worts and other fermented,
" fermentable, or other liquids, or melted or dissolved animal or
" vegetable substances, manufactured, made, or used by or in the
" processes of brewers, distillers, vinegar makers, soap makers,
" sugar refiners, chemists, or other manufacturers of articles of a
" similar nature or using similar processes." This consists of
a set of pumps which receive the liquor to be treated, and force
it into a main pipe in a chamber, from whence it passes through
a series of small tubes and falls down into a vessel at the bottom,
from which it runs by a zig-zag gutter which is surrounded with
cold water into a reservoir below, and, if necessary, it is again
passed by a pipe into the pump, and again treated as before. Or
the heated fluid is forced into a vessel at the top of the chamber,
and falls down through perforations in the vessel, or the perforations stopped up, it is conveyed by a pipe into the first main pipe,
and is forced through the above-mentioned series of small tubes.

[Printed, 6d. Drawing.]

A.D. 1809, May 15.—Nº 3236.

JOHNSON, WILLIAM. —"A new or improved process for
" heating fluids for the purposes of art and manufactures."
This consists, in reference to this subject, in apparatus for the

" boiling and evaporation of sugar " constructed as follows :—
A boiler is made of sufficient strength to bear a high pressure of
steam, on which there is a safety valve; from the top of this boiler
is a main pipe, having pipes branching from it, all of which have
cocks or valves "to be regulated by a weight, to govern the
" direction and quantity and temperature of the steam to be
" discharged." These branch pipes are connected to small
chambers in the bottom of other and larger vessels for the purpose
of conveying the high pressure steam into the same and so cause
the contents of the larger vessels to boil and be evaporated.

[Printed, 6d. Drawing. Rolls Chapel Report, 7th Report, p. 202.]

A.D. 1815, February 13.—N° 3884.

MOULT, WILLIAM.— "An improved method of sublimation
" and evaporation." This consists of a vessel with an outer
vessel or jacket, the space between is filled "with some fluid such
" as oil, sulphuric acid, mercury, or other similar fluid, which
" will boil or rise in vapour at that temperature which is the
" greatest that can be admitted for the process." "The interval
" between the two vessels is closed or shut up on all sides, and
" where it boils the steam or vapour cannot escape except by a pipe,
" which is provided to lead the vapour into a cooling apparatus
" or refrigeratory " so as to condense the vapour. The cooling
apparatus may be a worm pipe, a straight pipe in cold water, or a
flat metallic vessel in cold water, or two cylinders placed one
within the other, leaving a " space all round between the two,
" but this space is closed on all sides whilst the internal cylinder
" is open at each end; the vapour being admitted into the space,
" when the whole is immersed in cold water." " By the addition
" of this condensing apparatus the fluid of the bath " is condensed
and may be used again when required. For the purpose of
distilling spirits, &c. mercury is employed in the bath through
which the heat is transmitted, and may be also used for the
evaporation of sugar, &c.

[Printed, 4d. No Drawings. Repertory of Arts, vol. 28 (second series),
p. 134; Rolls Chapel Reports, 8th Report, p. 106.]

A.D. 1815, December 5.—N° 3965.

DIHL, CHRISTOPH.—(A communication.)--"Certain improve-
" ments in the method or apparatus for distilling." These are,
in reference to this subject, machines may be constructed to be

heated by steam, so as "to concentrate the sap of substances of
" any kind, particularly that of sugar cane, and to enable you
" to bake sugar. To procure this latter effect you must convey
" the steam that rises from the boiler into hollow platforms
" collected and connected together by means of tubes. The plat-
" forms being placed one above another, you cause to run over
" the surface of those platforms the juices whatever that you wish
" to concentrate." This juice drops from platform to platform.
Each of these platforms has a tube with a cock which communicates with a larger tube which conveys into the condenser "the
" produce of the distillation that has condensated into the interior
" of the platforms."

[Printed, 1s. Drawing. Rolls Chapel Reports, 8th Report, p. 111.]

A.D. 1816, May 25.—N° 4032.

TAYLOR, PHILIP.—" A new method of applying heat to liquors
" used in the processes of brewing, distilling, and sugar refining."
This consists in obtaining " the requisite ebullition and evapo-
" ration for" syrups and other liquors " employed in the above
" processes," which " liquids require a heat above that of boiling
" water to maintain them in such a state of perfect ebullition
" and evaporation as is necessary for the proper management of
" the different processes," employing reservoirs, the surface of
which are large compared with their capacity, and adapted " to the
" utensils or vessels employed in the said several processes in
" such manner that they may during the operation be immersed
" in the liquors to be heated." These reservoirs are charged and
supplied with high-pressure steam at pleasure "by means of
" proper apertures furnished with cocks or valves; other cocks
or valves letting off the air or superfluous steam together with
the water produced from the condensed steam. " In constructing
" my reservoirs containing the high-pressure steam, I prefer a
" combination of pipes or cylinders of small diameters and of
" considerable length (though other forms may be adopted), by
" which I obtain sufficient strength to resist the expansive force
" of the steam, thereby avoiding all danger from explosion, at
" the same time that the extension of the surface may be made
" to any required degree to enable the steam to communicate its
" heat rapidly to the surrounding fluid."

[Printed, 4d. No Drawings. Repertory of Arts, vol. 30 (second series),
pp. 193 and 257; Rolls Chapel Reports, 8th Report, p. 114.]

A.D. 1818, January 15.—N° 4197.

TAYLOR, PHILIP.—" A new method of applying heat in certain
" processes to which the same method hath not hitherto been
" applied, likewise for improvements in refrigerators." These
consists, in reference to this subject, as follows :—" Obtaining
" the requisite ebullition and evaporation " for syrups and other
liquors employed, which liquids " require a heat above that of
" boiling water to maintain them in such a state of perfect
" ebullition and evaporation as is necessary for the proper manage-
" ment of the different processes ;" employing reservoirs, the
surfaces of which are large compared to their capacity, and adapted
to the utensils or vessels employed in the said several processes
in such manner that they may during the operation be immersed
partially or completely in the liquors to be heated. These reservoirs are charged and supplied with high-pressure steam at
pleasure by means of proper apertures furnished with cocks and
valves ; other cocks and valves letting off the air or superfluous
steam together with the water produced from the condensed
steam. In constructing these reservoirs containing the high-
pressure steam, " I prefer a combination of pipes or cylinders
" of small diameters and of considerable length (though other
" forms may be adopted), by which I obtain sufficient strength
" to resist the expansive force of the steam, thereby avoiding
" all danger from explosion, at the same time that the extension
" of the surface may be made to any required degree to enable
" the steam to communicate its heat rapidly to the surrounding
" fluid," but these reservoirs may be variously constructed.
The mode " of applying heat in the process of making sugar and
" mollasses in the colonies is to adopt my steam reservoir above
" described to the clarifiers, rackers, boilers, teaches, or such
" vessels as are used in the various operations necessary to
" convert the cane juice into sugar." The remainder of the
Specification refers to the application of the above reservoirs to
other purposes than sugar making.

[Printed, 6d. No Drawings. Repertory of Arts, vol. 36 (second series),
p. 321 ; Rolls Chapel Reports, 8th Report, p. 124.]

A.D. 1822, July 27.—N° 4694.

PERRIER, SIR ANTHONY.—" Certain improvements in the
" apparatus for distilling, boiling, and concentrating by evapo-
" ration various sorts of liquids and fluids." These are, in

reference to this subject, in a boiler or evaporator for the evaporation of syrup, "the upper surface of its bottom divided by a ledge
" or ledges placed vertically, and of such height that the liquid
" shall never in case of ebullition pass their upper edge, the
" lower edge being attached by solder or otherwise to the
" bottom " so as to form a channel, which, whatever direction
" it is made to take, conducts the liquid so as to cause it cover
" in its course the whole surface of the bottom from the place of
" its falling in " at a given point from a tube from a reservoir
" until it arrives at the other end of the channel," when it is
discharged by a regulating tube arranged so as to regulate the
quantity allowed to flow. The fire may surround the bottom
and sides. The length of the course may be "augmented by
" placing upright divisions at suitable distances between the
" outside and ledge and the sides of the still attached to both,
" and an open left under the first and over the second throughout,
" so that the liquid should pass under the one and over the other
" in succession the whole round. This principle might occa-
" sionally be established generally over the whole channel." In
all cases the capacity of these evaporators for work may be
increased by placing layers of pipes "either in the furnace under
" the still, in the flue between the still and the chimney, or in
" the chimney itself, or in any other position more convenient
" where they shall experience the effect of the fire, so that by
" letting the liquid to be operated upon pass previously through
" these pipes, it arrives at the still in any state of forwardness
" sought."

[Printed, 10d. Drawings. London Journal (Newton's), vol. 6, p. 65;
Register of Arts and Sciences, vol. 1, p. 10; Engineers and Mechanics'
Encyclopædia, vol. 1, p. 437.]

A.D. 1822, August 17.—No 4696.

CLELAND, WILLIAM.—"An invention of an improved appa-
" ratus for the purpose of evaporating liquids." This consists,
in preference, of a square box or case made of metal or otherwise
" filled as completely as possible with steam pipes laying parallel
" to each other and equidistant." "A vessel like the rose of a
" garden pot or other suitable figure thickly perforated with small
" holes " is placed about a foot above the upper row of steam
pipes. This vessel is filled with the liquid to be evaporated by a
pump and distributes it "in showers or streamlets " upon the

pipes "which are by this means kept uniformly covered and wet
" with a succession of fresh liquid from the receiver." The
liquid falls from row to row of the pipes, and being arrived at the
bottom is discharged into the receiver outside, from whence it may
again be lifted by the pump. "A strong ascending current of
" heated air continually rushes into the evaporator near the
" bottom of the case, and passing with great rapidity over the
" wet surface of the steam pipes carries the aqueous vapour along
" with it." This apparatus is recommended "for the evaporation
" of syrups and other liquids to which a high temperature is
" prejudicial." In some cases rods or bars are substituted for
the pipes described above.

[Printed, 6d. Drawing. London Journal (*Newton's*), vol. 5, p. 91;
Repertory of Arts, vol. 4 (*third series*), p. 311.]

A.D. 1824, August 5.—N° 4997.

JOHNSON, WILLIAM.—"An apparatus for evaporating fluids,
" for the purpose of conveying heat into buildings, for manu-
" facturing, horticultural, and domestic uses, for heating liquids
" in distilling, brewing, and dyeing, and in making sugar and
" salt." This consists, first, making use of a "multiplied boiler"
with three evaporating vessels, one above each other, the lower
one being placed on the fire. Near the top of each vessel must be
pipes to carry off the steam, and to the lowest boiler should be
affixed a safety valve and steam gauge, and to all the vessels cocks.
The steam from this multiplied boiler may be conveyed into a
receptacle or receptacles in another vessel or vessels, as in a salt or
sugar pan, &c.

Second, "placing several close boilers separately over each
" other within one larger boiler, to be heated by the steam sur-
" rounding them generated in any liquid from such boiler, or
" they may be immersed in such liquid;" each vessel has its
charge of water for supplying steam. On the top of the inclosing
boiler a pan can be added for making sugar or salt, but this is not
claimed.

Third, "vessels formed the one above the other," "having
" openings for the steam to pass from water in a lower boiler to
" operate against distinct vessels formed above it, containing
" separate bodies of water and surfaces to evaporate from, by
" which means the heat of the steam or water contained in such
" boiler will act against every vessel above it with equal tempera-

" ture, and the steam from all may be discharged by one outer
" pipe."

Fourth, "employing the greater part of the steam generated in
" the vessels of a multiplied boiler to act under a side pan, whilst
" the remaining steam within them is employed evaporating fluids
" contained in an open vessel or pan placed upon the upper
" boiler," in making salt or sugar, &c.

[Printed, 6d. Drawing. Repertory of Arts, vol. 1 (*third series*), p. 443;
London Journal (*Newton's*), vol. 10, p. 295.]

A.D. 1826, February 11.—N° 5327.

GAMBLE, Josias Christopher.—" Certain apparatus for the
" concentration and crystallization of aluminous and other saline
" and crystalizable solutions, part of which apparatus may be
" applied to the general purposes of evaporation, distillation,
" inspissation, and desiccation, and especially to the generation
" of steam." These are, in reference to this subject, as follows :—
Two metallic vessels are fixed tightly the one within the other
with a space of two or three inches being at the bottom and sides.
" The outer vessel has a stroop at the bottom for drawing off the
" contents of the intermediate space. This boiler is set over a
" furnace in the usual way." A vessel or reservoir for holding
the fluid used to fill the space between the above vessels, or jacket,
is placed a little way from the double vessel, the lower part of such
vessel being on a level with the upper part of the double vessel.
One leg of a curved pipe is attached to the upper part of the above
jacket, the other being attached to the reservoir near the bottom,
thus maintaining a constant communication between the two ; an
air pipe proceeds from the highest part of the curved pipe, with a
stop cock a little way from the curve. In using this apparatus
" any liquid the boiling point of which is thirty-five degrees or
" more higher than the liquid to be boiled," is poured into the
reservoir ; it passes through the curved tube ; fills the jacket, and
expels the air therefrom by the air pipe, and the liquid to be
evaporated is placed in the pan. According to the nature of the
liquid to be evaporated so is the nature of the liquid in the jacket.
For " the boiling of sugar or for high-pressure steam not exceeding
" thirty pounds to the square inch I would recommend essential
" oils, especially oil of tar, on account of its great cheapness."

[Printed, 10d. Drawing. Repertory of Arts, vol. 3 (*third series*), p. 5;
London Journal (*Newton's*), vol. 14; p. 130.]

A.D. 1826, July 24.—N° 5394.

CLELAND, WILLIAM. — " Improvements in evaporation."
These are, " two separate parts of processes, the first of which may
" be used separately or conjointly with the second or not, as
" circumstances will permit or require." The first consists in
causing a rapid current of hot or cold air to pass over the surface
of a fluid or solution and so become charged or loaded with the
steam or vapour from the same. This consists of an evaporating
pan heated below as usual with a fire; the upper part of the pan
is closed in with wood or metal in the nature of a chest. In the
front is a passage to admit air and the same at the back for the
escape of the air into a flue or otherwise after it has performed its
office. On the top of the casing is a reservoir with a cullender
bottom, and a pump draws the hot liquor from the pan below
and delivers it into this reservoir, from which it runs through
the cullender into the pan below in showers, passing "through
" the perpetually changing sheet or current of air produced as
" aforesaid."

Second, making use " of such current of air after it has been
" so heated, and has combined with the hot vapour as aforesaid,
" for heating or raising the temperature of a second or other
" quantity of fluid for the purpose of evaporating it, or for any
" other purpose, by which means the caloric so obtained by the
" air is not wasted or dissipated, but is communicated to such
" second portion of the fluid, and is thus made to perform two
" or more successive, useful, and economical operations."

[Printed, 4d. No Drawings. Repertory of Arts, vol. 4 (third series), p. 311;
   London Journal (Newton's), vol. 1 (second series), p. 162; Engineers and
   Mechanics' Encyclopædia, vol. 1, p. 484.]

A.D. 1829, August 21.—N° 5837.

SHAND, WILLIAM.—"A certain improvement or improvements
" in distillation and evaporation." This consists, in reference to
this subject, of "a copper or sugar pan for evaporating syrups,
" sugar juice, and other substances." This is placed within a
vessel which is surrounded by a spiral channel. This vessel is
set in brickwork over a fire-place; into the vessel with the spiral
channel is put a quantity of oil, which when boiled comes into
close contact with the bottom of the pan. The heated oil
circulates several times around the still by means of the spiral

channel. The oil is partially condensed in its progress round the spiral, and is more effectually condensed by means of a water channel which lies upon the uppermost turn of the spiral channel. Water is caused to flow through this water channel. Any oil not condensed by these means passes from the spiral channel into a pipe, which pipe again passes into a condenser. In carrying out the above, "various kinds of essential oil may be " used as media for the communication of heat, but I prefer " spirit of turpentine for the purpose of evaporating syrups or " sugar juice and such like substances."

[Printed, 10d. Drawing. Repertory of Arts, vol. 11 (*third series*), p. 96; London Journal (*Newton's*), vol. 9 (*second series*), p. 183; Register of Arts and Sciences, vol. 4 (*new series*), p. 231.]

A.D. 1830, October 20.—N° 6012.

SHARP, JOSEPH BUDWORTH, and FAWCETT, WILLIAM.— " An improved mode of introducing air into fluids for the " purpose of evaporation." This consists of a copper or pan for boiling sugar, containing the liquid to be evaporated. Two iron pillars, fixed on each side of the masonry or brickwork enclosing the above pan, support two cross frames of iron placed one above the other. Through these cross frames is passed a vertical hollow shaft or cylinder, at the end of which is a hollow brass bulb, and in which are two or more small radiating pipes fixed nearly horizontally, but the extremities rather depressed and reaching within a short distance of the bottom of the pan. A loose pinion wheel rests upon the lower cross frame, and passing round it, but not attached to the hollow shaft. A horizontal shaft, in preference, of wood, is connected by a crank movement to a toothed quadrant which works into the above loose pinion. To this horizontal shaft a reciprocating motion is given by a crank or otherwise. The reciprocating motion given to it is not necessarily communicated to the vertical hollow shaft, but when desired this is accomplished by means of a coupling box. This arrangement of hollow shaft and pipes is suspended in the liquid over pulleys, and balanced by a weight, so as to be raised and lowered as desired. " By giving motion to the air pipes under the surface " of the fluid to be evaporated in such manner as to cause a " vacuum in the fluid to fill which, the air rushes through the " pipes into the flue, and it evaporates it," the fluid.

[Printed, 6d. Drawing. Repertory of Arts, vol. 11 (*third series*), p. 281; London Journal (*Newton's*), vol. 8 (*conjoined series*), p. 360; Register of Arts and Sciences, vol. 6 (*new series*), p. 79.]

A.D. 1830, October 20.—N° 6014.

URE, Andrew.—" An apparatus for regulating the temperature
" in vaporization, distillation, and other processes." This
apparatus is said to operate " on the well-established physical
" principle that when two rulers or flat bars of two different
" solids or metallic alloys, which differ in the rates of their
" expansibilities by heat, are firmly united together, with their
" flat surfaces in contact, and fastened by rivets, solder, or other
" means, a change of temperature in such a compound bar will
" occasion it to bend laterally, with a sensible movement in
" flexure, provided that neither of the two rulers be so thick and
" rigid as to resist and counteract the physical force impressed
" on it by the other." This " apparatus may be composed of
" several such compound bars, which are combined, in order to
" augment and accumulate their motion in flexure, so as to
" operate on moveable stopcocks, air ventilators, or registers,
" or other means of opening or closing apertures, in order to
" regulate the temperature of the media in which the said
" combinations of compound bars are immersed." It is said
that " the applications of the above-described thermostatic
" apparatus for regulating temperature in vaporization, distilla-
" tion, and other processes are very numerous;" among these is
a double pan for boiling by a saline solution at a uniform
temperature;" this pan, it is said, is excellently adapted for
concentrating, among other solutions named, " cane juice,
" syrups," &c. liable to be injured by boiling them over a naked
fire. The space between the two pans communicates by pipes
with two cisterns, a small one with a small pipe with a
thermostatic cock or valve, to admit the water to flow into the
saline solution when the bath concentrates and the heat requires
to be lowered. The other communicates with a bath to receive
the overflow of the solution whenever it happens to boil furiously
from too great a dilution with water or is suddenly subjected to
too strong a heat from the fire.

[Printed, 1s. Drawing. London Journal (*Newton's*), vol. 8, p. 307; Register of Arts and Sciences, vol. 6 (*new series*), p. 60; Mechanics' Magazine,
vol. 15, p. 246.]

A.D. 1830, October 20.—N° 6016.

URE, Andrew.—" An air stove apparatus for the exhalation
" and condensation of vapours." A chamber is constructed of
" a rectangular or other form, generally oblong, of which the

" height is much less than the breadth," of " stone, brick, wood,
" metal, or other suitable materials, and its sole may be heated
" from beneath by a fire flue, by steam, hot water, or other hot
" liquids. Into one end of the chamber a sliding door is fitted,
" having its bottom close to the sole, and into the other end,
" immediately beneath the roof or cover, an opening is made for
" the discharge of the vapours, to which opening is attached an
" exhausting and forcing pump, piston, bellows, or fanner, for
" first drawing off the air and vapours, and then propelling them
" into a cistern of condensation." " In working with this
" apparatus the door plate must never be slid quite down to the
" bottom, but a stratum of air, more or less thick," must " be
" admitted to favour the exhalation." This apparatus " is
" peculiarly adapted for drying raw sugar without injury to its
" grain."

[Printed, 6d. Drawing. Repertory of Arts, vol. 11 (*third series*), p. 278;
London Journal (*Newton's*), vol. 1 (*conjoined series*), p. 418 ; Register of
Arts and Sciences, vol. 6 (*new series*), p. 74; Engineers and Mechanics'
Encyclopædia, vol. 2, p. 790.]

A.D. 1830, November 29.—N° 6041.

CHURCH, WILLIAM.—" Certain improvements in apparatus
" applicable to propelling boats and driving machinery by the
" agency of steam, parts of which improvements are applicable to
" the purposes of evaporation." These are, " the employment
" of apparatus whatever may be its form for raising a current of
" air, gas, or vapour to abstract heat from vapour generated in
" boilers, stills, or other vessels used in raising substances by
" means of heat to a gaseous vaporous or elastic state, and making
" use of such air, gas, or vapour (when and after it shall have
" thus abstracted heat) for the purposes of evaporation." No
claim is made to any particular " form of apparatus, as the form
" may be varied in the construction of the steam engine or the
" boilers or vessels of the brewer, chymist, distiller, rectifier, salt
" or soap manufacturers or refiners of sugar, and so forth, as
" may be found most convenient in their different processes of
" vaporisation." No example is given of the application of the
above to a pan for evaporation, but its application to a still is as
follows :—The still being heated by steam into a jacket or other.
means, the vapour generated from the wash passes into a pipe
and from thence into a spiral passage in a vessel, from whence it
passes into a worm tub. At the same time that the hot vapour

is passing along the above spiral passage "atmospheric or other " air, gas, or vapour which supports combustion," is driven into a corresponding spiral passage which travels in an opposite direction, and from thence is conducted by a tube "in a heated state " to supply fuel with air or gas for its combustion."

[Printed, 3s. 4d. Drawings. London Journal (*Newton's*), vol. 8 (*second series*), page 1; Register of Arts and Sciences, vol. 6 (*new series*), p. 106; Rolls Chapel Reports, 7th Report, page 134.]

A.D. 1831, January 15.—N° 6061.

PARKER, WILLIAM. — " Certain improvements in preparing " animal charcoal." These are, in " revivifying animal charcoal " after it has been used in the process of refining sugar; it is first washed, dried by air or otherwise, sifted, charged into iron crucibles, which are "closely packed and rammed, and until the " crucible be completely filled ;" these crucibles are then placed one above the other, covering the upper crucible of each set or column with a close fitting cover " luted with loom or clay," having a small hole to allow the escape of the gas evolved during theprocess. " Fish bones or any animal matter, or fatty or oily " matter, or any resinous or bituminous substances," are used " in admixture with the deteriorated material aforesaid," but preferring, "the former on account of greater economy." The operation of baking or burning is continued until the flame from the top of the crucibles is nearly extinct, when the firing is stopped, as also any supply of air by closing the doors of the ashpits. When the crucibles are of a dull red color the operation of revivifying is complete, the furnace is thrown open to cool the crucibles, when their contents are " ground, sifted, and prepared " for use in the usual or most convenient way suitable for its " intended use."

[Printed, 10d. Drawing. Repertory of Arts, vol. 11 (*third series*), p. 342; London Journal (*Newton's*), vol. 7 (*conjoined series*), p. 357; Mechanics' Magazine, vol. 15, p. 349; Register of Arts and Sciences, vol. 6 (*new series*), p. 101.]

A.D. 1831, July 30.—N° 6146.

PERKINS, ANGIER MARCH.—"Certain improvements in the " apparatus or method of heating the air in buildings, heating " and evaporating fluids, and heating metal." These are, " circulating water in tubes or pipes which are closed in all parts, " allowing a sufficient space for the expansion of the water which

" is contained within the apparatus, by which means the water
" will at all times be kept in contact with the metal, however
" high the degree of heat, such apparatus may be submitted
" to, and yet at the same time there will be no danger of
" bursting the apparatus in consequence of the water having
" sufficient space to expand." In reference to this subject,
namely, in " the boiling of syrup in the making or refining of
" sugar," "it will be seen that the heated water is made to cir-
" culate through a series of tubes, and give off its heat to the
" fluid contained in the boiler, or these tubes may be made to
" pass into steam or other boilers, in a similar manner, and will
" cause the fluid contained in such boilers to become heated and
" evaporated."

[Printed, 1s. Drawings. Repertory of Arts, vol. 13 (*third series*), p. 12); London Journal (*Newton's*), vol. 2 (*conjoined series*), p. 14; Mechanics' Magazine, vol. 22, pp. 58, 101, and 197; also vol. 23, p. 461; Register of Arts and Sciences, vol. 7 (*new series*), pp. 36 and 70; Webster's Reports, vol. 2, pp. 6, 15, and 17. Extended for 5 years (see No. 10,778.]

A.D. 1831, August 27.—N° 6154.

PERKINS, JACOB. — " Certain improvements in my former
" patent, dated the Second day of July, one thousand eight
" hundred and thirty-one, making the same applicable to the
" evaporating and boiling fluids for certain purposes." These
are, "in the application of certain apparatus or machines," on
the following principle, " for the purpose of boiling worts, solu-
" tions of sugar and salt or other substances." In a copper or
boiler heated by a furnace or flues is an apparatus of a similar
shape to the bottom of the boiler used, having a space all round
between the apparatus and boiler for the flow of the liquid; but
it is open at the top, and there is also a space at the bottom, and
it is supported by two small rods or legs. By this arrangement
it is said that when heat is applied, the liquor will travel up
between the pan and the apparatus to the surface, while at the same
time there will be a constant flow of the liquid down the centre
of the apparatus to become heated.

[Printed, 8d. Drawings. Repertory of Arts, vol. 13 (*third series*), p. 198; London Journal (*Newton's*), vol. 11 (*conjoined series*), p. 41; Mechanics' Magazine, vol. 17, pp. 91, 103, 158, and 316; also vol. 24, pp. 387 and 459; Engineers and Mechanics' Encyclopædia, vol. 1, pp. 213 and 482.]

A.D. 1831, September 22.—N° 6165.

URE, ANDREW. — "An improved apparatus for evaporating
" syrups and saccharine juices." This is said to consist, first, in

" the combinations of apparatus," afterwards described for the above purposes, " by the heat of a saline bath, composed of a " strong solution of chloride of calcium or acetate of soda, which " boils at a temperature considerably above the boiling point of " water."

Second, " the application of one or other of the said saline " solution baths to sugar pans with corrugated bottoms or " corrugated bottoms and sides."

A corrugated pan for evaporating the syrup has an outer vessel or jacket made of iron or other suitable metal for containing the saline solution, a pipe is led from the top of the jacket and passes in a zig-zag form through a cistern above, in which there is water. This pipe is terminated above the cistern by a safety valve. The water from the condensed steam flows into the bottom of the jacket by a small tube in which are a series of apertures. There is another tube which issues immediately from the bottom of the cistern named above, which is intended to supply water to the saline bath through a stop-cock or valve into a perforated distribution tube near the bottom of the pan. The action of that stop cock or valve of supply is regulated by what the Patentee calls a " thermostat," see No. 6014, Old Law. The manner in which the thermostat is applied is as follows: —There are two pairs of compound thermostatic bars acting in concert, the undermost bar of which is fastened firmly by the middle to a bracket, bolted to the side of the iron pan, and the uppermost bar is connected at the middle to a rod passing upwards through a stuffing box in the top " moves up and down by the motion in " flexure of the said thermostatic bars, correspondent with the " increase and diminution of the temperature of the bath," and by this means moves the stop cock or valve so as to admit water into the saline bath when required. A mercurial thermometer is in " the saline bath ; it serves as a cheque on the thermostat," &c. There is a rake covered with a glove to assist the final discharge of the syrup along the corrugations of the pan.

[Printed, 10d. Drawing. Repertory of Arts, vol. 14 (*third series*), p. 149; London Journal (*Newton's*), vol. 1 (*conjoined series*), p. 1; Register of Arts and Sciences, vol. 7 (*new series*), p. 105; Rolls Chapel Reports, 7th Report, p, 136.]

A.D. 1833, June 20.—N° 6439.

URE, ANDREW. — " An improved apparatus for evaporating " syrups and saccharine juices, which is also applicable to other

" purposes." This is said to consist, first, " the application of a " strong solution of caustic alkali as a bath to an evaporating " pan or still with a corrugated bottom." The above solution is put between the two pans; the inner pan has its bottom or bottom and sides "amplified by corrugation with angular or curvilinear " ridges and furrows."

Second, the bath space of the double pan should be provided with a safety tube for discharging any redundant steam; this tube should terminate in a water trap cistern charged with milk of lime.

Third, " the application of the said alkaline solution, either as a " bath to the external surface of a plane pan, as conducted from a " boiler, through pipes and conduits in order to supply a regu- " lated heat to various purposes."

Fourth, " the application of sulphuric and phosphoric acids or " mixtures of them as baths or mediums for transmitting heat " at definite temperatures."

Both the alkaline and acid baths are furnished with safety tubes, thermometers, and water supplying pipes for dilution.

[Printed, 4d. No Drawings. London Journal (*Newton's*), vol. 3 (*conjoined series*), p. 285; Mechanics' Magazine, vol. 24, p. 465; Rolls Chapel Report, 7th Report, p. 145.]

A.D. 1834, April 8.—N° 6590.

CROSLEY, HENRY. — "An improved method or process, " arrangement, and combination of apparatus with certain agents " used or employed therewith, whereby evaporation of fluids and " solutions may be effected advantageously, and also for other " beneficial purposes to which the said method or process is " applicable or can be applied." The apparatus is at follows :— Open pans heated by steam or otherwise are charged with "the " fluid or liquid or solution to be evaporized;" "the air pumps, " blowing cylinders, or other proper apparatus for forcing or " propelling air," heated or otherwise, are put in motion by any power, natural or artificial, and air forced into a condensing vessel or reservoir, whence it passes through one or more branches into pipes connected with a metallic hollow wheel or wheels immersed in the fluid, &c. in the open pans. The air finds egress through " perforations, holes, or apertures in the central hollow box or " nave and in the hollow spokes of the wheel;" the air descends, then ascends through the fluid, &c. and carries off the aqueous parts thereof and accelerates evaporation, which is further increased by the motion of the wheel. Scrapers may be affixed to one or more spokes of the wheel. Two wheels may be worked in an oblong pan. For "melting or dissolving sugar for clearance" steam is forced into the mass and knives or scrapers may be attached to the spokes of the wheel.

[Printed, 1s. Drawing. London Journal (*Newton's*), vol. 19 (*conjoined series*), p. 30.]

A.D. 1835, August 17.—N° 6883.

BOWMAN, FREDERICK.—"An improvement in the process of " renewing the virtues of animal charcoal when exhausted or " impaired." This is, in place of heating the charcoal for the above purpose in large bulks as in retorts as is now done, " exposing the animal charcoal in detail, either by spreading it in " thin layers, constantly moved or otherwise kept in motion, to " the heat of a succession of heating surfaces, disposed so that as " the charcoal is brought on to each successive surface it receives " more heat than from the preceding one, the hottest of such surfaces " being maintained at or a little above a red heat, and of keeping " the said animal charcoal as much as possible in motion while " being dried and carbonized ;" and "never suffering any part " of it to obtain a red heat if it is sufficiently recarbonized before " attaining that heat, or to retain such red heat longer than is " necessary for producing complete carbonization." Above a furnace is a brick arch, on the crown of which is a platform on which is a plate of iron upon hinged joints, the fire rises and goes horizontally away under a series of iron plates to the chimney. The plates nearer to the chimney are for drying the charcoal, upon which it is spread and raked at a depth of about four inches. It is gradually moved by rakes to the plate next the platform named above, where the carbonization is completed, when it is cleared from this plate into receivers or it is raked on to the platform and cleared from it.

[Printed, 10d. Drawing. London Journal (*Newton's*), vol. 14 (*conjoined series*), p. 234.]

A,D. 1836, June 13.—N° 7115.

BERRY, MILES.—(*A communication.*)—"An improved apparatus " for torrifying, baking, and roasting vegetable substances which " with certain modifications and additions is also applicable to the

" evaporation and concentration of saccharine juices and other
" liquids." These are, in reference to this subject, as follows :—
Neither " the application of an oil or other bath to the boiler, nor
" the corrugated or fluted bottom of the boiler or evaporating
" vessel, nor the evaporating of the aqueous parts of the solutions
" or syrups by passing them over the surface of shallow trays from
" one to another," are claimed, " as they are all old contrivances
" and have been before carried into effect. But what I consider
" novel under this application of the apparatus is, the manner of
" effecting the vacuum by the two upper condensing vessels in
" connection with the refrigerator or condenser and its ventilators
" or rotary fans, and also the general arrangement and con-
" struction of the apparatus." "The manner of producing the
" vacuum :—The steam or vapour from the evaporating vessel
" rises into the upper part of the serpent or condenser, and
" travelling along all the fluted chambers arrives at the left-hand
" one of two condensing vessels," "which is provided with two
" cocks, the one for the introduction of steam and the other to
" allow the exit of the air driven out by the steam. The con-
" densing vessel is air-tight, and has no opening except at the
" bottom, and is plunged in a tub or tank filled with water" and
when the above vessel "is full of steam the air pipe is closed, and
" the cock of another pipe placed in the lower part of the con-
" denser is opened ; and a stream of water from a reservoi .
" placed above the condensing vessels being ejected into the vessel
" filled with steam, a condensation takes place, and the condensed
" steam descends from the upper part of the refrigerator into the
" lower part and falls through a pipe into a recipient together
" with all the air it may have carried with it ; thus a vacuum is
" produced in the apparatus," &c. &c.

[Printed, 1s. 2d. Drawing. London Journal (*Newton's*), vol. 10 (*conjoined series*), p. 257 ; Rolls Chapel Reports, 7th Report, p. 176.]

A.D. 1839, June 22.—N° 8123.

PARKER, FREDERICK.—" Improvements in revivifying or re-
" burning animal charcoal." These are, so conducting the above
process "that the retort, oven, or vessel shall not be required
" to be cooled down, and yet at the same time the drawing and
" cooling process shall not cause the charcoal to pass into the
" atmosphere when in a red hot' condition or in a state to be

" injured thereby." The charcoal to be revivified is placed in a vertical retort or vessel of iron or otherwise, which is surrounded by the flues of the fire-place or furnace below. A hopper or chamber above is kept filled with animal charcoal to be reburned; the cooling vessel which is below is connected to the lower part of the retort, oven, or vessel by a sand-joint. The cooler is, in preference, of thin sheet iron and of considerable extent. The lower end of the cooler is closed by a slide, "there being a series of openings " in the bottom and slide, so that when it is desired to draw the " charcoal, the openings are made coincident in the slide and " bottom ; at other times they are to be slided." There "is an " apparatus for measuring the charcoal as it comes from the " cooler ;" this " apparatus has a slide and perforated bottom " similar to that above described to the cooler."

[Printed, 8d. Drawing. London Journal (*Newton's*), vol. 23 (*conjoined series*), p. 28 ; Inventors' Advocate, vol. 1, p. 179.]

A.D. 1852, March 8.—N° 14,015.

VAN KEMPEN, Peter.—(*A communication from Gerrit Abraham Cramer.*)—"An improved refrigerator to be used in brewing, " distilling, and other similar useful purposes." This consists, first, " of a long continuous trough, along which the wort to be " cooled is caused to flow." The length depends on the average temperature of the worts, the degree of temperature to which it may be necessary to reduce them, and the cold water or liquor employed in connection with the apparatus. The apparatus consists of an external trough "of any kind of metal or any other " suitable material containing the cooling medium, and through " which it passes in an opposite direction to that in which the " wort flows." The inner trough is of thin copper tinned. The inner and outer troughs are composed of parts of the same length with vulcanized india-rubber packing. "The refrigerator is " placed at an incline of not less than one foot in two hundred " feet, and of an uniform inclination throughout." The water or cooling liquid is introduced at the lower end from a head of the same.

Second, " preventing atmospheric influences taking effect on the " wort when in the refrigerator. For this purpose a strip of " zinc is placed throughout the entire length of the refrigerator," and covered by the fluid in it. Both ends of the strip are placed

in connection with two plates " deposited in the ground, one of
" which, of copper, attached by a copper strip reaching to the
" refrigerator forms the positive pole, while the other earth plate
" is of zinc, and forms the negative pole. Any atmospheric
" electricity which during thunderstorms is liable to turn wort
" sour is carried off by this arrangement.

[Printed, 1s. 2d. Drawings.]

APPENDIX B.

A.D. 1802, March 24.—N° 2599.

TREVITHICK, RICHARD, and VIVIAN, ANDREW.—" Methods
" for improving the construction of steam engines, and the
" application thereof for driving carriages, and for other pur-
" poses." These are, in reference to this subject, an " engine with
" rollers for pressing or crushing sugar canes, moved by a steam
" engine." A case, in the form of a drum or cylinder, suspended
upon two strong trunnions or pivots, its flat ends standing
upright; within this iron case is fixed a boiler, so as to leave a
small vacant space between itself and the case, and within the
boiler is fixed a fire-place, having its grate above the ash hole;
the heated vapour and smoke rise, at the inner extremity and pass
out by two flues, through the boiler and case which join above
in a chimney which is more closely applied, slung between two
centres in a ring. The working cylinder with its piston steam
pipe, nozzle, and cock, are inserted in the upper part of the
boiler. The piston rod drives the fly, upon the arbor of which
is fixed a small wheel, which drives a great wheel upon the
axis of the middle roller. " In such cases or constructions as
" may render it more desirable to fix the boiler with its chimney
" and other apparatus, and to place the cylinder out of the
" boiler; the cylinder itself may be suspended for the same pur-
" pose upon trunnions or pivots in the same manner, one or both
" of which trunnions or pivots may be perforated so as to admit·
" the introduction and escape of the steam, or its condensation."
To allow of no vibratory motion of the boiler or cylinder the

S.   D D

same may be fixed by guides. " The steam which escapes in this
" engine is made to circulate in the case round the boiler, where
" it prevents the external atmosphere from affecting the tempera-
" ture of the included water, and affords, by its partial conden-
" sation, a supply for the boiler itself, and is or may be directed
" to useful purposes."

[Printed, 1s. Drawing. Repertory of Arts, vol. 4 (*second series*), p. 241 ;
Mechanics' Magazine, vol. 12, p. 162 ; and vol. 13, p. 313 ; Register of Arts
and Sciences, vol. 4, pp. 316 and 332 ; Engineers' and Architects' Journal,
vol. 2, p. 93 ; and vol. 10, p. 113 ; Engineers' and Mechanics' Encyclopediæ.
vol. 2, pp. 386 and 734 ; Stuart's History of the Steam Engine, pp. 162 and
183 ; Stuart's Anecdotes of Steam Engines, vol. 2, p. 455 ; Rolls Chapel
Reports, 6th Report, p. 151.]

A.D. 1837, January 11.—N° 7276.

GOODLET, GEORGE.—" A new and improved mode of distilling
" spirits from wash and other articles, also applicable to general
" purposes of rectifying, boiling, and evaporating or concentrating."
These are, " causing the fluids to be operated on for the purpose
" of evaporation, to circulate through pipes or other vessels,
" acted on by steam, hot water, oil, or other suitable heated fluid
" contained in proper vessels." In the application of this principle " to the evaporation of worts, syrups, the juice of the sugar
" cane, of mangle worzel, and other articles, the process may be
" advantageously conducted in the following manner :—Suppose
" a reservoir, which may be of wood or any other suitable material,
" with a spout or vent to suit the premises, placed above the said
" reservoir, allowing the steam to get quickly away; this reser-
" voir being charged with the fluid to be operated upon, and
" a connection being formed with it and the circulating pipes in
" the heated medium, the fluid, with the aid of a force pump or
" otherwise, is made to pass through the said circulating pipes
" and returned by the ejection pipe entering at the under part of
" said steam spout or vent. The steam immediately ascends
" with great force, and the concentrated fluid falls into the reser-
" voir to be recirculated until the process is finished; or the
" heated fluid from the circulating pipe may enter below the
" surface of the fluid in the reservoir."

[Printed, 4d. No Drawings. Repertory of Arts, vol. 8 (*new series*),
p. 238.]

ADDENDUM.

A.D. 1859, February 18.—N° 451.

GARTON, CHARLES.—"An improved method of treating cane
" sugar, in order to render it fitter to be employed in brewing,
" distilling, and wine and vinegar making." This consists "in
" assimilating the properties of cane sugar to those of malt
" saccrum (or more properly saccharum) and fruit sugar," as
follows :—The sugar is liquified by boiling water, the solution is
mechanically agitated " for about forty-eight hours, more or less,
" at the same time raising the temperature as quickly as possible
" to about 160° Fahrenheit. When the agitation has proceeded
" for about six hours I add acid to the solution, and after ceasing
" the agitation I neutralize the acid by chalk or other suitable
" agent; I then separate the precipitate from the saccharine
" solution thus obtained by subsidence or filtration, when it will
" be ready for use; or I liquify the sugar, apply heat thereto,
" and subject the solution to mechanical agitation as before
" described, raising the temperature of the solution to about
" 180° Fahrenheit, and continuing the agitation for about sixty
" hours, more or less, but without adding acid." In both processes, " glucose of any kind, in the proportion of about one-
" twentieth part of the whole, may with advantage be added to
" the solution at the commencement of the process." To each
hundredweight of sugar, "eight ounces of sulphuric acid, 184°
" gravity, diluted with twenty-four ounces of water," may be
used.

[Printed, 4d. No Drawings.]

S.

A.D. 1850, February 18.—N° 151.

GARTON, CHARLES.—"An improved method of treating cane
" sugar, in order to render it fitter to be employed in brewing,
" distilling, and wine and vinegar making." This consists "in
" assimilating the impurities of cane sugar to those of malt
" saccrum (or more properly saccharum) and fruit sugar," as
follows :—The sugar is liquified by boiling with the solution is
mechanically agitated " for about forty-eight hours, more or less,
" at the same time raising the temperature as quickly as possible
" to about 160° Fahrenheit. When the agitation has proceeded
" for about six hours I add acid to the solution; and after ceasing
" .... I then separate the precipitate from the ....
" and subject the solution to mechanical agitation as before
" described, raising the temperature of the solution to about
" 160° Fahrenheit, and continuing the agitation for about sixty
" hours, more or less, but without adding acid." In both processes, " alumina, of any kind, in the proportion or about one-
" .... of the whole, may with advantage be added to
" the solution at the commencement of the process." To each
hundredweight of sugar, " eight ounces of sulphuric acid, 1·4,
" figure ......ated with twenty-four ounces of water," may be
used.

[Printed, 2d. No Drawing.]

INDEX OF SUBJECT MATTER.

Charcoal—*cont.*

Warner, 241.
Green, 261.
De Lisle, 263.
Gilbee (*Pesier*), 273.
Clark (*De Gemini, E. T. and E. O.*), 296.
Newton (*Fryatt*), 299.
Haseltine (*Goessling*), 358.
Leon, Tessimond, and Kissack, 366.
Léon (*Leplay*), 371.

Making;
Constant, 15.
Martineau, P. and J., 21.
Derosne, 46.
Pertins, 49.
Parker, 408 (*Appendix*).
Crosley and Stevens, 70.
Derosne, 77.
Champion, 81.
Sievier, 94.
De Cavaillon, 108.
Birkmyre, 115.
Macfie, 121.
Gwynne, 130.
Ebingre, 148.
Brandeis, 149.
Knab, 162.
Brandeis, 164.
Gwynne, 173.
Maumené, 175.
Challeton, 177.
Dimsdale, 177.
Taylor and Brown, J. and J., 182.
Way and Paine, 183.
Duquesne, 188.
Oxland, 195.
Chantrell, 199.
Lodge and Marshall, 200.
Ellis, 201.
Stenhouse, 203.
Brooman (*Barry*), 210.
Stenhouse, 216.
Ziegler, 227.
Galy-Cazalet and Hibbard, 231.
Botturi, 231.
Warner, 241.
Spencer, 242.
Lichtenstadt and Duff, 243.
Bruère, 244.
Henry (*Layard*), 246.
Chantrell, 248.
Chantrell, 252.
Belton, 280.
Richardson and Prentice, 283.
Townsend, 286.
Williams, 289.
Robertson, 300.
Torr, 312.
Robertson, 312.
Henry (*Société Coignet Père et Fils et Compagnie*), 313.
Beanes, 340.
Paterson, 348.

Charcoal—*cont.*

Carey, 351.
Bringes, 362.
Müller, Weld, and Powell, 364.
Clark (*Gaade*), 370.
Clark (*Gaade*), 373.
Cormack, 374.
Cormack, 382.
Ensley, 387.

Revivifying;
De Cavaillon, 24.
Pertins, 49.
Parker, 408 (*Appendix*).
Bowman, 412 (*Appendix*).
Oliver, 60, 63.
Parker, 413 (*Appendix*).
Bancroft and MacInnes, 65.
Champion, 81.
Bowman, 93.
Sievier, 94.
Gwynne, 130.
Brooman, 137.
Torr, 139.
Nash, 151.
Hills, F. C. and G., 163.
Scott, 165.
Picciotto, 174.
Chantrell, 176.
Coste, 185.
Oxland, 195.
Chantrell, 199.
Bryant, 211.
Anderson, 217.
Ziegler, 227.
Finzel and Bryant, 233.
Parsons, 234.
Bringes and Collins, 235.
Warner, 241.
Spencer, 242.
Bensen, 245.
Henry (*Layard*), 246.
Chantrell, 248.
Chantrell, 252.
Green, 261.
Cowan, J. and P., 270.
Duncan, Scott, and Dawson, 272.
Brearley, 278.
Williams, 289.
Cowan, 294.
Carey and Pierce, 297.
Duncan, 299.
Cowan, 304.
Brooman (*Leplay and Cuisinier*), 306.
Mac Kirdy, 307.
Paterson (*Drummond*), 307.
Torr, 312.
Brooman (*Leplay and Cuisinier*), 318.
Gits, 325.
Mackirday, 335.
Beanes, 340.
Brooman (*Trolliet*), 345.
Brinjes, 346.
Beanes, 347.

Charcoal—*cont.*
Fryer, 348.
Davies (*Jacquier and Danek*), 352.
Brinjes, 355.
White, 363.
Ogston, 364.
Müller, Weld, and Powell, 364.
Fletcher, 368.
Clark (*Gaade*), 370.
Beanes, 372.
Clark (*Gaade*), 373.
Cormack, 374.
Norman, 377.
Elliott, 378.
Robey, 379.
Gordon, 380.
Cormack, 382.
Jasper, 383.
Knaggs, 384.
Patrick, 385.
Cormack, 389.
Gordon, 390.

Charcoal substitutes. *See* Charcoal making.

Charcoal substitutes revivifying. *See* Charcoal, revivifying.

Chlorine, use of:
Newton, 110.
Nash, 151.
Herapath, 320.

Chloroform, use of:
Clark (*Lion*), 274.

Clay. *See* Alumina.

Claying sugar methods of:
Murray, 5.
Vaughan, 12.
Drake, 22.
Bates, 40.
Garnett, 45.
Rotch, 119.
Bessemer. 170.

Coals, use of. *See also* Bituminous substances.
Martineau, P. and J., 21.
Crosley and Stevens, 70.
De Cavaillon, 108.
Gwynne, 173.
Challeton, 177.
Oxland, 195.
Spencer, 242.
Heusner, 327.
Burq, 360.

Coke, use of. *See also* Bituminous substances.
Martineau, P. and J., 21.
Crosley and Stevens, 70.
Spencer, 242.
Burq, 360.

Colouring sugar:
Green, 261.
Ware, 296.

Coolers. *See* Pans.

Cold producing or using:
Tickell, 397 (*Appendix*).
Nesmond, 160.
Vion, 161.
Bessemer, 172.
Green, 261.
Garton and Hill, 335.

Coral. *See* Alkaline earths.

Creosote:
Newton, 110.

Crushing or cutting sugar. *See* Nippers.

Crystallizing vessels. *See* Pans.

Cylinders. *See* Pans.

Dialysis, application of:
Brooman (*Dubrunfaut*), 329.
Gedge (*Minchin*), 376.
Kohn (*Minchin*), 393.

Diastase, use of:
Manbré, 260.
Manbré, 266.

Diffusion, use of. *See* Dialysis.

Drying sugar, &c.:
Wyatt, 9.
Ingram, 10.
Vaughan, 12.
Ure, 406 (*Appendix*).
Stokes, 35.
Gutteridge and Stevens, 53.
Crosley, 60.
Hardman, 72.
Constable, 73.
Gye, 84.
Newton, 98.
Brooman, 109.
Shears, 122.
Herring, 129.
Macintosh, 145.
Macintosh, 148.
Bessemer, 170.
Archbald, 186.
Delabarre, 197.
Aspinall, 198.

Vacuum pans, making or using —cont.

Crosley and Stevens, 70.
Cooper, 72.
Robinson, 75.
Borrie, 76.
Gadesden, 78.
Johnston, 79.
St. Clair, 82.
Pearse and Child, 83.
Britten, 84.
Richardson, 89.
Johnston, 93.
Steiner, 95.
Claypole, 97.
Murdoch, 108.
Finzel, 112.
Gwynne, 116.
Shears, 122.
Bessemer, 127.
Herring, 129.
Gwynne, 130.
Varillat, 135.
Brooman, 137.
Bessemer, 140.
Walker, 142.
Bessemer, 144.
Brooman, 145.
Macintosh, 148.
Aspinall, 149.
Brown, 150.
Young, 153.
Bessemer, 156.
Brooman, 157.
Nesmond, 160.
Pidding, 162.
Greenwood, 166.
Finzel, 167.
Bessemer, 179.
Beanes, 180.
Collette, 181.
Aspinall, 183.
Wright (*Reid*), 184.
Aspinall, 187.
Fairrie, 193.
Fairrie, 193.
Delabarre, 197.
Schiele, 201.
Leitch, 205.
Golding, 207.
Blackwell, 209.
Allman and Bethune, 215.
Lewsey, 220.
Longbottom, 220.
Edwards, 223.
Travis and Casartelli, 229.
Bensen, 232.
Bensen, 233.
Manbré, 239.
Bensen, 245.
Cameron, 247.
Newton (*Bertholemey*), 250.
Wagner, 254.
Aspinall, 255.
Manbré, 260.
Clark (*Lion*), 274.

Vacuum pans, making or using —cont,

Fletcher, 276.
Newton (*Fryatt*), 278.
Green, 284.
Field, 291.
Davies (*Lavignac*), 293.
Duncan, 299.
Schwartz, 302.
Patrick, 307.
Paterson (*Drummond*), 309.
Peek, 322.
Newman, 328.
Kissler, 330.
Reid, 333.
Doddrell, 336.
Vanderfeesten, 337.
Beanes, 347.
Lebaudy, 357.
Beanes and Finzel, 358.
Haseltine (*Goessling*), 358.
Haseltine (*Goessling*), 359.
Fletcher, 369.

Valves, making or using :

Shannon, 396 (*Appendix*).
Johnson, 397 (*Appendix*).
Taylor, 399 (*Appendix*).
Taylor, 400 (*Appendix*).
Johnson, 402 (*Appendix*).
Ure, 409 (*Appendix*).
Ronald, 78.
Steiner, 95.
Scoffern, 115.
Shears, 122.
Varillat, 135.
Brooman, 137.
Macintosh, 148.
Bessemer, 156.
Banfield, 156.
Greenwood, 166.
Mayelston, 169.
Leitch, 205.
Blackwell, 209.
Lewsey, 220.
Travis and Casartelli, 229.
Fletcher, 316.
Newton (*De Villeneuve*), 321.
Carey, 351.
Fletcher, 368.
Campbell, 370.

Vapours, exhausting. *See* Vacuum pans.

Wassama, use of :

Shears, 122.

Water, use of :

Shannon, 396 (*Appendix*).
Tickell, 397 (*Appendix*).
Constant, 15.
Howard, 17.
Taylor, 22.
Hague, 23.

LONDON:

Printed by GEORGE E. EYRE and WILLIAM SPOTTISWOODE,
Printers to the Queen's most Excellent Majesty.

PATENT LAW AMENDMENT ACT, 1852.

LIST OF WORKS printed by order of THE COMMIS-SIONERS OF PATENTS FOR INVENTIONS, and sold at the PATENT OFFICE, 25, Southampton Buildings, Chancery Lane, London.

I.

1. SPECIFICATIONS of PATENTS for INVENTIONS, DIS-CLAIMERS, &c., enrolled under the Old Law, from A.D. 1617 to Oct. 1852, comprised in 13,561 Blue Books, or 691 thick vols. imp. 8vo. Total cost price about 600*l.*

2. SPECIFICATIONS of INVENTIONS, DISCLAIMERS, &c., deposited and filed under the Patent Law Amendment Act from Oct. 1, 1852, to Dec. 31, 1870, comprised in 59,025 Blue Books, or 1,842 thick vols. imp. 8vo. Total cost price, about 1,848*l.*

II.

1. CHRONOLOGICAL INDEX of PATENTS of INVENTION from A.D. 1617 to Oct. 1852. 2 vols. (1554 pages). Price 30*s.* By Post, 33*s.* 2*d.*

ALPHABETICAL INDEX for the above period. 1 vol. (647 pages). Price 20*s.* By Post, 21*s.* 5*d.*

SUBJECT-MATTER INDEX for the above period. 2 vols. (907 pages). Second Edition. 1857. Price 2*l.* 16*s.* By Post, 2*l.* 18*s.* 8*d.*

REFERENCE INDEX for the above period, pointing out the Office in which each enrolled Specification may be consulted; the Books in which Specifications, Law Proceedings connected with Inventions, &c. have been noticed. 1 vol. (710 pages). Second Edition. 1862. Price 30*s.* By Post, 31*s.* 5*d.*

APPENDIX to REFERENCE INDEX, containing abstracts from such of the early Patents and Signet Bills as describe the nature of the Invention. 1 vol. (91 pages). Price 4*s.* By Post 4*s.* 6*d.*

2. CHRONOLOGICAL INDEXES of APPLICATIONS for PATENTS and PATENTS GRANTED from Oct. 1 to Dec. 31. 1852, and for the year 1853. 1 vol. (258 pages). Price 11s, By Post, 12s.

ALPHABETICAL INDEXES for the above periods. 1 vol. (181 pages). Price 13s. By Post, 13s. 8d.

SUBJECT-MATTER INDEX for 1852. 1 vol. (132 pages). Price 9s. By Post, 9s. 7d.

SUBJECT-MATTER INDEX for 1853. 1 vol. (291 pages). Price 16s. By Post, 16s. 11d.

3. CHRONOLOGICAL INDEX for 1854. 1 vol. (167 pages). Price 6s. By Post, 6s. 7d.

ALPHABETICAL INDEX for 1854. 1 vol. (119 pages). Price 7s. By Post, 7s. 7d.

SUBJECT-MATTER INDEX for 1854. 1 vol. (311 pages). Price 16s. 6d. By Post, 17s. 6d.

4. CHRONOLOGICAL INDEX for 1855. 1 vol. (188 pages). Price 6s. 6d. By Post, 7s. 2d.

ALPHABETICAL INDEX for 1855. 1 vol. (129 pages). Price 7s. 6d. By Post, 8s. 1d.

SUBJECT-MATTER INDEX for 1855. 1 vol. (311 pages). Price 17s. By Post, 17s. 11d.

5. CHRONOLOGICAL INDEX for 1856. 1 vol. (189 pages). Price 6s. 6d. By Post, 7s. 1d.

ALPHABETICAL INDEX for 1856. 1 vol. (143 pages). Price 8s. By Post, 8s. 7d.

SUBJECT-MATTER INDEX for 1856. 1 vol. (335 pages). Price 18s. 6d. By Post, 19s. 7d.

6. CHRONOLOGICAL INDEX for 1857. 1 vol. (196 pages). Price 6s. 6d. By Post, 7s. 2d.

ALPHABETICAL INDEX for 1857. 1 vol. (153 pages). Price 8s. By Post, 8s. 8d.

SUBJECT-MATTER INDEX for 1857. 1 vol. (367 pages). Price 19s. 6d. By Post, 20s. 8d.

7. CHRONOLOGICAL INDEX for 1858. 1 vol. (188 pages). Price 6s. By Post, 6s. 8d.

ALPHABETICAL INDEX for 1858. 1 vol. (148 pages). Price 8s. By Post, 8s. 7d.

SUBJECT-MATTER INDEX for 1858. 1 vol. (360 pages). Price 19s. 6d. By Post, 20s. 6d.

3. CHRONOLOGICAL INDEX for 1859. 1 vol. (196 pages). Price 6s. 6d. By Post, 7s. 1d.

ALPHABETICAL INDEX for 1859. 1 vol. (188 pages). Price 10s. By Post, 10s. 7d.

SUBJECT-MATTER INDEX for 1859. 1 vol. (381 pages). Price 20s. By Post, 20s. 11d.

2

9. CHRONOLOGICAL INDEX for 1860. 1 vol. (209 pages).
   Price 7s. By Post, 7s. 7d.
   ALPHABETICAL INDEX for 1860. 1 vol. (203 pages).
   Price 10s. 6d. By Post, 11s. 1d.
   SUBJECT-MATTER INDEX for 1860. 1 vol. (405 pages).
   Price 22s. By Post, 23s.

10. CHRONOLOGICAL INDEX for 1861. 1 vol. (215 pages).
   Price 7s. By Post, 7s. 7d.
   ALPHABETICAL INDEX for 1861. 1 vol. (222 pages).
   Price 10s. 6d. By Post, 11s. 2d.
   SUBJECT-MATTER INDEX for 1861. 1 vol. (442 pages).
   Price 23s. By Post, 24s. 1d.

11. CHRONOLOGICAL INDEX for 1862. 1 vol. (237 pages).
   Price 7s. 6d. By Post, 8s. 2d.
   ALPHABETICAL INDEX for 1862. 1 vol. (240 pages).
   Price 11s. 6d. By Post, 12s. 2d.
   SUBJECT-MATTER INDEX for 1862. 1 vol. (465 pages).
   Price 23s. By Post, 24s. 1d.

12. CHRONOLOGICAL INDEX for 1863. 1 vol. (220 pages).
   Price 7s. By Post, 7s. 7d.
   ALPHABETICAL INDEX for 1863. 1 vol. (218 pages).
   Price 11s. By Post, 11s. 8d.
   SUBJECT-MATTER INDEX for 1863. 1 vol. (432 pages).
   Price 22s. By Post, 23s.

13. CHRONOLOGICAL INDEX for 1864. 1 vol. (222 pages).
   Price 7s. By Post, 7s. 7d.
   ALPHABETICAL INDEX for 1864. 1 vol. (220 pages).
   Price 11s. By Post, 11s. 8d.
   SUBJECT-MATTER INDEX for 1864. 1 vol. (446 pages).
   Price 23s. By Post, 24s. 1d.

14. CHRONOLOGICAL INDEX for 1865. 1 vol. (230 pages).
   Price 7s. By Post, 7s. 7d.
   ALPHABETICAL INDEX for 1865. 1 vol. (236 pages).
   Price 11s. 6d. By Post, 12s. 2d.
   SUBJECT-MATTER INDEX for 1865. 1 vol. (474 pages).
   Price 23s. By Post, 24s. 1d.

15. CHRONOLOGICAL INDEX for 1866. 1 vol. (239 pages).
   Price 7s. By Post, 7s. 8d.
   ALPHABETICAL INDEX for 1866. 1 vol. (243 pages).
   Price 11s. 6d. By Post, 12s. 2d.
   SUBJECT-MATTER INDEX for 1866. 1 vol. (465 pages).
   Price 23s. By Post, 24s. 4d.

16. CHRONOLOGICAL INDEX for 1867. 1 vol. (254 pages).
   Price 7s. 6d. By Post, 8s. 2d.
   ALPHABETICAL INDEX for 1867. 1 vol. (258 pages).
   Price 12s. By Post, 12s. 8d.

3

SUBJECT-MATTER INDEX for 1867. 1 vol. (508 pages). Price 25s. By Post, 26s. 2d.

'DESCRIPTIVE INDEX (Abridgments of Provisional and Complete Specifications) for 1867.
  a. Quarter ending 31st March. 1 vol. (228 pages). Price 1s. 8d. By Post, 2s. 1d.
  b. Quarter ending 30th June. 1 vol. (224 pages). Price 1s. 8d. By Post, 2s. 1d.
  c. Quarter ending 30th September. 1 vol. (196 pages). Price 1s. 8d. By Post, 2s.
  d. Quarter ending 31st December. 1 vol. (232 pages). Price 1s. 8d. By Post, 2s. 1d.

17. CHRONOLOGICAL INDEX for 1868. 1 vol. (274 pages). Price 8s. By Post, 8s. 8d.

ALPHABETICAL INDEX for 1868. 1 vol. (291 pages). Price 13s. By Post, 13s. 10d.

SUBJECT MATTER INDEX for 1868. 1 vol. (632 pages). Price 30s. By Post, 31s. 5d.

DESCRIPTIVE INDEX (Abridgments of Provisional and Complete Specifications) for 1868.
  a. Quarter ending 31st March. 1 vol. (236 pages). Price 1s. 8d. By Post, 2s. 1d.
  b. Quarter ending 30th June. 1 vol. (218 pages). Price 1s. 8d. By Post, 2s. 1d.
  c. Quarter ending 30th September. 1 vol. (194 pages). Price 1s. 8d. By Post, 2s.
  d. Quarter ending 31st December. 1 vol. (224 pages). Price 1s. 8d. By Post, 2s. 1d.

18. CHRONOLOGICAL AND DESCRIPTIVE INDEX (containing the Abridgments of Provisional and Complete Specifications) for 1869.
  a. Quarter ending 31st March. 1 vol. (226 pages). Price 1s. 8d. By Post, 2s. 1d.
  b. Quarter ending 30th June. 1 vol. (234 pages). Price 1s. 8d. By Post, 2s. 1d.
  c. Quarter ending 30th September. 1 vol. (200 pages). Price 1s. 8d. By Post, 2s. 1d.
  d. Quarter ending 31st December. 1 vol. (212 pages). Price 1s. 8d. By Post, 2s. 1d.

ALPHABETICAL INDEX for 1869. 1 vol. (272 pages). Price 13s. By Post, 13s. 9d.

SUBJECT MATTER INDEX for 1869. 1 vol. (587 pages). Price 28s. By Post, 29s. 2½d.

19. CHRONOLOGICAL AND DESCRIPTIVE INDEX (containing the Abridgments of Provisional and Complete Specifications) for 1870.
  a. Quarter ending 31st March. 1 vol. (222 pages). Price 1s. 8d. By Post, 2s. 1d.

4

b. Quarter ending 30th June. 1 vol. (218 pages). Price 1*s.* 8*d.* By Post, 2*s.* 1*d.*

c. Quarter ending 30th September. 1 vol. (168 pages). Price 1*s.* 8*d.* By Post, 2*s.* 1*d.*

ALPHABETICAL INDEX for 1870. 1 vol. (242 pages). Price 12*s.* By Post, 12*s.* 8*d.*

20. CHRONOLOGICAL AND DESCRIPTIVE INDEX (containing the Abridgments of Provisional and Complete Specifications) for 1871, with Indexes of Names and Subject Matter. Published in weekly numbers, price 4*d.* each.*

III.

ABRIDGMENTS (in Classes and Chronologically arranged) of SPECIFICATIONS of PATENTED INVENTIONS, from the earliest enrolled to those published under the Act of 1852.

These books are of 12mo. size, and each is limited to inventions of one class only. They are so arranged as to form at once a Chronological, Alphabetical, Subject-matter, and Reference Index to the class to which they relate. Inventors are strongly recommended, before applying for Letters Patent, to consult the classes of Abridgments of Specifications which relate to the subjects of their inventions, and by the aid of these works to select the Specifications they may consider it necessary to examine in order to ascertain if their inventions are new.

The following series of Abridgments do not extend beyond the end of the year 1866. From that date the Abridgments have not been published in classes, but will be found in chronological order in the " Chronological and Descriptive Index " (*see* Section II. of this List of Works). It is intended, however, to publish these Abridgments in classes as soon as the Abridgments of all the Specifications from the earliest period to the end of 1866 have appeared in a classified form. Until that takes place the Inventor (by the aid of the Subject Matter Index for each year) can continue his examination of the Abridgments relating to the subject of his invention in the Chronological and Descriptive Index.

The classes already published are,—

1. DRAIN TILES AND PIPES, price 4*d.*, by post 5*d.*
2. SEWING AND EMBROIDERING, price 6*d.*, by post 7*d.*
3. MANURE, price 4*d.*, by post 5*d.*
4. PRESERVATION OF FOOD, Part I., A.D. 1691-1855, price 4*d.*, by post 5*d.*—Part II., A.D. 1856-1866, price 6*d.*, by post 7*d.*
5. MARINE PROPULSION, price 1*s.* 10*d.*, by post 2*s.* 2*d.*
6. MANUFACTURE OF IRON AND STEEL, Parts I., II., & III., A.D. 1621-1857, price 1*s.* 6*d.*, by post 1*s.* 9*d.*—Part IV., A.D. 1857-1865, price 2*s.* 6*d.*, by post 2*s.* 8*d.*
7. AIDS TO LOCOMOTION, price 6*d.*, by post 7*d.*
8. STEAM CULTURE, price 8*d.*, by post 10*d.*

---

* *See* Notice on page 16.

2. APPENDIX to the SPECIFICATIONS of ENGLISH PA-
TENTS for REAPING MACHINES. By B. WOODCROFT, F.R.S.
Price 6s. 6d.   By Post, 6s. 11d.

3. INDEX to ALL INVENTIONS PATENTED in ENGLAND
from .1617 to 1854 inclusive, arranged under the greatest
number of heads, with parallel references to INVENTIONS and
DISCOVERIES described in the scientific works of VARIOUS
NATIONS, as classified by Professor Schubarth.   By B. WOOD-
CROFT, F.R.S.   Price 1s.   By Post, 1s. 1d.
The foreign works thus indexed form a portion of the Library of
the Commissioners of Patents, where they may be consulted.

4. EXTENSION of PATENTS to the COLONIES.—Abstract of
Replies to the Secretary of State's Circular Despatch of January 2,
1853, on the subject of the Extension of Patents for Inventions
to the Colonies.   Second Edition, with Revised Table.   1861.
Price 2s.   By Post, 2s. 2d.

5. SUPPLEMENT to the SERIES of LETTERS PATENT and
SPECIFICATIONS. from A.D. 1617 to Oct. 1852; consisting
for the most part of Reprints of scarce Pamphlets, descriptive of
the early patented Inventions comprised in that Series.

CONTENTS.

1. Metallica; or the Treatise of Metallica, briefly comprehending the doctrine
of diverse new metallical inventions, &c.   By SIMON STURTEVANT. (Letters Patent, dated 29th February 1611.)   Price 1s. 4d.; by post, 1s. 5d.

2. A Treatise of Metallica, but not that which was published by Mr. Simon
Sturtevant, upon his Patent, &c.   By JOHN ROVENZON. (Letters Patent
granted A.D. 1612.)   Price 4d; by post, 4½d.

3. A Commission directed to Sir Richard Wynne and others to inquire upon
oath whether NICHOLAS PAGE or Sir NICHOLAS HAISE was the first inventor of certaine kilnes for the drying of malt, &c. &c. (Letters Patent,
Nos. 83 and 85, respectively dated 8th April 1626, and 23rd July 1635.)
Price 2d.; by post, 2½d.

4. DUD DUDLEY's Metallum Martis; or iron made with pit-coale, sea-coale,
&c. (Letters Patent, Nos. 18 and 117, respectively dated 22nd February
1620, and 2nd May 1638.)   Price 8d.; by post, 9d.

5. Description of the nature and working of the Patent Waterscoop Wheels
invented by WILLIAM WHELER, as compared with the raising wheels
now in common use.   By J. W. B.   Translated from the Dutch by
Dr. Tolhausen. (Letters Patent, No. 127, dated 24th June 1642.)   Price 2s.;
by post, 2s. 1½d.

6. An exact and true definition of the stupendous Water-commanding Engine
invented by the Right Honourable (and deservedly to be praised an
admired) EDWARD SOMERSET, Lord Marquis of WORCESTER, &c. &c
(Stat. 15 Car. II. c. 12. A.D. 1663.)   Price 4d.; by post, 4½d.

7. Navigation improved; or the art of rowing ships of all rates in calms with
a more easy, swift, and steady motion than oars can. By THOMAS SAVERY.
(Letters Patent, No. 347, dated 10th January 1696.)   Price 1s.; by post,
1s. ½d.

8. The Miner's Friend; or an engine to raise water by fire, described, &c
By THOMAS SAVERY. (Letters Patent, No. 356, dated 25th July 1698, and
Stat. 10 & 11 Will. III. c. 31, A.D. 1699.)   Price 1s.; by post, 1s. 1d.

9. Specimina Ichnographica; or a brief narrative of several new inventions
and experiments, particularly the navigating a ship in a calm, &c. By JOHN
ALLEN, M.D. (Letters Patent, No. 513, dated 7th August 1729.)   Price 8d.;
by post, 8½d.

10. A description and draught of a new-invented Machine for carrying vessels or ships out of or into any harbour, port, or river against wind and tide, or in a calm, &c. By JONATHAN HULLS. (*Letters Patent, No. 556, dated 21st December* 1736.) Price 8d.; by post, 9d.

11. An historical account of a new method for extracting the foul air out of ships, &c., with the description and draught of the machines by which it is performed, &c. By SAMUEL SUTTON, the Inventor. To which are annexed two relations given thereof to the Royal Society by Dr. Mead and Mr. Watson. (*Letters Patent, No. 602, dated 16th March* 1744.) Price 1s.; by post, 1s. 1d.

12. The letter of Master WILLIAM DRUMMOND for the construction of machines, weapons, and engines of war for attack or defence by land or sea, &c. Dated the 29th Septomper 1626. (*Scotch Patent, temp. Car. II.*) Price 4d.

A FREE LIBRARY and READING ROOMS are open to the Public daily, from 10 till 4 o'clock, in the Office of the Commissioners of Patents, 25, Southampton Buildings, Chancery Lane. In addition to the printed Specifications, Indexes, and other publications of the Commissioners, the Library includes a Collection of the leading British and Foreign Scientific Journals, and text-books in the various departments of science and art.

Complete sets of the Commissioners of Patents' publications (each set including more than 2,700 volumes and costing for printing and paper nearly £2,600) have been presented to the authorities of the most important towns in the kingdom, on condition that the works shall be rendered daily accessible to the public, for reference or for copying, free of all charge. The following list gives the names of the towns, and shows the place of deposit, so far as ascertained, of each set of the works thus presented:—

Aberdeen (*Mechanics' Institution*).
Belfast (*Queen's College*).
Beverley (*Guildhall*).
Birmingham (*Central Free Library—Reference Department, Ratcliff Place*).
Blackburn (*Free Library and Museum, Town Hall Street*).
Bolton-le-Moors (*Public Library, Exchange Buildings*).
Bradford, Yorkshire (*Borough Accountant's Office, Corporation Buildings, Swain Street*).
Brighton (*Town Hall*).
Bristol (*City Library, King Street*).
Burnley (*Office of the Burnley Improvement Commissioners*).
Bury.
Carlisle (*Public Free Library, Police Office*).
Chester (*Town Hall, Northgate St.*)
Cork (*Royal Cork Insta, Nelson Place*).
Crewe (*Railway Station*).
Darlington (*Mechanics' Institute, Skinnergate*).
Derby (*Free Public Library*).
Dorchester.
Drogheda.
Dublin (*Royal Dublin Soy, Kildare St.*)
Dundalk (*Free Library*).

Falmouth (*Public Libr, Church St.*)
Gateshead (*Mechanics' Institute*).
Gorton (*Railway Station*).
Glasgow (*Stirling's Libr, Miller St.*)
Grimsby, Great (*Mechanics' Institution, Victoria Street*).
Halifax.
Hanley, Staffordshire Potteries (*Town Hall*).
Hertford (*Free Public Library, Town Hall*).
Huddersfield (*Improvement Commissioners' Offices, South Parade*).
Hull (*Mechanics' Inst., George St.*)
Ipswich (*Museum Library, Museum Street*).
Keighley (*Mechanics' Inst., North St.*)
Kidderminster (*Public Free Library, Public Buildings, Vicar Street*).
King's Lynn, Norfolk (*Stanley Library, Athenæum*).
Lancaster (*Mechanics' Institute, Market Street*).
Leamington Priors (*Public Library, Town Hall*).
Leeds (*Public Library, Infirmary Buildings*).
Leicester (*Free Library, Wellington Street*).

9

Limerick (*Town Hall*).
Liverpool (*Free Public Library, William Brown Street*).
London (*British Museum*).
———— (*Society of Arts, John Street, Adelphi*).
Macclesfield (*Useful Knowledge Society*).
Maidstone (*Free Library*)
Manchester (*Free Lib, Camp Field*).
Montrose (*Free Library*).
Newark, (*Mechanics' Institute, Middle Gate*).
Newcastle-upon-Tyne (*Literary and Philosophical Society*).
Newport, Monmouth (*Commercial Room, Town Hall*).
Northampton.
Norwich (*Free Library, St. John's, Maddermarket*).
Nottingham (*Free Library*).
Oldham (*School of Arts and Sciences, Lyceum*).
Oxford (*Public Free Library, Town Hall.*)
Paisley (*Government School of Design, Gilmour Street*).
Plymouth (*Mechanics' Institute, Princess Square*).
Preston, Lancashire (*Dr. Shepherd's Library, the Institution, Avenham*).

Reading (*Literary, Scientific, and Mechanics' Institution, London St.*)
Rochdale (*Commissioners' Rooms, Smith Street*).
Rotherham (*Board of Health Offices, Howard Street*).
Salford (*Royal Museum and Library, Peel Park*).
Sheffield (*Free Public Library, Surrey Street*).
Shrewsbury (*Public Museum, College Street*).
Southampton (*Hartley Institution*).
Stirling (*Burgh Library, Town House, Broad Street*).
Stockport (*Museum, Vernon Park*).
Sunderland (*Corporation Museum, Athenæum, Fawcett Street*).
Wakefield (*Mechanics' Institution, Barstow Square*).
Warrington (*The Museum and Library*).
Waterford (*Town Hall, The Mall*).
Wexford (*Mechanics' Institute, Crescent Quay*).
Wigan.
Wolverhampton (*School of Practical Art, Darlington Street*).
Wolverton (*Railway Station*).
York (*Lower Council Chamber, Guildhall*).

The Commissioners' publications have also been presented to the following Public Offices, Seats of Learning, Societies, British Colonies, and Foreign States :—

Public Offices, &c.

Admiralty—Director of Works' Department.
    Chief Constructor's Department.
    Chatham Dockyard.
    Sheerness ditto.
    Portsmouth ditto.
    Devonport ditto.
    Pembroke ditto.
Artillery Institute, Woolwich.
Board of Trade, Whitehall.

Ordnance Office—Pall Mall.
    Small Arms Factory, Enfield.
War Office, Pall Mall.
India Office.
Royal School of Mines, &c., Jermyn Street, Piccadilly.
Dublin Castle, Dublin.
Record and Writ Office, Chancery, Dublin.
Office of Chancery, Edinburgh.

Seats of Learning and Societies.

Cambridge University.
Trinity College, Dublin.

Queen's College, Galway.
Incorporated Law Society, Chancery Lane, London.

British Colonies.

Antigua.
Barbados.
British Guiana.
Canada—Library of Parliament, Ottawa.
    Bureau of Agriculture, Toronto.
    Board of Arts and Manufactures, Montreal.
Cape of Good Hope.
Ceylon.

India—Bengal.
    Bombay.
    Madras.
    N.-W. Provinces.
Jamaica.
Malta.
Mauritius.
New Brunswick.
Newfoundland.
New South Wales.
New Zealand.
Nova Scotia.

Prince Edward Island.
South Australia—Colonial Institute, Adelaide.
Tasmania.
Trinidad.
Victoria—Parliamentary Library, Melbourne.
    Patent Office, Melbourne.
    Public Library, Melbourne.

Foreign States.

Argentine Republic—Buenos Ayres.
Austria—Handels Ministerium, Vienna.
Belgium—Ministère de l'Intérieur, Brussels.
        Musée de l'Industrie, Brussels.
France—Bibliothèque Nationale
      Conservatoire des Arts et Métiers, } Paris.
      Hôtel de Ville,
Germany—Alsace—Société Industrielle, Mulhouse.
      Bavaria—Königliche Bibliothek, Munich
      Gotha—Ducal Friedenstein Collection.
      Prussia—Gewerbe-Akademie, Berlin.
         Königliche Bibliothek, Berlin.
         Königliche Polytechnische Schule, Hanover.
      Saxony—Polytechnische Schule, Dresden.
      Wurtemberg—Bibliothek des Musterlagers, Stuttgart.
Italy—Ufficio delle Privative, Florence.
Netherlands—Harlem.
Russia—Bibliothèque Impériale, St. Petersburg.
Spain—Madrid.
Sweden—Teknologiska Institutet, Stockholm.
United States—Patent Office, Washington.
      Astor Library, New York.
      State Library, Albany.
      Franklin Institute, Philadelphia.
      Free Library, Boston.
      Library Company, Philadelphia.
      Library Association, Chicago.
      Peabody Institute, Baltimore.
      Historical Society, Madison, Wisconsin.
      Cornell University, Ithaca, N.Y.
      Mercantile Library, St. Louis.

Grants of complete series of Abridgments of Specifications have been made to the undermentioned Mechanics, Literary, and Scientific Institutions:—

Aberystwith (*Literary and Working Men's Reading Room*).
Alnwick (*Scientific and Mechanical Institution*).
Altrincham (*Altrincham and Bowdon Literary Institution*).
Ashby-de-la-Zouch (*Mutual Improvement Society*).
Bacup (*Mechanics' Institution*).
Ballymoney (*Town Hall*).
Banbridge (*Literary and Mutual Improvement Society*).
Banbury (*Mechanics' Institution*).
Barnstaple (*Literary and Scientific Institution*).
Bath (*Athenæum*).
Batley (*Mechanics' Institution*).
Battle (*Young Men's Christian Association*).
Belfast (*Athenæum*).
Berkhampstead, Great (*Working Men's College*).
Birmingham (*Birmingham Heath and Smethwick Working Men's Club and Institute*).
——— (*Bloomsbury Institution*).
——— (*Central Lending Library*).
——— (*Deritend Working Men's Association*).

Birmingham (*Graham Street Institution*).
Bodmin (*Literary Institution*).
Bolton (*Mechanics' Institute*).
——— (*School of Art*).
Bradford, near Manchester (*Bradford Working Men's Club*).
———, Yorkshire (*Library and Literary Society*).
——— (*Mechanics' Institute*).
Brampton, near Chesterfield (*Local Museum and Literary Institute*).
Breage, Cornwall (*Breage Institution*).
Bristol (*Athenæum*).
——— (*Institution*).
——— (*Law Library Society*).
——— (*Library*).
Broomsgrove (*Literary and Mechanics' Institute*).
Burnley (*Mechanics' Institution*).
——— (*Literary Institution*).
Burslem (*Wedgwood Institute*).
Bury St. Edmund's (*Mechanics' Institution*).
Canterbury (*Working Men's Club*).
Cardiff (*Free Library and Museum*)
Cardigan (*Mechanics' Institute*).
Carharrack (*Literary Institute*).

11

Carmarthen (*Literary and Scientific Institution*).
Cheltenham (*Literary and Philosophical Society*).
——— (*Permanent Library*).
——— (*Working Men's Club*).
Chertsey (*Literary and Scientific Institution*).
Chester (*City Library and Reading Room*).
Chesterfield (*Mechanics' Institution*).
Chichester (*Literary and Philosophical Society*).
——— (*Literary Society and Mechanics' Institute*).
Coalbrookdale(*Literary and Scientific Institution*).
Cockermouth (*Mechanics' Institution*).
Colchester (*Literary Institution*).
——— (*Young Men's Christian Association*).
Compstall (*Athenæum*).
Coventry (*Free Library*).
——— (*Institute*).
——— (*School of Art*).
Crediton (*Working Men's Club*).
Dartmouth (*Mutual Improvement Society*).
Deal (*Deal and Walmer Institute*).
Denton (*Denton and Haughton Mechanics' Institution*).
Derby (*Mechanics' Institution*).
Devonport (*Mechanics' Institute*).
Dewsbury (*Mechanics' Institution*).
Doncaster (*Free Library*).
——— (*Great Northern Mechanics' Institute*).
Dorchester (*County Museum and Library*).
——— (*Working Men's Institute*).
Dudley (*Mechanics' Institution*).
Dukinfield (*Mechanics' Institute*).
——— (*Village Library and Reading Room*).
Dumbarton (*Philosophical and Literary Society*).
Dumfries (*Mechanics' Institution*).
Durham (*Mechanics' Institute*).
Eagley, Bolton-le-Moors (*Library and Institute*).
Earlestown, Newton-le-Willows (*Mutual Improvement Society*).
Edinburgh (*Horological Society*).
——— (*Mechanics' Library*).
——— (*Philosophical Institution*).
——— (*Royal Society of Arts*).
——— (*Royal Scottish Society of Arts*).
——— (*Subscription Library*).
——— (*Watt Institution and School of Art*).
——— (*Working Men's Club*).
Egham (*Literary Institute*).
Egremont (*Mechanics' Institution*).
Exeter (*Devon and Exeter Institution*).
Faversham (*Institute*).
Frome (*Literary and Scientific Institution*).
12

Gainsborough (*Literary, Scientific and Mechanics' Institute*).
Garforth, near Leeds (*Working Men's Club*).
Glasgow (*Athenæum*).
——— (*Central Working Men's Club and Institute*).
——— (*Institution of Engineers in Scotland*).
——— (*Mechanics' Institution, Bath Street*).
——— (*Philosophical Society*).
Grantham (*Public Literary Institution*).
Gravesend (*Gravesend and Milton Library and Reading Rooms*).
Greenwich (*Working Men's Institute*).
Guildford (*Mechanics' Institute*).
Halesworth (*Mechanics' Institute*).
Halifax (*Literary and Philosophical Society*).
——— (*Mechanics' Institute*).
——— (*Working Men's College*).
Haslingdon (*Institute*).
Hastings (*Literary and Scientific Institute*).
Hawarden (*Literary Institution*).
Hebden Bridge, near Todmorden (*Mechanics' Institution*).
Helston (*Reading Room and Library*).
Hereford (*Natural History, Philosophical, Antiquarian, and Literary Society*).
Hertford (*Literary and Scientific Institution*).
Heywood (*Mechanics' Institute*).
Holbeck (*Mechanics' Institution*).
Hollingwood (*Working Men's Club*).
Holywell Green (*Mechanics' Institution*).
Huddersfield(*Mechanics' Institution*).
Hull (*Church Institute*).
—— (*Literary, Scientific and Mechanics' Institute*).
—— (*Lyceum Library*).
—— (*Royal Institution, Albion Street*).
—— (*Young People's Institute*).
Huntingdon (*Literary and Scientific Institution*).
Kendal (*Christian and Literary Institute*).
——— (*Working Men's Institute*).
Kidderminster (*Mechanics' Institute*).
Lancaster (*Mechanics' Institute and School of Science*).
Leeds (*Church Institute*).
—— (*Library*).
—— (*Mechanics' Institution and Literary Society*).
—— (*Philosophical and Literary Society*).
—— (*Working Men's Institute*).
—— (*Young Men's Christian Association*).
Leighton Buzzard (*Working Men's Mutual Improvement Society*).
Leith (*Mechanics' Subscription Library*).
Lewes (*Mechanics' Institute*).
—— (*School of Science and Art*).
Lincoln (*Mechanics' Institute*).

Liverpool (*Institute*).
———— (*Mechanics' Institute*).
———— (*Medical Institution*).
———— (*Polytechnic Society*).
Llanelly (*Chamber of Commerce and Reading Room*).
London (*Athenæum Club, Pall Mall*).
———— (*Beaumont Institute, Mile End*).
———— (*Bedford Working Men's Institute, Spitalfields*).
———— (*Birkbeck Institution, Southampton Buildings, Chancery Lane*).
———— (*Bow Common Working Men's Club, Devon's Road, Bow Common*).
———— (*Christchurch Working Men's Club. New Street, Lark Hall Lane, Clapham*).
———— (*Clerkenwell Club, Lower Rosoman Street*).
———— (*Holloway Working Men's Club and Institute, Holloway Road*).
———— (*Literary and Scientific Society, Wellington Street, Islington*).
———— (*Literary and Scientific Institution, Walworth*).
———— (*St. James and Soho Working Men's Club, Rupert Street, Soho*).
———— (*St. Mary Charterhouse Working Men's Club, Golden Lane*).
———— (*South London Working Men's College, Blackfriars Road*).
———— (*Southwark Working Men's Club, Broadwall, Stamford Street*).
———— (*Spring Vale Institution, Hammersmith*).
———— (*Working Men's Club, Brixton Hill*).
———— (*Working Men's Club, St. Mark's, Victoria Docks*).
———— (*Working Men's Club and Institute. Battersea*).
———— (*Working Men's Club and Institute Union, Strand*).
———— (*Working Men's College, Great Ormond Street*).
Loughborough (*Working Men's Club and Institute*).
Madeley (*Anstice Memorial, Workmen's Club and Institute*).
Manchester (*Ancoats Branch Free Library*).
———— (*Athenæum*).
———— (*Campfield Free Lending Library*).
———— (*Chorlton and Ardwick Branch Free Library*).
———— (*Hulme Branch Free Library*).
———— (*Law Library*).
———— (*Mechanics' Institution*).
———— (*Natural History Museum, Peter Street*).
———— (*Portico Library, Moseley Street*).
———— (*Rochdale Road Branch Free Library*).
———— (*Royal Exchange Library*).
Mansfield (*Co-operative Industrial Society*).

Mansfield (*Mechanics', Artizans', and Apprentices' Library*).
Melksham (*Mutual Improvement Society*).
Merthyr-Tydfil (*South Wales Institute of Engineers*).
Middlesborough (*Mechanics' Institution*).
Modbury (*Mechanics' Institution*).
Mossley (*Mechanics' Institute*).
Newark (*Mechanics' Institute*).
Newcastle-upon-Tyne (*Mechanics' Institution*).
———— (*Working Men's Club*).
New Mills, near Stockport (*Mechanics' Institute*).
Newport, Isle of Wight (*Young Men's Society and Reading Room*).
Northampton (*Mechanics' Institute*).
Nottingham (*Free Library*).
———— (*Mechanics' Institution*).
———— (*Subscription Library, Bromley House*).
Oldham (*Analytic Literary Institution*).
———— (*Mechanics' Institution, Werneth*).
Ormskirk (*Public Library*).
Oswestry (*Institute*).
Patricroft (*Mechanics' Institution*).
Pembroke Dock (*Mechanics' Institute*).
Pendleton (*Mechanics' Institution*).
Penryn (*Working Men's Club and Reading Room*).
Perth (*Mechanics' Library, High Street*).
Peterborough (*Mechanics' Institution*).
Plymouth (*Working Men's Institute*).
Poole (*Literary and Scientific Institution*).
———— (*Mechanics' Institute*).
Portsea (*Athenæum and Mechanics' Institution*).
Preston (*Avenham Institution*).
———— (*Society of Useful Knowledge*).
Rawtenstall (*Mechanics' Institution*).
Richmond (*Working Men's College*).
Rotherham (*Rotherham and Masbro' Literary and Mechanics' Institute*).
Royston (*Institute*).
Ryde, Isle of Wight (*Philosophical and Scientific Society*).
Saffron Walden (*Literary and Scientific Institution*).
St. Just (*Institution*).
St. Leonard's (*Mechanics' Institution*).
Salford (*Working Men's Club*).
Saltaire (*Literary Institute*).
Selby (*Mechanics' Institute*).
Sheffield (*Branch Free Library*).
———— (*Literary and Philosophical Society, School of Arts*).
Skipton, Yorkshire (*Mechanics' Institute*).
Southampton (*Hartley Institution*).
———— (*Polytechnic Institution*).
Southport (*Athenæum*).

South Shields (*Working Men's Institute and Club*).
Spalding (*Mechanics' Institute*).
—————— (*Christian Young Men's Association*).
Staines (*Literary and Scientific Institution*).
—————— (*Mechanics' Institute and Reading Room*).
Stamford (*Institution*).
Stourbridge (*Church of England Association*).
—————— (*Iron Works Reading Room and Library*).
—————— (*Mechanics' Institution*).
—————— *Working Men's Institute*).
Stratford (*Working Men's Hall*).
Sunderland (*Working Men's Club*).
Swansea (*Royal Institution of South Wales*).
—————— (*Working Man's Institute*).
Tavistock (*Mechanics' Institute*).
—————— (*Public Library*).
Thornton, near Bradford (*Mechanics' Institute*).
Thornton Heath, Croydon (*Workmen's Club*).
Todmorden (*Mechanics' Institution*).
Truro (*Cornwall County Library*).
—————— (*Institution*).
—————— (*Royal Institution of Cornwall*).
Tunbridge Wells (*Mechanics' Institution*).

Tunbridge Wells (*Society of Literature and Science*).
Turton near Bolton (*Chapel Town Institute*).
Ulverston (*Temperance Hall*).
Uttoxeter (*Mechanics' Literary Institute*).
Wakefield (*Mechanics' Institute*).
Watford (*Literary Institute*).
Wells, Somerset (*Mechanics' Institution, Grove Lane*).
—————— (*Young Men's Society*).
Whaleybridge (*Mechanics' Institute*).
Whitby (*Institute*).
—————— (*Museum*).
—————— (*Subscription Library*).
Whitehaven (*Mechanics' Institute*).
—————— (*Working Men's Reading Room*).
Whitstable (*Institute*).
Wisbeach (*Mechanics' Institute*).
Wolverhampton (*Library*).
Wolverton (*Institute*).
Woodbridge (*Literary and Mechanics' Institute*).
—————— (*Working Men's Hall*).
Worcester (*Railway Literary Institute*).
—————— (*Workman's Hall*).
Workington (*Mechanics' Institution*).
York (*Church Institute*).
—————— (*Institute of Popular Science, &c.*)
—————— (*Railway Library*).

Presentations of portions of the Works, published by order of the Commissioners of Patents, have been made to the following Libraries :—

Armagh (*Town Clerk's Office*).
Aylesbury (*Mechanics' Institution and Literary Society, Kingsbury*).
Birmingham (*Institution of Mechanical Engineers, Newhall Street*).
Boston, Lincolnshire (*Public Offices, Market Place*).
Cambridge (*Free Library, Jesus Lane*).
Chester (*Mechanics' Institute, St. John Street*).
Coalbrookdale (*Literary and Scientific Institution*).
Coventry (*Watchmakers' Association*).
Darwen, Over (*Free Public Library*).
Dublin (*Dublin Library, D'Olier Street*).
Edinburgh (*Horological Society*).
Ennis (*Public Library*).
Gloucester (*Working Men's Institute, Southgate Street*).
Ipswich (*Mechanics' Institute, Tavern Street*).
Kew (*Library of the Royal Gardens*).

Kington, Herefordshire (*Reading Institute*).
Leominster (*Literary Institute*).
London (*House of Lords*).
—————— (*House of Commons*).
—————— (*Hon. Soc. of Gray's Inn*).
—————— (*Hon. Soc. of Inner Temple*).
—————— (,, ,, *Lincoln's Inn*).
—————— (,, ,, *Middle Temple*).
—————— (*Aeronautical Society*).
—————— (*British Horological Institute*).
—————— (*General Post Office*).
—————— (*Institution of Civil Engineers*).
—————— (*Odontological Society*).
—————— (*Royal Society*).
—————— (*United Service Museum*)
Manchester (*Literary and Philosophical Society, George Street*).
—————— (*Mechanics' Institution, David Street*).

Newcastle-upon-Tyne (*North of England Institute of Mining Engineers*).
Oxford (*Bodleian Library*).
Stretford, near Manchester (*Mechanics' Institute*).

Swindon, New (*Mechanics' Institute*)
Tamworth (*Library and Reading Room, George Street*).
Yarmouth, Norfolk (*Public Library, South Quay*).

British Colonies and Foreign States.

British Columbia—Mechanics' Institute, Victoria.
——————————— Public Library, New Westminster.
France—Academy of Science, Paris.
Netherlands—Bibliothéque de l'Ecole Polytechnique de Delft.
Russia—Imperial Technological Institute, St. Petersburg.
Smyrna—Literary and Scientific Institute.

United States—American Academy of Arts and Sciences, Boston.
——————————American Institute, New York.
——————————— American Society of Civil Engineers, New York.
——————————— Odd Fellows' Library Association, San Francisco.
——————————— Smithsonian Institute, Washington.

PATENT OFFICE MUSEUM, SOUTH KENSINGTON.

THIS Museum is open to the public daily, free of charge. The hours of admission are as follows :—
   Mondays, Tuesdays, and Saturdays, 10 A.M. till 10 P.M.
   Wednesdays, Thursdays, and Fridays, from 10 A.M. till 4, 5, or 6 P.M., according to the season.

If any Patentee should be desirous of exhibiting a model of his invention in London, he may avail himself of this Museum, which has been visited since its opening on the 22nd June 1857 by more than 2,150,000 persons. The model will be received either as a gift or loan ; if deposited as a loan, it will be returned on demand. Before sending a model, it is requested that the size and description of it shall first be given to the Superintendent of the Patent Office Museum.

GALLERY OF PORTRAITS OF INVENTORS, DISCO-VERERS, AND INTRODUCERS OF USEFUL ARTS.—This Collection, formed by Mr. Woodcroft, and first opened to public view in 1853, is now exhibited in the Patent Office.
   Presentations or loans of Portraits, Medallions, Busts, and Statues, in augmentation of the Collection; are solicited. They will be duly acknowledged in the Commissioners of Patents' Journal, and included in the next edition of the Catalogue.

All communications relating to the Patent Office, or to the Museum and Portrait Gallery, to be addressed to B. WOODCROFT, Clerk to the Commissioners of Patents and Superintendent of the Patent Office Museum, at the Patent Office, 25, Southampton Buildings, Chancery Lane, London, W.C.

NOTICE.

The Abridgments delivered at the Patent Office by the Applicants for Letters Patent will in future be published weekly (commencing on Friday, July 14), with Indexes of Persons and Subjects. In the body of the work the Abridgments of the Provisional and Complete Specifications will be published in regular numerical order at the expiration of the term of six months from the date of application. But each weekly number will have an appendix, containing the Abridgments open to public inspection before the expiration of the term of six months, in consequence of the Patentees having filed their Final Specifications, and also the Abridgments of Complete Specifications just received. These Abridgments will be subsequently printed in the body of the work in their proper places, in order to preserve the numerical and chronological arrangement of the book. In the indexes of each successive number all the previous indexes will be incorporated until the end of the year; and then the last indexes only should be retained to bind with the fifty-two weekly parts in one volume for the year.

B. WOODCROFT.

July 10, 1871.

*** The work referred to in the above notice is published (under the title of "Chronological and Descriptive Index of Patents," &c.) on Friday in each week, and is forwarded, post free, to subscribers. Terms 22s. per annum. Subscriptions received at the Sale Room of the Patent Office, 25, Southampton Buildings, Holborn, where also single copies, at 4d. each, may be obtained. Post Office Orders to be made payable at the Post Office, Holborn, to Mr. Bennet Woodcroft, Clerk to the Commissioners of Patents.